CW00591626

The Orbis Pocket Encyclopedia of the World

The Orbis Pocket Encyclopedia of the World

Orbis Publishing, London

Published in Great Britain
by Orbis Publishing Limited
First Edition 1981 — Revised 1982

Maps, text and index prepared
and printed by Kartografie,
Prague – Czechoslovakia

Cartographic Editor:
RNDr. Jiří Novotný

Technical Editor:
Marie Pánková

© KARTOGRAFIE, PRAGUE, 1981

All rights reserved. No part of this publication may be reproduced, stored in a retrieval system, or transmitted, in any form or by any means, electronic, mechanical, photocopying, recording or otherwise, without the prior permission of the publishers. Such permission, if granted, is subject to a fee depending of the nature of the use.

ISBN 0-85613-339-6

CONTENTS

EXPLANATIO

CITIES AND TOWNS

British Isles

⊖	LONDON	over 1,000,000 inhabitants
◉	GLASGOW	500,000 - 1,000,000 inhabitants
◎	Belfast	100,000-500,000 inhabitants
⊙	Bath	50,000-100,000 inhabitants
○	Perth	20,000-50,000 inhabitants
○	Douglas	less than 20,000 inhabitants

Other Maps

⊖	NEW YORK	over 1,000,000 inhabitants
◉	BRISBANE	500,000-1,000,000 inhabitants
◎	Utrecht	100,000-500,000 inhabitants
⊙	Calais	50,000-100,000 inhabitants
○	Gibraltar	less than 50,000 inhabitants

Physical and World Maps

○ Ottawa

Canberra National Capitals

CANADA

Victoria

St. Helena
(U.K.)

SYMBOLS

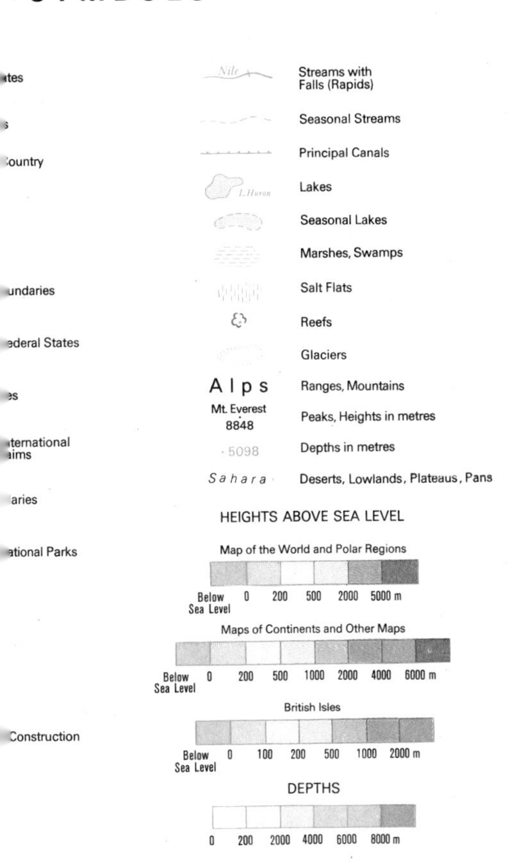

ites
s
ountry
undaries
ederal States
es
ternational
aims
aries
ational Parks
Construction
Nile
Streams with Falls (Rapids)
Seasonal Streams
Principal Canals
L. Huron
Lakes
Seasonal Lakes
Marshes, Swamps
Salt Flats
Reefs
Glaciers
Alps
Ranges, Mountains
Mt. Everest 8848
Peaks, Heights in metres
5098
Depths in metres
Sahara
Deserts, Lowlands, Plateaus, Pans
HEIGHTS ABOVE SEA LEVEL
Map of the World and Polar Regions
Below Sea Level 0 200 500 2000 5000 m
Maps of Continents and Other Maps
Below Sea Level 0 200 500 1000 2000 4000 6000 m
British Isles
Below Sea Level 0 100 200 500 1000 2000 m
DEPTHS
0 200 2000 4000 6000 8000 m

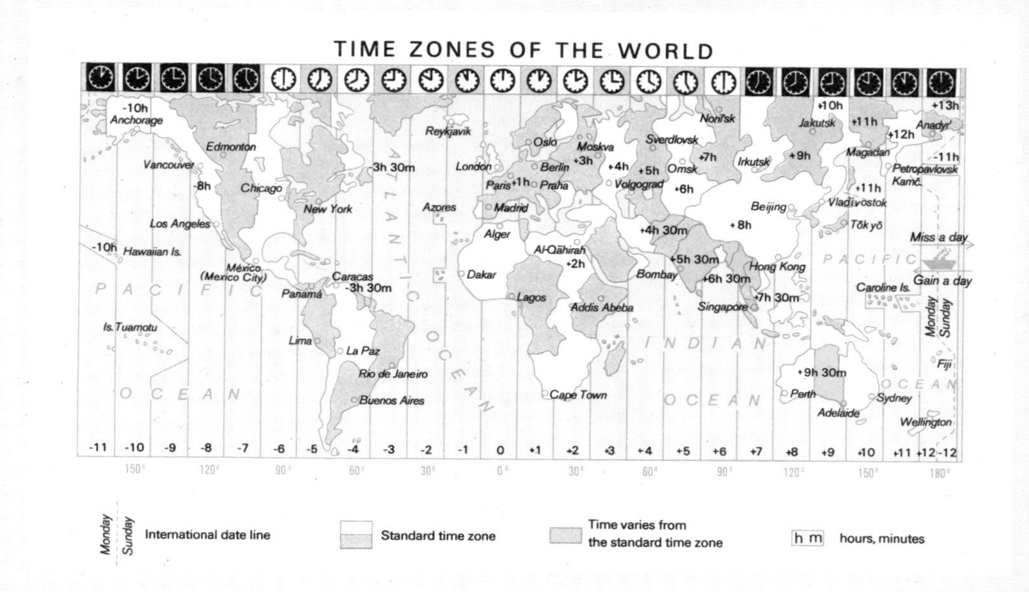
TIME ZONES OF THE WORLD
-10h
Anchorage
Edmonton
Vancouver
-8h
Chicago
New York
-3h 30m
Reykjavik
London
Paris
+1h
Praha
Madrid
Azores
Oslo
Berlin
Moskva
+3h
+4h
+5h
Volgograd
Sverdlovsk
Omsk
+6h
+7h
Noril'sk
Irkutsk
+9h
Jakutsk
+10h
+11h
Magadan
+12h
Anadyr'
+13h
-11h
Petropavlovsk
Kamč.
+11h
Vladivostok
Tōkyō
Beijing
+8h
Los Angeles
-10h
Hawaiian Is.
México
(Mexico City)
Caracas
-3h 30m
Panamá
Alger
Al-Qāhirah
+2h
Dakar
Lagos
Addis Abeba
+4h 30m
+5h 30m
Bombay
+6h 30m
Hong Kong
+7h 30m
Singapore
Caroline Is.
Miss a day
Gain a day
Monday
Sunday
Fiji
Is. Tuamotu
Lima
La Paz
Rio de Janeiro
Buenos Aires
Cape Town
+9h 30m
Perth
Adelaide
Sydney
Wellington
PACIFIC OCEAN
ATLANTIC OCEAN
INDIAN OCEAN
PACIFIC OCEAN
-11 -10 -9 -8 -7 -6 -5 -4 -3 -2 -1 0 +1 +2 +3 +4 +5 +6 +7 +8 +9 +10 +11 +12 -12
150° 120° 90° 60° 30° 0° 30° 60° 90° 120° 150° 180°
Monday
Sunday
International date line
Standard time zone
Time varies from the standard time zone
h m hours, minutes

The Orbit of the Earth around the Sun

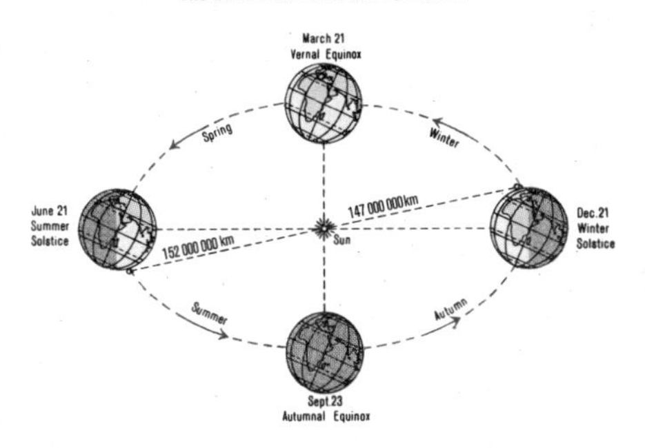

PLANETS

Planet	Mean distance from the Sun /astronomical units/	Mean distance from the Sun /million km/	Sidereal period /years/	Mean orbit velocity / km/s /	Equatorial diameter /Earth=1/	Equatorial diameter /km/	Mass /Earth=1/	Mean density / km/m³/	Sidereal rotation period /equatorial/	Number of satellites /R Rings/
Mercury	0.387	57.9	0.24	47.9	0.38	4 850	0.0558	5 600	58.65 days	0
Venus	0.723	108.2	0.61	35.0	0.95	12 140	0.8150	5 200	243 days	0
Earth	1.000	149.6	1.00	29.8	1.00	12 756	1.0000	5 518	23 hrs 56 mins 4.1s	1
Mars	1.524	227.9	1.88	24.1	0.53	6 780	0.1074	3 933	24 hrs 37 mins 22.7s	2
Jupiter	5.203	778.3	11.86	13.1	11.23	143 200	317.893	1 314	9 hrs 50 mins	14 + R
Saturn	9.539	1 427.0	29.46	9.6	9.41	120 000	95.147	704	10 hrs 14 mins	11 + R
Uranus	19.182	2 869.6	84.01	6.8	4.15	52 900	14.54	1 210	24 hrs ?	5 + R
Neptune	30.058	4 496.6	164.79	5.4	3.88	49 500	17.23	1 670	15-18hrs	2
Pluto	39.44	5 900	247.7	4.7	0.22	2 800	0.0024	1 240	6.4 days ?	1

Astronomical units of distance:

Astronomical unit = mean Sun-Earth distance = AU = $1.495\ 978\ 70 \times 10^{11}$ m = 149.6 mn. km

Parsec / pc / = distance from which the astronomical unit appears at a visual angle of 1″ = $3.085\ 678 \times 10^{16}$ m =
= 206 264.8 AU = 3.26 light years

Light year = distance covered by light in one year = 9.46×10^{15} m = 9.5 billion km

THE NORTHERN SKY

THE MOON

Distance from the Earth: perigee 364 000 km
mean 384 400 km
apogee 406 700 km
Orbital velocity 3 680 km/h (1.02 km/s)
Diameter 3 476 km (0.27 Earth's diameter)
Mass 0.0123 Earth's mass
Mean density 3 340 kg/m^3 (0.60 mean Earth's density)
Surface temperature: daytime +130°C
nightime −150°C
Surface gravity 1.62 m/s^2 (0.165 Earth's gravity)
circular velocity (at the surface) 1.7 km/s
parabolic velocity (at the surface) 2.4 km/s
Sidereal month 27.321 661 days
Synodic month 29.530 588 days
Extreme libration: in latitude 6°50′, in longitude 7°54′

Almost 59 % of the Moon's surface can be observed from the Earth; 18 % is visible only at certain times and 41 % can never be seen from the Earth. Nevertheless, the whole of the Moon has been mapped (except for 1% of its surface) by means of space probes

THE MOON'S PHASES

The Moon does not shine by its own light; one of its hemispheres is illuminated by the Sun / a day /, the opposite is in a shadow / a night /. We can see only the part of the illuminated hemisphere which belongs to the Near Side. All Moon's phases change through 1 synodic month.

THE ROTATION OF THE MOON

The Moon's period of rotation on its axis is equal to its period of revolution around the Earth. Therefore the Moon always presents the same face to an observer on the Earth.

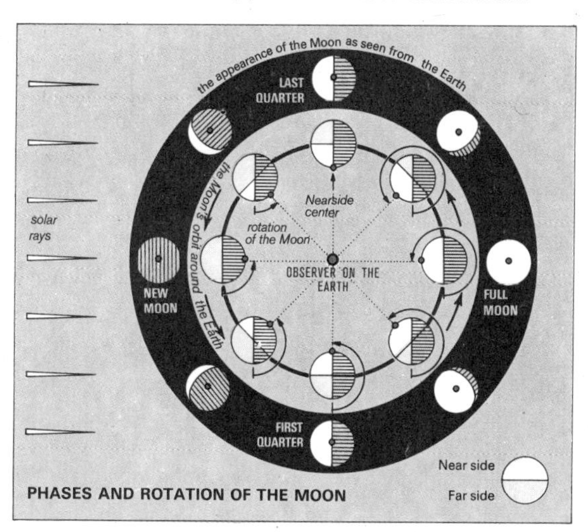

PHASES AND ROTATION OF THE MOON

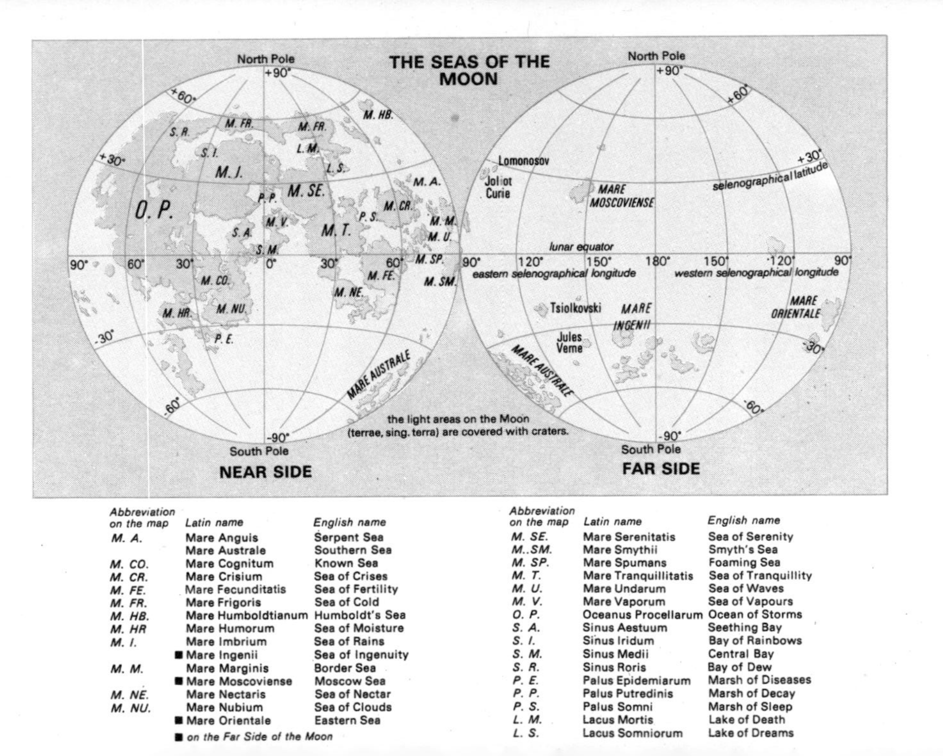

Abbreviation on the map	Latin name	English name
M. A.	Mare Anguis	Serpent Sea
	Mare Australe	Southern Sea
M. CO.	Mare Cognitum	Known Sea
M. CR.	Mare Crisium	Sea of Crises
M. FE.	Mare Fecunditatis	Sea of Fertility
M. FR.	Mare Frigoris	Sea of Cold
M. HB.	Mare Humboldtianum	Humboldt's Sea
M. HR	Mare Humorum	Sea of Moisture
M. I.	Mare Imbrium	Sea of Rains
	■ Mare Ingenii	Sea of Ingenuity
M. M.	Mare Marginis	Border Sea
	■ Mare Moscoviense	Moscow Sea
M. NE.	Mare Nectaris	Sea of Nectar
M. NU.	Mare Nubium	Sea of Clouds
	■ Mare Orientale	Eastern Sea

■ *on the Far Side of the Moon*

Abbreviation on the map	Latin name	English name
M. SE.	Mare Serenitatis	Sea of Serenity
M..SM.	Mare Smythii	Smyth's Sea
M. SP.	Mare Spumans	Foaming Sea
M. T.	Mare Tranquillitatis	Sea of Tranquillity
M. U.	Mare Undarum	Sea of Waves
M. V.	Mare Vaporum	Sea of Vapours
O. P.	Oceanus Procellarum	Ocean of Storms
S. A.	Sinus Aestuum	Seething Bay
S. I.	Sinus Iridum	Bay of Rainbows
S. M.	Sinus Medii	Central Bay
S. R.	Sinus Roris	Bay of Dew
P. E.	Palus Epidemiarum	Marsh of Diseases
P. P.	Palus Putredinis	Marsh of Decay
P. S.	Palus Somni	Marsh of Sleep
L. M.	Lacus Mortis	Lake of Death
L. S.	Lacus Somniorum	Lake of Dreams

map 2b

MARE	...... sea
LACUS	...... lake
SINUS	...... bay
DEPRESSIO	...... depression
REGIO	...... landscape
PALUS	...... marsh
FRETUM	...... strait

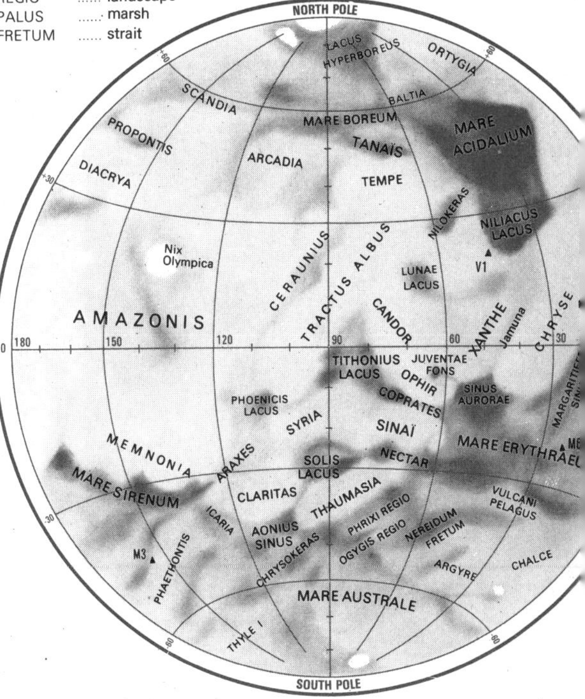

The appearance of albedo or white features, their colour, intensity, and shape, change in accordance with Mars's seasons and the influence of atmospherical conditions (clouds, fogs, dust storms, etc.).

Atmosphere of Mars: CO_2 95%, N_2 2.7%, Ar 1.6%, O_2 0.15% and further small amount of CO, Xe, Kr, H_2O.

Atmospherical pressure at the zero level: 610 Pa /like about 30 km above the Earth/

Surface temperature: maximum on the equator about +30°C

minimum on poles about −120°C

The surface of the planet: the southern hemisphere of Mars is thickly covered by craters like the similar areas of so-called continents on the Moon. Extensive lava flows with volcanic features /e.g. shield volcanoes of which Olympus Mons 24 km high with the base diameter of 500 km is the highest /predominate on the northern hemisphere.

The polar caps of Mars are formed by the layer of frozen CO_2 and water ice; remnants of these caps which do not melt even in summer probably consist of a thick layer of ice.

MARS

ale 1 : 110 000 000

quivalent projections

rawn by: Ing. A. Růkl

On the map there are shown the albedo features on Mars visible from t Earth by telescopes / i.e. bright and dark areas with different capability reflect a light /. The names of albedo features correspond to the recommendation of the I. A. U. from 1958.

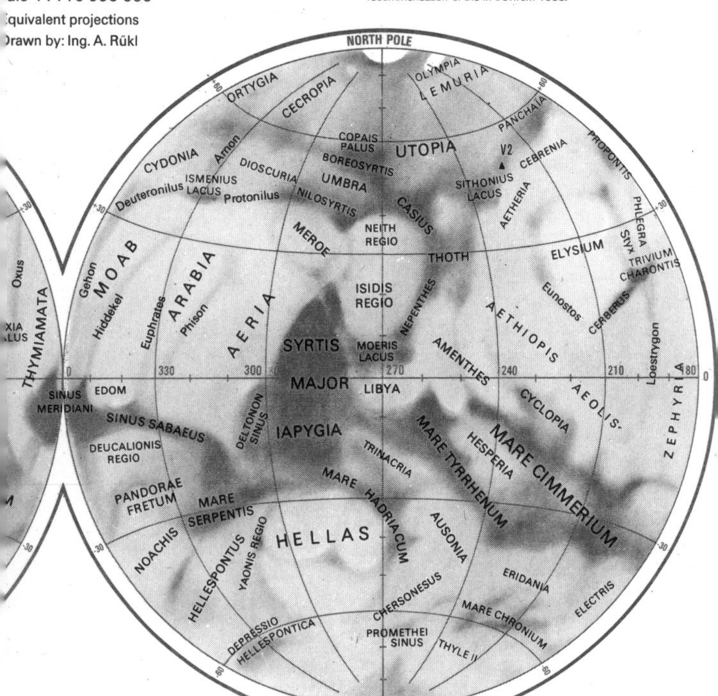

Successful probes to Mars

Probe	Date of reaching Mars		
Mariner 4 /U.S.A./	15 July	1965	Encounter, the first 21 photographs of the surface.
Mariner 6 /U.S.A./	31 July	1969	Encounter, 76 photographs, the survey of the atmosphere and surface.
Mariner 7 /U.S.A./	5 Aug.	1969	Encounter, 126 photographs, the survey of the atmosphere and surface of the plane
Mariner 9 /U.S.A./	14 Nov.	1971	The first artificial satellite of Mars. Mapping of the planet: 7 329 photographs.
Mars 2 /U.S.S.R./	27 Nov.	1971	Satellite of Mars. Crash landing of a landing capsule on the planet.
+Mars 3 /U.S.S.R./	2 Dec.	1971	Satellite of Mars. The first soft landing of a capsule.
Mars 5 /U.S.S.R./	12 Feb.	1974	Satellite of Mars. Photographs of the surface, the survey of the atmosphere.
+Mars 6 /U.S.S.R./	12 March	1974	The first direct measuring in the atmosphere, landing of the capsule on Mars.
+Viking 1 /U.S.A./	18 June	1976	Satellite of Mars. Viking Lander 1 landed 20 July 1976. Extensive survey of the plane from the orbit and surface.
+Viking 2 /U.S.A./	7 Aug.	1976	Satellite of Mars. Viking Lander 2 landed 3 Sept. 1976. Further longterm survey of th planet and its moons. The existence of life on Mars was not proved.

+... landing points marked on the map

DIMENSIONS OF THE EARTH

	According to F. N. Krasovskij's ellipsoid	According to Hayford's ellipsoid
Radius of the Equator (a)	6,378,245.0 m	6,378,388.0 m
Radius of the axis (b)	6,356,863.0 m	6,356,911.9 m
Flattening of the Earth $(\frac{a-b}{b})$	$\frac{1}{298.3}$	$\frac{1}{297.0}$
Circumference of the Equator	40,075,704 m	40,076,594 m
Circumference of one of the meridians	40,008,548 m	40,009,152 m
Total surface area of the Earth	510,083,000 sq.km	510,100,900 sq.km
Area of dry land (29.2%)	148,628,000 sq.km	149,460,000 sq.km
Area of sea (70.8%)	361,455,000 sq.km	360,641,000 sq.km
Volume of Earth	1,083,319,780,000 cub.km	
Weight of Earth	5.978×10^{24} kg	
Area of tropical zone (39.7%)	202,505,000 sq.km	
Area of temperate zones (52%)	265,418,000 sq.km	
Area of polar zones (8.3%)	42,160,000 sq.km	

Chemical composition of the Earth's surface: oxygen 49.13%, silicon 26%, aluminium 7.45%, iron 4.2%, calcium 3.25%, sodium 2.4%, potassium 2.35%, hydrogen 1%, titanium 0.61%, carbon 0.35%, chlorine 0.20%, phosphorus 0.12%, manganese 0.10%, sulphur 0.10%, fluorine 0.08%, barium 0.05%, nitrogen 0.04% etc.

WATER

Chemical composition of sea water: oxygen 85.82%, hydrogen 10.72%, chlorine 1.89%, sodium 1.056%, magnesium 0.14%, sulphur 0.088%, calcium 0.041%, potassium 0.038%, bromine 0.0065%, carbon 0.002% etc.

Oceans and Seas

Name	Area in 1,000 sq.km	Volume in 1,000 cub.km	Greatest depth in m	Average depth in m
PACIFIC OCEAN	179,680	723,699	11,034	3,780
Philippine Sea	5,726	23,522	10,830	4,188
Coral Sea	4,791	11,470	9,174	2,243
South China Sea	3,537	3,622	5,559	1,024
Tasman Sea	3,336	10,960	5,944	3,285
Bering Sea	2,315	3,796	4,191	1,640
Sea of Okhotsk	1,603	1,316	3,916	821
Sea of Japan	1,062	1,630	3,699	1,535
ATLANTIC OCEAN	91,655	329,700	8,648	3,597
Weddell Sea	2,910	8,375	6,820	2,878
Caribbean Sea	2,776	6,745	7,491	2,429
Mediterranean Sea	2,505	3,603	5,121	1,438
Gulf of Mexico	1,554	2,366	4,376	1,522
Labrador Sea	841	1,596	4,316	1,898
North Sea	565	49	725	87
Black Sea	422	555	2,210	1,315
Baltic Sea	419	215	470	51
INDIAN OCEAN	76,170	282,600	7,725	3,710
Arabian Sea	4,592	14,514	5,803	3,160
Bay of Bengal	2,191	5,664	3,835	2,585
Arafura Sea	1,017	157	3,680	154
Red Sea	460	182	3,039	396
ARCTIC OCEAN	13,950	17,100	5,450	1,328
Barents Sea	1,405	322	610	254
Norwegian Sea	1,383	2,408	4,487	1,724
Greenland Sea	1,205	1,740	4,848	1,497
East Siberian Sea (Vostočno-Sibirskoje More)	936	42	155	45
Hudson Bay	822	92	258	128
Baffin Sea	780	593	2,414	861

Greatest ocean depth: in Pacific Ocean – Mariana Trench 11,034 m (11°21' N.Lat., 142°12' E.Long.); in Atlantic Ocean – Puerto Rico Trench 8,648 m (19°35' N.Lat., 68°17' W.Long.); in Indian Ocean – Java Trench 7,725 m (10°15' S.Lat., 109° E.Long. – approx.); in Arctic Ocean – Eurasia Basin 5,450 m (82°23' N.Lat., 19°31' E.Long.).

1 : 150 000 000

0	1000	2000	3000	4000 Km
0	500	1 000	1 500	2 000	2 500 Mi

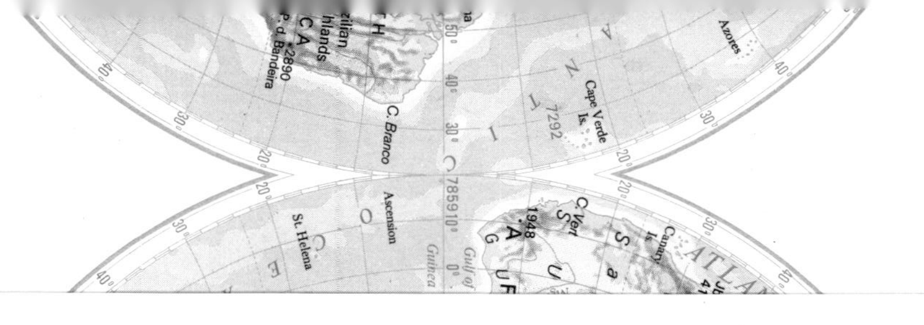

add page 26

Longest Rivers of the World

	Name	Length in km	River basin in sq.km
1.	Amazonas (-Ucayali-Apurímac)	7,025	7,050,000
2.	Nile-Kagera	6,671	2,881,000
3.	Mississippi-Missouri	6,212	3,250,000
4.	Changjiang /Yangtze/	5,520	1,942,000
5.	Ob' (-Irtyš)	5,410	2,975,000
6.	Huanghe /Yellow/	4,845	772,000
7.	Congo /Zaïre/ -Lualaba	4,835	3,822,000
8.	Mekong /Lancangjiang/	4,500	810,000
9.	Amur (-Šilka, -Onon)	4,416	1,855,000
10.	Lena	4,400	2,490,000

Largest Islands of the World

	Name (Continent)	Area in sq.km
1.	Greenland (N.America)	2,175,600
2.	New Guinea (Oceania)	785,000
3.	Borneo /Kalimantan/ (Asia)	746,546
4.	Madagascar (Africa)	587,041
5.	Baffin I. (N.America)	507,414
6.	Sumatera (Asia)	433,800
7.	Honshū (Asia)	227,414
8.	Victoria I. (N.America)	217,274
9.	Great Britain (Europe)	216,325
10.	Ellesmere I. (N.America)	196,221

Highest Waterfalls of the World

	Name (Country)	Height in m
1.	Salto Angel (Venezuela)	979
2.	Tugela (Natal, South Africa)	948
3.	Yosemite (Cal., U.S.A.)	739
4.	Cuquenán (Venezuela)	610
5.	Sutherland (New Zealand)	579
6.	Takakkaw (B.C., Canada)	503
7.	Giétroz (Switzerland)	498
8.	Ribbon (Cal., U.S.A.)	491
9.	King George VI (Guyana)	488
10.	Della (B.C., Canada)	440

Largest Lakes of the World

	Name (Continent)	Area in sq.km	Greatest depth in m		Name (Continent)	Area in sq.km	Greatest depth in m
1.	Caspian Sea (Asia)	371,000	1,025	6.	L. Michigan (N.America)	58,016	281
2.	L. Superior (N.America)	82,414	393	7.	L.Tanganyika (Africa)	32,880	1,435
3.	L. Victoria (Africa)	68,800	125	8.	O. Bajkal (Asia)	31,500	1,620
4.	Aral'skoje More (Asia)	64,115	67	9.	Great Bear Lake (N.America)	31,328	137
5.	L. Huron (N.America)	59,596	226	10.	Great Slave Lake (N.America)	28,570	140

LAND SURFACE

Continent	Area in 1,000 sq.km	% of Land surface	Altitude in metres highest	aver.	lowest	Length of the coast in km	Population in millions (1978)
Europe	10,527	7.04	4,810	340	−28	37,900	628
Asia	44,413	29.72	8,848	960	−394	69,000	2,476
Africa	30,319	20.28	5,895	750	−173	30,500	436
North America	24,247	16.22	6,194	720	−86	75,600	360
South America	17,834	11.93	6,959	580	−40	28,700	236
Australia and Oceania	8,511	5.98	5,030	350	−16	19,700	23
Antarctica	13,209	8.83	5,140	2,280		30,000	
WORLD	149,460	100.00	8,848	840	−394	291,400	4,159

Highest point on the Earth surface: Mount Everest 8,848 m.
Lowest point on the Earth surface: The Dead Sea 394 m below sea-level.
Highest active volcano: Volcan Guallatiri (in Chile) 6,060 m (eruption in 1960).
Largest island: Greenland 2,175,600 sq.km.
Largest peninsula: Arabian Peninsula 2,780,000 sq.km.
Longest mountain chain: Rocky Mountains Range – Andes (Cordilleras), length 15,000 km.
Largest lowlands: Amazonian Lowlands (S.America) approx. 5,000,000 sq.km.
Largest desert: Sahara (Africa) 7,750,000 sq.km.
Largest glacier (excl. Antarctica): Greenland, area 1,830,000 sq.km, volume 2,700,000 cub.km.
Largest lake: Caspian Sea 371,000 sq.km.
Deepest lake: O. Bajkal 1,620 m.
Longest river: Amazonas (-Ucayali, -Apurímac) 7,025 km.
Largest river basin: Amazonas (Amazon) 7,050,000 sq.km.
Highest average flow: Amazonas 120,000 cub.m per sec.
Highest waterfalls: Salto Angel (in Venezuela) 979 m.
Deepest cave system: Réseau de la Pierre-Saint-Martin (in French Pyrenees) depth 1,332 m.
Longest cave system: Flint Ridge-Mammoth Cave (Kentucky, U.S.A.) 297,080 m.
Highest absolute temperature: Al-Azīzīyah (in Libya) + 58°C.
Highest average annual temperature: Dalol (in Ethiopia) + 34.4°C.
Lowest absolute temperature: Vostok (3,488 m high, permanent Soviet base in Antarctica) −88.3°C.
Lowest average annual temperature: Pole of Cold (in Antarctica) −57.8°C.
Highest average annual precipitation: Waialeale (Kauai – Hawaii, U.S.A.) 11,684 mm.
Lowest average annual precipitation: Arica (in Chile) 0.8 mm.
Largest national park: Wood Buffalo National Park, area 44,807 sq.km.

Countries of the World according to area

Country	Area in sq.km
1. U.S.S.R.	22,274,900
2. Canada	9,976,139
3. China	9,560,980
4. U.S.A.	9,363,166
5. Brazil	8,511,965
6. Australia	7,686,848
7. India	3,287,590
8. Argentina	2,780,092
9. Sudan	2,505,813
10. Algeria	2,381,740
11. Zaïre	2,345,409
12. Greenland (Den.)	2,175,600
13. Saudi Arabia	2,149,690
14. Mexico	1,972,546
15. Indonesia	1,919,270
16. Libya	1,759,540
17. Iran	1,648,100
18. Mongolia	1,565,000
19. Peru	1,285,216
20. Chad	1,284,000
21. Niger	1,266,995
22. Angola	1,246,700
23. Mali	1,239,710
24. Ethiopia	1,221,900
25. South Africa	1,221,037

Countries of the World according to population
Mid-year estimates 1978 (*) 1979)

Country	Population
1. China	879,250,000
2. India	634,200,000
3. U.S.S.R.	260,100,000
4. U.S.A.	220,000,000
5. Indonesia	145,350,000
6. Brazil	115,397,000
7. Japan	115,100,000
8. Bangladesh	84,655,000
9. Nigeria	80,627,000
10. Pakistan	76,770,000
11. Mexico	69,381,000*)
12. Fed. Rep. of Germany	59,398,000+
13. Italy	56,696,000
14. United Kingdom	56,010,000
15. France	53,241,000
16. Vietnam	49,120,000
17. Philippines	46,351,000
18. Thailand	45,100,000
19. Turkey	43,210,000
20. Egypt	40,287,000*)
21. Spain	37,109,000
22. Rep. of Korea	37,019,000
23. Iran	35,213,000
24. Poland	35,032,000
25. Burma	32,334,000

+excluding West Berlin

THE UNITED NATIONS (UN)

The most important international organization in the world. The Charter of the United Nations was signed by 50 states at the San Francisco Conference held from 25 April to 26 June 1945, and it came into force on 24 October 1945 (United Nations Day). The Preamble to the Charter lays down the purposes and principles of the UN: to maintain international peace and security; to develop friendly relations among nations; to cooperate internationally in solving international economic, social, cultural and humanitarian problems and in promoting respect for human rights and fundamental freedoms; to support social progress and improve standards of living and to coordinate the fundamental principles of the member nations so as to attain these common ends.

Members of the UN are the sovereign states that established the organization and those admitted by the General Assembly upon recommendation of the Security Council. The UN had 157 members on 12 November 1981. Original member states (since 1945): Argentina, Australia, Belgium, Bolivia, Brazil, Byelorussian S.S.R. Canada, Chile, China, Colombia, Costa Rica, Cuba, Czechoslovakia, Denmark, Dominican Republic, Ecuador, Egypt, El Salvador, Ethiopia, France, Greece, Guatemala, Haiti, Honduras, India, Iran, Iraq, Lebanon, Liberia, Luxembourg, Mexico, Netherlands, New Zealand, Nicaragua, Norway, Panama, Paraguay, Peru, Philippines, Poland, Saudi Arabia, South Africa, Syria, Turkey, Ukrainian S.S.R., United Kingdom of Great Britain and Northern Ireland, Uruguay, Union of Soviet Socialist Republics, United States of America, Venezuela and Yugoslavia. Other states admitted as members: in 1946 – Afghanistan, Iceland, Sweden, Thailand; 1947 – Pakistan, Yemen; 1948 – Burma; 1949 – Israel; 1950 – Indonesia; 1955 – Albania, Austria, Bulgaria, Finland, Hungary, Ireland, Italy, Jordan, Kampuchea, Laos, Libya, Nepal, Portugal, Romania, Spain, Sri Lanka; 1956 – Japan, Morocco, Sudan, Tunisia; 1957 – Ghana, Malaysia; 1958 – Guinea; 1960 – Benin, Cameroon, Central African Republic, Chad, Congo, Cyprus, Gabon, Ivory Coast, Madagascar, Mali, Niger, Nigeria, Senegal, Somalia, Togo, Upper Volta, Zaïre; 1961 – Mauritania, Mongolia, Sierra Leone, Tanzania; 1962 – Algeria, Burundi, Jamaica, Rwanda, Trinidad and Tobago, Uganda; 1963 – Kenya, Kuwait; 1964 – Malawi, Malta, Zambia; 1965 – Gambia, Maldives, Singapore; 1966 – Barbados, Botswana, Guyana, Lesotho; 1967 – South Yemen; 1968 – Equatorial Guinea, Mauritius, Swaziland; 1970 – Fiji; 1971 – Bahrain, Bhutan, Oman, Qatar, United Arab Emirates; 1973 – Bahamas, German Democratic Republic, Federal Republic of Germany; 1974 – Bangladesh, Grenada, Guinea-Bissau; 1975 – Cape Verde, Comoros, Mozambique, Papua New Guinea, Surinam, São Tomé and Principe; 1976 – Angola, Samoa, Seychelles; 1977 – Djibouti, Vietnam; 1978 – Dominica, Solomon Islands; 1979 – Saint Lucia; 1980 – Zimbabwe, Saint Vincent; 1981 – Belize, Vanuatu, Antigua.

The principal organs of the UN: 1. The General Assembly – consists of all member states; each of them has juridically an equal position irrespective of size, power or importance; 2. The Security Council – bears the primary responsibility for the maintenance of peace and security (it has 5 permanent and 10 elected members); 3. The Economic and Social Council (54 elected members); 4. The Trusteeship Council (5 members); 5. The International Court of Justice (with its seat at 's-Gravenhage); 6. The Secretariat – carries out all administrative functions in the UN. It is headed by the Secretary-General, who is appointed by the General Assembly on the recommendation of the Security Council. **Secretary-General:** Javier Pérez de Cuellar (since 1 Jan. 1982). **Headquarters of the UN:** New York. **Official languages:** Arabic, Chinese, English, French, Spanish and Russian.

The Economic and Social Council has 5 regional economic commissions: the Economic Commission for Europe (ECE; H.Q.: Genève), the Economic Commission for Asia and the Far East (ECAFE; H.Q.: Krung Thep), the Economic Commission for Latin America (ECLA; H.Q.: Santiago), the Economic Commission for Africa (ECA; H.Q.: Addis Abeba), and the Economic Commission for Western Asia (ECWA; H.Q.: Bayrūt); functional commissions (Statistical Commission, Population Commission, Social Development Commission, Commission on Human Rights, Commission on the Status of Women, Commission on Narcotic Drugs, Commission on International Raw Materials Trade). **Special related agencies:** United Nations Children's Fund (UNICEF), Office of the United Nations High Commissioner for Refugees (UNHCR; H.Q.: Genève), United Nations Conference on Trade and Development (UNCTAD; H.Q.: Genève), United Nations Development Programme (UNDP), United Nations Industrial Development Organization (UNIDO; H.Q.: Wien), United Nations Institute for Training and Research (UNITAR), World Food Council (WFC; H.Q.: Rome), United Nations University (UNU; H.Q.: Tōkyō), as well as other commissions and agencies.

International specialized organizations in relationship with the UN (14): International Atomic Energy Agency (IAEA; H.Q.: Wien), International Labour Organization (ILO; H.Q.: Genève), Food and Agriculture Organization (FAO; H.Q.: Rome), United Nations Educational, Scientific and Cultural Organization (UNESCO; H.Q.: Paris), World Health Organization (WHO; H.Q.: Genève), International Bank for Reconstruction and Development (IBRD – World Bank; H.Q.: Washington), International Development Association (IDA; H.Q.: Washington), International Finance Corporation (IFC; H.Q.: Washington), International Monetary Fund (IMF; H.Q.: Washington), International Civil Aviation Organization (ICAO; H.Q.: Québec), Universal Postal Union (UPU; H.Q.: Bern), International Telecommunication Union (ITU, H.Q.: Genève), World Meteorological Organization (WMO; H.Q.: Genève), Inter-Governmental Maritime Consultative Organization (IMCO; H.Q.: London), General Agreement on Tariffs and Trade (GATT; H.Q. Genève), World Intellectual Property Organization (WIPO; H.Q.: Genève).

INTERNATIONAL ORGANIZATIONS

The Council of Europe – established on 5 May 1949 in London (H.Q.: Strasbourg). Membership: 21 European countries; founder members: United Kingdom, France, Italy, Belgium, the Netherlands, Luxembourg, Ireland, Denmark, Norway, Sweden. Turkey and Greece joined in 1949, Iceland in 1950, Fed. Rep. of Germany in 1951, Austria in 1956, Cyprus in 1961, Switzerland in 1963, Malta in 1965, Portugal in 1976 and Spain in 1977. Finland and the Vatican City participate in the work of certain bodies. Control organs: the Committee of Ministers (usually 21 Ministers of Foreign Affairs), the Joint Committee (21 members), the Consultative Assembly (consists of 154 parliamentary representatives from 21 member countries) and the Secretariat. The Council's aim is to achieve a greater unity between its members and to co-ordinate policies on economic, social, cultural, legal, scientific and administrative matters. **European Communities** – short name for 3 communities established by 6 countries of Western Europe to integrate economic policies and move towards political unity. The original members are France, Italy, the Fed. Rep. of Germany, Belgium, the Netherlands and Luxembourg. Other members since 1973 are the United

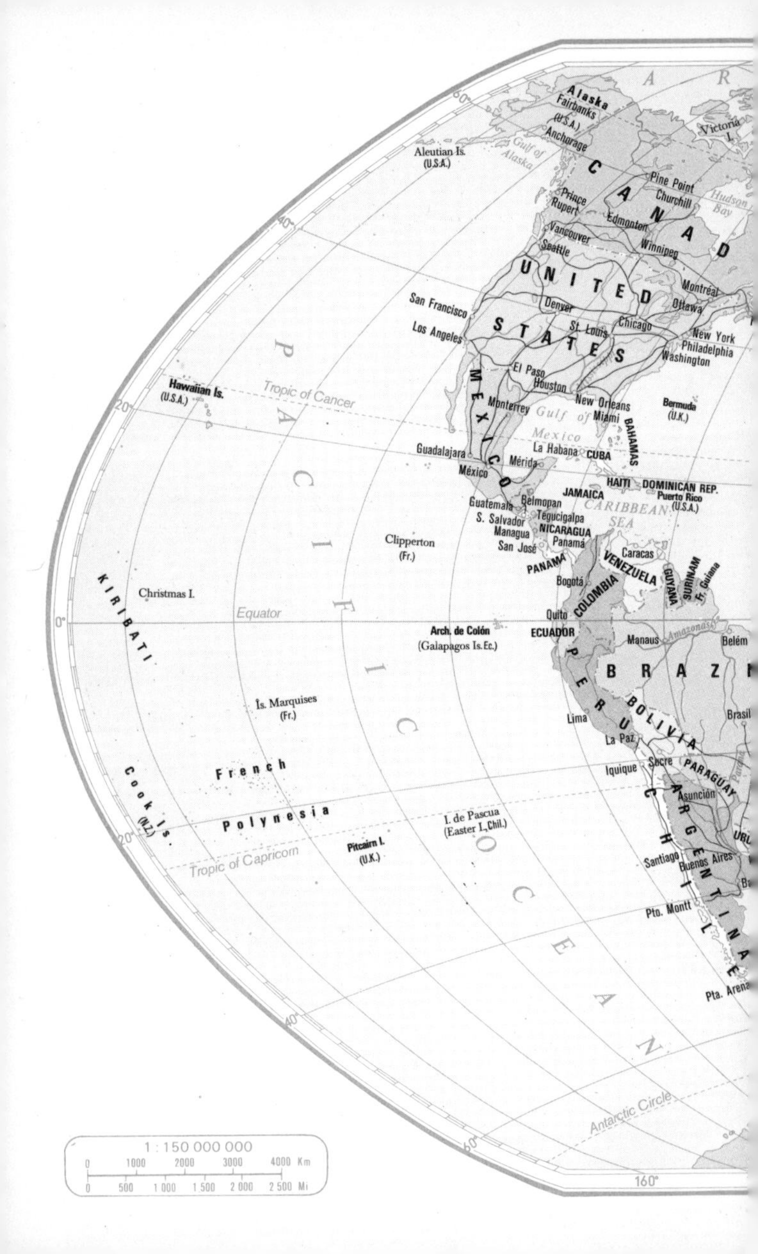

Alaska
Fairbanks
(U.S.A.)
Anchorage
Aleutian Is.
(U.S.A.)
Gulf of Alaska
Victoria I.
CANADA
Pine Point
Churchill
Hudson Bay
Prince Rupert
Edmonton
Vancouver
Seattle
Winnipeg
UNITED STATES
Montréal
Ottawa
San Francisco
Denver
Chicago
Los Angeles
St. Louis
New York
Philadelphia
Washington
El Paso
Houston
MEXICO
Hawaiian Is.
(U.S.A.)
Tropic of Cancer
Monterrey
Gulf of Mexico
New Orleans
Miami
Bermuda
(U.K.)
BAHAMAS
PACIFIC OCEAN
Guadalajara
La Habana
CUBA
Mérida
México
HAITI
DOMINICAN REP.
JAMAICA
Puerto Rico
(U.S.A.)
Guatemala
Belmopan
CARIBBEAN SEA
S. Salvador
Tegucigalpa
Managua
NICARAGUA
Clipperton
(Fr.)
San José
Panamá
Caracas
PANAMA
VENEZUELA
GUYANA
SURINAM
Fr. Guiana
KIRIBATI
Bogotá
COLOMBIA
Christmas I.
Equator
Quito
Arch. de Colón
(Galapagos Is.Ec.)
ECUADOR
Manaus
Amazonas
Belém
BRAZIL
PERU
Îs. Marquises
(Fr.)
BOLIVIA
Brasil
Lima
La Paz
French Polynesia
Iquique
Sucre
PARAGUAY
Cook Is.
(N.Z.)
Asunción
CHILE
I. de Pascua
(Easter I., Chil.)
ARGENTINA
Pitcairn I.
(U.K.)
Santiago
Buenos Aires
Tropic of Capricorn
Pto. Montt
Pta. Arena
Antarctic Circle
60°
40°
20°
0°
160°
1 : 150 000 000
0 1000 2000 3000 4000 Km
0 500 1 000 1 500 2 000 2 500 Mi

The General Agreement on Tariffs and Trade (GATT; H.Q.: Genève) came into force in 1948. It deals with problems of international trade and lays down a code of conduct in international relations. 102 countries were associated in this organization at the beginning of 1975.

The World Council of Churches (office: Genève) was formally constituted in Amsterdam on 23 August 1948. In 1978, the Council had 293 members from more than 100 countries and territories.

MILITARY PACTS

NATO – the North Atlantic Treaty Organization (H.Q.: Bruxelles). The treaty was signed in Washington on 4 April 1949. A military and political grouping of western countries: Belgium, Canada, Denmark, France, Iceland, Italy, Luxembourg, the Netherlands, Norway, Portugal, the United Kingdom of Great Britain and Northern Ireland and the United States of America; Greece and Turkey since 1952, the Federal Republic of Germany since 1955. France withdrew from NATO in 1966 and Greece in 1974.

The Warsaw Pact (H.Q.: Moskva), a military and political union of the European socialist countries for defensive purposes, signed in Warszawa on 14 May 1955. Members: Bulgaria, Czechoslovakia, the German Democratic Republic, Hungary, Poland, Romania and U.S.S.R; Albania withdrew in 1962.

ANZUS – The Pacific Security Treaty, a military pact signed at San Francisco on 1 Sept. 1951 between Australia, New Zealand and the U.S.A.

ANZUK – a military treaty which came into being in 1971. Members: Australia, Malaysia, New Zealand, Singapore and the United Kingdom of Great Britain and Northern Ireland.

SEATO – the South East Asia Treaty Organization (H.Q.: Krung Thep) – a military and political treaty signed in 1954. Members: Australia, France, New Zealand, the Philippines, Thailand, the United Kingdom of Great Britain and Northern Ireland and the United States of America.

Largest cities of the World
(Population of the city proper)

Name (Country)	Population	(Year)
1. Shanghai (China)	11,500,000	(1977)
2. México (Mex.)	8,988,230	(1978)
3. Beijing (China)	8,600,000	(1977)
4. Tōkyō (Jap.)	8,392,425	(1976)
5. Moskva (U.S.S.R.)	7,911,000	(1978)
6. New York (U.S.A.)	7,420,000	(1978)
7. São Paulo (Braz.)	7,198,608	(1975)
8. London (U.K.)	7,028,200	(1976)
9. Tianjin (China)	7,000,000	(1977)
10. Sŏul (Rep. of Korea)	6,889,470	(1975)
11. Chongqing (China)	6,000,000	(1977)
12. Bombay (India)	5,970,575	(1971)
13. Al-Qāhirah (Egypt)	5,414,000	(1978)
14. Guangzhou (Canton, China)	5,000,000	(1977)
15. Rio de Janeiro (Braz.)	4,857,716	(1975)

Largest cities of the World
(Population of the urban agglomeration)

Name (Country)	Population	(Year)
1. New York (U.S.A.)	17,040,000	(1978)
2. México (Mex.)	13,993,866	(1978)
3. Tōkyō (Jap.)	11,683,613	(1976)
4. Shanghai (China)	11,500,000	(1977)
5. London (U.K.)	11,175,000	(1976)
6. Moskva (U.S.S.R.)	10,925,000	(1977)
7. São Paulo (Braz.)	10,000,000	(1975)
8. Paris (Fr.)	9,878,524	(1975)
9. Los Angeles (U.S.A.)	9,200,000	(1978)
10. Calcutta (India)	9,100,000	(1975)
11. Buenos Aires (Arg.)	8,925,000	(1974)
12. Sŏul (Rep. of Korea)	8,625,000	(1975)
13. Beijing (China)	8,600,000	(1977)
14. Al-Qāhirah (Egypt)	8,539,000	(1978)
15. Rio de Janeiro (Braz.)	8,300,000	(1975)

Kingdom, Denmark and Ireland. **The European Parliament** consists of 410 members who are elected directly within each of the 9 member countries (seat: Strasbourg and Luxembourg). **The European Economic Community** (**EEC** or "the Common Market"; H.Q.; Bruxelles), established on 25 March 1957 in Rome with the task of achieving a customs union and to co-ordinate the economic policies of member states. 52 independent developing countries of Africa, the Caribbean and the Pacific are affiliated to the Community. – **The European Coal and Steel Community** (**ECSC;** H.Q.: Luxembourg) established on 18 April 1951 in Paris with a task of contributing to the economic development and rising standard of living in member countries by establishing a common market for coal, coke, iron ore and steel. – **The European Atomic Energy Community** (**EAEC,** "Euratom"; H.Q.: Bruxelles) founded in Rome on 25 March 1957 to promote the nuclear energy industry and nuclear research.

The European Free Trade Association (**EFTA;** H.Q.: Genève) was set up in Stockholm on 3 May 1960. Its task was to eliminate tariffs and quantitative restrictions on the import and export of goods between member countries. Members: Norway, Sweden, Switzerland, Iceland, Portugal and Austria; associated member Finland since 1961.

The Organization for Economic Co-operation and Development (**OECD;** H.Q.: Paris) was set up on 30 Sept. 1961 to further economic growth rises in the standard of living, to contribute to economic co-operation between member and non-member countries and to contribute to the expansion of world trade. Member countries are: Australia, Austria, Belgium, Canada, Denmark, the Federal Republic of Germany, Finland, France, Greece, Iceland, Ireland, Italy, Japan, Luxembourg, the Netherlands, New Zealand, Norway, Portugal, Spain, Sweden, Switzerland, Turkey, the United Kingdom of Great Britain and Northern Ireland and the United States of America; associated member Yugoslavia.

The Council for Mutual Economic Assistance (**CMEA;** H.Q.: Moskva) was established on 8 Jan. 1949 by the socialist countries to coordinate economic planning, the implementation of joint provisions for the expansion of industry and agriculture on the basis of an international socialist division of labour, specialization and co-operation in production, the construction of fuel and power and transport systems and the exchange of scientific and technological knowledge. This international socialist organization is based on the principle of a voluntary choice of criteria and on mutual collaboration. Member countries: Bulgaria, Czechoslovakia, Hungary, Poland, Romania, the Union of Soviet Socialist Republics; since 1950 – the German Democratic Republic, 1962 – Mongolia, 1972 – Cuba and 1978 – Vietnam (10 members). Albania withdrew in 1961. Observers: Yugoslavia, the Democratic People's Republic of Korea, Laos, Angola; China has not taken part since 1961. Special Status of co-operation: Finland, Iraq, Mexico. **The International Bank for Economic Cooperation** (H.Q.: Moskva) came into being in 1963; it is used by member countries of CMEA for accountancy purposes in trade and economic development.

The Organization of American States (**OAS,** formerly "Pan-American Union"; secretariat: Washington) was formed at a conference in Bogotá on 30 April 1948 to work towards international peace, to promote American solidarity and to coordinate economic, social, scientific, technological and cultural policies. Membership: 27 independent countries (Cuba was expelled in 1962 after U.S. intervention). **The International Bank of Development** was established in 1960 by 25 Latin American countries (excluding Cuba and Guyana), Canada and the United States of America.

The Latin American Free Trade Association (**ALALC = LAFTA;** secretariat Montevideo) has aimed at removing trade restrictions and tariffs since 1961. Membership: Argentina, Brazil, Chile, Colombia, Ecuador, Mexico, Paraguay, Peru, Uruguay; Venezuela since 1966 and Bolivia since 1967. **The Andian Group** (Grupo Andino) is an association of Bolivia, Chile, Colombia, Ecuador and Peru since 1969; Venezuela is an observer. Its aims are economic control and the regulation of foreign investments. **"The Agreement of La Plata Countries"** – signed by Argentina, Bolivia, Brazil, Paraguay, Uruguay. **The Latin American Economic System** (**SELA;** secretariat: Caracas) came into being in Panamá (1975) – it is an inter-Governmental organization of 25 Latin American countries, including Cuba.

The Caribbean Community (**CARICOM**) and the **Caribbean Common Market (CCM)** – 14 small island countries in the Caribbean (incl. Belize and Guyana) have been associated since 1973 for the purposes of coordinating foreign trade and economic development and restricting the influence of foreign business in the economic life of member countries. H.Q.: Georgetown.

The Organization of African Unity (**OAU;** H.Q.: Addis Abeba) was established on 20 May 1963. Its chief aims are the furtherance of African unity and solidarity, the coordination of the political, economic, cultural, health, scientific and defence policies; the elimination of colonialism in Africa, common defence and the independence of the member countries as well as the development of international collaboration according to the Charter of the United Nations. There were 45 members in 1975. **The African Bank of Development** (H.Q.: Abidjan) was founded in 1963 to contribute to economic and social development of member countries.

The Arab League (H.Q.: formerly Cairo, now Tunis) was founded in Cairo on 22 March 1945 with the purpose of promoting collaboration and unity in political, economic, military, financial and cultural matters. Original members: Egypt, Iraq, Jordan, Lebanon, Saudi Arabia, Syria, Yemen. There were 22 Arab member countries, and the Palestine Liberation Organization, in 1978. **The Arab Common Market** was founded in 1965.

The Asian and Pacific Council (**ASPAC;** H.Q.: Canberra) is a regional political and economic body. Members: Australia, Japan, New Zealand, the Philippines, the Republic of Korea and Taiwan; observers: Indonesia and Laos; Malaysia withdrew in 1973.

The Association of South East Asian Nations (**ASEAN;** H.Q.: Krung Thep), came into being in 1967 to stimulate the economic, social and cultural development of this region. Members: Malaysia, Indonesia, the Philippines, Singapore and Thailand.

The Organization of Petroleum Exporting Countries (**OPEC;** H.Q.: Wien), was established in 1960 as an association of the developing countries that extract and export petroleum to protect their interests particularly against the international oil monopolies. Members: Algeria, Gabon, Indonesia, Iran, Iraq, Kuwait, Libya, Nigeria, Qatar, Saudi Arabia, Syria, the United Arab Emirates and Venezuela.

EUROPE

Europe lies in the temperate belt of the northern hemisphere, and although it is connected to the western side of Asia it is regarded as a separate continent for both historical and cultural reasons. Its name derives from the Accadian word "ereb", meaning "evening twilight or sunset", i.e "the land where the sun sets", and the ancient Greeks adapted this Semitic word to the form in which it has come down to us.

Europe covers an **area of 10,527,000 sq.km**, which is 7.04% of the world's land surface (incl. the European part of the U.S.S.R.). It has **628 mn. inhabitants** (1978) and a population density of 60 persons per sq.km. **Geographical position:** northernmost point of the mainland: cape Nordkinn in Norway 71°08′ N.Lat. (of the entire continent: cape Mys Fligeli on island O. Rudol'fa in Zeml'a Franca Iosifa (Fr. Joseph Land) 81°51′ N.Lat.); southernmost point: Punta de Tarifa in Spain in the Strait of Gibraltar 35°59′ N.Lat. (island Gávdhos off the southern coast of Crete 34°48′ N.Lat.); westernmost point: Cabo da Roca in Portugal 9°29′ W.Long (Tearaght I. off the west coast of Ireland 10°39′ W.Long.); easternmost point: the eastern foothills of the Pol'arnyj Ural in the U.S.S.R. 67°20′ E.Long. The eastern continental boundary between Europe and Asia is over 3,500 km long and leads along the eastern foothills of the Ural Mts. (Ural'skije Gory), further along the river Emba to the Caspian Sea, from there along the Kuma-Manyč Depression (along the rivers Kuma and Manyč) to the mouth of the river Don on the Sea of Azov (Azovskoje More).

Europe's coastline, the longest of all the continents, is extremely varied and measures 37,900 km (excl. the coastlines of all large and small offshore islands). Largest peninsulas: Scandinavia (area 762,500 sq.km), Iberian (581,000 sq.km), Balkan (496,700 sq.km), Appennine (251,000 sq.km). Islands are situated mainly to the North-West, where the largest are the British Isles (area 315,000 sq.km) and Iceland (102,829 sq.km); further to the North Novaja Zeml'a (82,180 sq.km) and Svalbard (62,050 sq.km), and in the South in the Mediterranean Sea there are Sicily (25,426 sq.km) and Sardinia (23,813 sq.km).

The land surface of Europe varies considerably, although it has the lowest average height of all the continents. Plains rising no higher than 200 m make up 57% of the total area of the continent. The distribution of lowland and mountain regions is determined by both geological structure and geomorphological evolution. The oldest core of the continent is the lowlands of Eastern Europe ("Fennosarmatia") with the vast East European Plain. The lowest point is the Caspian Sea (−28 m below sea level). To the East rises the longest mountain system, the Ural Mountains (Ural'skije Gory, 1,894 m) stretching over 2,500 km. The Scandinavian mountains and their numerous glaciers date from the Caledonian age with typical deep fjords. The Central European Plain, which bears the marks of glaciation, and the German-Bohemian ranges (1,602 m) date from the Hercynian age, like the French Massif Central (1,886 m). The highest and most typical European mountains, the Alps (Mt. Blanc, 4,810 m), arose during the Tertiary folding as did the Carpathian Mts. (2,655 m) and the Dinara (2,751 m). Most of the Southern European peninsulas are mountainous, the highest ranges (of Tertiary age) are the Pyrenees (3,404 m) and the Appennines (2,914 m). This is a volcanic region (Etna 3,340m) with earthquakes.

Europe's relatively moist climate has resulted in a dense **network of rivers**. The mean annual discharge is 2,560 cub.km and roughly 80% of its rivers drain into the marginal seas of the Atlantic Ocean. About 20% drain into the Caspian Sea, which has no outlet and is fed by Europe's longest river, the Volga (length 3,531 km, river basin 1,360,000 sq.km, mean annual flow 8,220 cub.m per sec.).

The Eastern European rivers are at their fullest in spring and early summer and at their lowest in autumn and winter. The Central and Northern European rivers are at their fullest in spring and their lowest in early autumn. The Western European rivers, however, are at their greatest flow in winter, and at their smallest in the summer. The Southern European rivers are filled mostly by rain water in winter and some of them dry up completely in summer. The rivers of the high mountain regions have their maximum discharge (as much as 85%) in summer. The majority of **European lakes** are of glacial origin and the largest is Lake Ladoga (Ladožskoje Oz., 18,400 sq.km).

Europe spreads across 4 **climatic belts** in the Northern hemisphere. In the Far North there is the Arctic and Subarctic zone (Svalbard, Novaja Zeml'a and the northern shores of the U.S.S.R.). The main part of the continent lies in the belt of temperate climate. Western Europe has the oceanic type with prevailing westerly winds (mild winters and summers); Central Europe has a transitional climate (the summer is warm with precipitation, the winter has permanent snow cover) and Eastern Europe has the continental type (long cold winters and hot summers). Highest absolute temperature: Sevilla (Spain) 47.8°C; lowest: Pustozersk (U.S.S.R.) −55.2°C; maximum rainfall Crkvice (Yugoslavia) 4,624 mm a year; minimum Almeria (Spain) 218 mm.

Mean January and July temperatures in °C (annual rainfall in mm) are as follows: Vardø −5.0 and 9.1 (597), Reykjavík −1.2 and 10.9 (870), Bergen 1.7 and 16.3 (2,002), Stockholm −2.9 and 17.8 (555), Perm' −15.1 and 18.1 (570), Aberdeen 3.3 and 13.5 (748), Hamburg −0.3 and 17.0 (734), Warszawa −3.6 and 18.9 (531), Moskva −10.5 and 18.3 (694), London 4.2 and 17.0 (612), Paris 3.6 and 19.3 (645), Davos −7.0 and 12.1 (959), Praha −0.8 and 19.9 (491), Budapest −2.3 and 20.9 (647), Genova 7.5 and 24.4 (1184), Dubrovnik 9.0 and 24.5 (1391), Madrid 4.4 and 23.6 (419), Valletta 12.8 and 25.5 (513), Athínai 8.8 and 26.5 (402).

The soil, **flora and fauna** vary with the climatic conditions. From North to South there are the following soil and vegetation zones: tundra (moss, lichen), taiga with coniferous trees, mixed broad-leaved and coniferous forest, broad-leaved woodland of the temperate belt, tree steppe, steppe (dry grasses), semi-desert (in the Caspian region), evergreen maquis and broad-leaved Mediterranean forest. Alpine flora varies according to the altitude at which it is growing. Europe has widespread cultivated steppe with fertile arable soil. The fauna includes reindeer, wolf, brown bear, lynx, fox, deer, hare, hedgehog, various species of birds, reptiles, amphibious animals, fishes and insects.

ATLANTIC OCEAN
NORWEGIAN SEA
NORTH SEA
Iceland
Reykjavik
Askja
1510
2119
1491
Hvannadalshnúkur
Denmark Str.
Rifstangi
Arctic Circle
3921
3970
3062
33
Faeroe Is.
Shetland Is.
Orkney Is.
Duncansby Hd.
I. of Lewis
Hebrides
Ben Nevis
1343
Grampian Mts.
Great Britain
Glasgow
North Chan.
Ireland
Shannon
IRISH SEA
Dublin
Slea Hd.
1041
978
Pennines
14
1085
Severn
St. George's Chan.
CELTIC SEA
Land's End
London
Thames
Str. of Dover
English Channel
Channel Is.
Brest
Pte. du Raz
Bretagne
Normandie
Le Havre
Paris
Seine
Loire
Nantes
Bay of Biscay
5080
5463
5100
5053
Trondheim
2472
Glittertinden
Bergen
Oslo
Glåma
Vänern
44
725
Lindesnes
Skagerrak
Kattegat
Göteborg
Jylland
Fyn
København
Sjælland
Kiel
Hamburg
Elbe
Berlin
Amsterdam
-5
Rotterdam
Central
Bruxelles
Maas
694
Ardennes
Rhein
Frankfurt a. M.
Main
1142
Praha
Bohemian
1456
Donau
Inn
Strasbourg
Rhine
1493
Plain of France
Massif Central
Puy de Sancy
1886
Jura
Bern
4274
4049
3797
Grossglockner
Alps
Lyon
4810
Mt. Blanc
Milano
Venezia
Trieste
Les Ecrins
4103
Rhône
Po
Genova
Marseille
LIGURIAN SEA
ADRIA
Appennini
Tevere
Corno Grande
2914
Roma
Vesuvi
Napoli
Mte. Cinto
2710
Corse (Corsica)
Str. of Bonifacio
Sardegna (Sardinia)
1834
Cagliari
3830
TYRRHENIAN SEA
G. du Lion
Bordeaux
Garonne
La Coruña
C. de Finisterre
2648
Cord. Cantábrica
Porto
Iberian Peninsula
Duero
Sist. Iberico
2316
Pico de Aneto
3404
Pyrenees
Ebro
Barcelona
Sist. Central
2592
Madrid
Tagus (Tajo)
C. da Roca
Lisboa
Guadiana
Sa. Morena
Guadalquivir
Valencia
Turia
Islas Baleares
Menorca
Ibiza
Mallorca
Sa. Nevada
3478
Mulhacén
Sevilla
Almeria
Pta. de Tarifa
Str. of Gibraltar
Tanger
MEDITERRANEAN
2887
Alger
2308
O. Chéliff
Atlas Tellien
R. Ben Sekka (C. Blanc)
Tunis
Plateau of the Shotts
Dj. Chélia
2328
3340
Messina
Etna
Sicilia (Sicily)
Str. of Sicily
Malta
Lofoten
1 : 25 000 000
0 200 400 600 Km
0 100 200 300 400 Mi
A B C D
1 2 3 4 5
60° 50° 40°
30° 20° 10° 0° 10°

BARENTS SEA
Nordkinn
Hammerfest
Murmansk
Kol'skij Pol.
Lappland
Inari
Oz. Imandra
BELOJE MORE
Češskaja Guba
Pol. Kanin
Timanskij Kraž
Pustozersk
Pečora
Narodnaja
Ural'skije Gory
Zapadno-Sibirskaja Nizmen.
Chanty-Mansijsk
Sos'va
Ob'
Mezen'
Archangel'sk
Sev. Dvina
Vyčegda
Kemijoki
Oulu
Oulujärvi
Bothnia
Finland
Vaasa
Oz. Vygozero
Onega
Onežskoje Oz.
Kotlas
Suchona
Kama
Kamskoje Vdchr.
Perm'
Sverdlovsk
Tavda
Tobol
Ladožskoje Oz.
Päijänne
Saimaa
Helsinki
Leningrad
G. of Finland
Tallinn
Neva
Svir'
Oz. Beloje
Rybinskoje Vdchr.
East European Plain
Vjatka
Belaja
Ufa
G. Jamantau
Ural
Gor'kovskoje Vdchr.
Gor'kij
Volga
Kazan'
Kujbyševskoje Vdchr.
Kujbyšev
Orenburg
Čudskoje Oz.
Volchov
Valdajskaja Vozvyšen.
Saaremaa
Rīga
Moskva
Oka
Zap. Dvina
Dnepr
Srednerusskaja Vozvyšennost'
Privolžskaja Vozvyšen.
Neman
Minsk
Narew
Warszawa
Polesje
Pripjat'
Desna
Don
Volgogradskoje Vdchr.
Ural
Prikaspijskaja Nizmen.
Emba
Gurjev
Volgograd
Char'kov
Kijev
Wisla
Donec
Cimlanskoje Vdchr.
Astrachan'
CASPIAN SEA
Carpathian Mts.
Rostov-n.-D.
Manyč
Juž. Bug
Dnestr
Prut
Tisza
Odessa
AZOVSKOJE MORE
Kuban'
Kuma
Terek
Krymskij Pol.
Kerč'
Bol'šoj Kavkaz
G. El'brus
G. Kazbek
Baku
Kura
Tbilisi
Oz. Sevan
Szeged
Moldoveanu
Carpaţii Mer.
Mureşul
Izmail
Wallachia
Bucureşti
Danube
M. Saryč
Jalta
BLACK SEA
Beograd
Morava
Stara Planina
Sofija
Botev
Burgas
Karadeniz Boğ.
Musala
Rhodope Mts.
Korab
Thessaloniki
İstanbul
SEA OF MARMARA
Sakarya
Ankara
Kizil
Çanakkale Boğ.
Baba Burnu
Asia Minor
Ólimbos
Pindhos
Lésvos
Tuz G.
AEGEAN SEA
Évvoia
İzmir
Athinai
Toros Dağ.
Pelopónnisos
Kikládhes
Dhodhekanisos
Ródhos
Cyprus
A. Tainaron
Kriti (Crete)
5a Gibraltar
1 : 2 500 000
SPAIN
San Roque
La Línea
Los Barrios
Algeciras
Facinas
Gibraltar
Gibraltar (U.K.)
Tarifa
Pta. de Tarifa
MEDITERRANEAN SEA
ATLANTIC OCEAN
Strait of Gibraltar
Benzu
Ksar-es-Seghir
Musa
Ceuta (Sp.)
Tanger
Restinga
R. Sabartil
MOROCCO
Mdiq
El-Borj
Regaia
Martil
Aakba
El Fendek
Tétouan

LONGEST RIVERS

Name	Length in km	River basin in sq.km
Volga	3,531	1,360,000
Danube /Donau, Dunaj, Duna, Dunav, Dunărea/	2,850	817,000
Ural	2,428	231,000
Dnepr	2,201	503,000
Kama	2,032	522,000
Don	1,870	423,000
Pečora	1,809	322,000
Oka	1,480	245,000
Belaja	1,420	142,000
V'atka	1,367	129,000
Dnestr	1,352	72,000
Rhine /Rhein, Rhin, Rijn/	1,326	224,400
Severnaja Dvina (-Suchona)	1,302	367,000
Desna	1,187	89,000
Labe /Elbe/	1,165	144,055
Vyčegda	1,070	123,000
Wisła	1,047	194,424
Loire	1,020	115,000
Zapadnaja Dvina	1,020	88,000
Tisa /Tisza/	997	157,000
Meuse /Mass/	950	49,000

LARGEST LAKES

Name	Area in sq.km	Greatest Depth in m	Altitude in m
Ladožskoje Oz.	18,400	225	4
Onežskoje Oz.	9,616	120	32
Vänern	5,585	93	44
Saimaa	4,400	58	76
Čudskoje Oz. -Pskovskoje Oz.	3,650	14	30
Vättern	1,912	120	88
IJselmeer	1,250	6	0
Oz. Vygozero	1,159	40	29
Mälaren	1,140	64	1
Oz. Beloje	1,125	11	110
Päijänne	1,090	93	78
Inari	1,000	60	114
Oz. Il'men'	982	11	18
Oulujärvi	980	38	124
Oz. Topozero	910	56	109
Kallavesi	900	102	85
Oz. Imandra	880	67	126
Pielinen	850	48	94
Balaton	591	11	106
L. Geneva	582	310	372

LARGEST ISLANDS

Name	Area in sq.km	Name	Area in sq.km	Name	Area in sq.km
Great Britain	216,325	Novaja Zeml'a (South I.)	33,275	Crete /Kriti/	8,259
Iceland	102,829	Sicily /Sicilia/	25,426	Sjælland	7,015
Ireland	83,849	Sardinia /Sardegna/	23,813	O. Kolgujev	5,250
Novaja Zeml'a (North I.)	48,905	Nordaustlandet	14,530	Évvoia	3,654
Vestspitzbergen	39,044	Corsica /Corse/	8,681	Mallorca	3,411

HIGHEST MOUNTAINS

Name (Country)	Height in m	Name (Country)	Height in m
Mont Blanc /Mte. Bianco/ (Fr.-It.)	4,810	La Meije (Fr.)	3,987
Dufourspitze /Mte. Rosa/ (Switz.-It.)	4,634	Eiger (Switz.)	3,970
Dom (Switz.)	4,545	Mt. Pelvoux (Fr.)	3,946
Weisshorn (Switz.)	4,506	Ortles (It.)	3,899
Matterhorn /Mte. Cervino/ (Switz.-It.)	4,478	Monte Viso (It.)	3,841
Dent Blanche (Switz.)	4,357	Grossglockner (Aust.)	3,797
Grand Combin (Switz.)	4,314	Wildspitze (Aust.)	3,774
Finsteraarhorn (Switz.)	4,274	Grossvenediger (Aust.)	3,674
Aletschhorn (Switz.)	4,195	Tödi (Switz.)	3,614
Jungfrau (Switz.)	4,158	Adamello (It.)	3,554
Les Ecrins (Fr.)	4,103	Mulhacén (Sp.)	3,478
Gran Paradiso (It.)	4,061	Pico de Aneto (Sp.)	3,404
Piz Bernina (Switz.)	4,049	Monte Perdido (Sp.)	3,355

ACTIVE VOLCANOES

Name (Country)	Altitude in m	Latest eruption
Etna (Sicilia, It.)	3,340	1979
Beerenberg (Jan Mayen, Nor.)	2,278	1971
Askja (Ice.)	1,510	1961
Hekla (Ice.)	1,491	1970
Vesuvio (It.)	1,277	1949
Pico Gorda (Azores, Port.)	1,021	1968
Stromboli (Ie. Eolie, It.)	926	1975

FAMOUS NATIONAL PARKS

Name (Country)	Area in sq.km
Pečoro-Ilyčskij Zap. (U.S.S.R.)	12,000
Sareks (Sweden)	5,350
Lake District N.P. (U.K.)	2,251
North Wales (Snowdonia) N.P. (U.K.)	2,176
Šumava-Böhmerwald (Czech.- F.R.Ger.)	1,860
Białowieski P.N. (Pol.- U.S.S.R.)	850
Gran Paradiso (It.)	800

EUROPE

Country	Area in sq.km	Population	Year	Density per sq.km
Albania	28,748	2,616,000	1977	91
Andorra	453	30,584	1977	68
Austria	83,853	7,512,000	1978	90
Belgium	30,521	9,837,113	1978	322
Bulgaria	110,912	8,822,000	1977	80
Channel Islands (U.K.)	194	126,000	1977	649
Czechoslovakia	127,876	15,184,323	1979	119
Denmark	43,075	5,105,423	1978	119
Faeroe Islands (Den.)	1,399	41,575	1977	30
Finland	337,032	4,760,000	1979	14
France	543,965	53,241,000	1978	98
German Democratic Republic	108,178	16,757,857	1978	155
Germany, Federal Republic of	248,097	59,398,000	1978	239
Gibraltar (U.K.)	6	28,275	1977	4,713
Greece	131,944	9,284,000	1977	70
Hungary	93,032	10,698,000	1978	115
Iceland	102,829	222,000	1977	2
Ireland	70,282	3,221,000	1978	46
Isle of Man (U.K.)	588	61,000	1977	104
Italy	301,260	56,696,000	1978	188
Liechtenstein	157	25,000	1978	159
Luxembourg	2,586	356,000	1977	138
Malta	316	336,000	1978	1,063
Monaco	1.8	25,000	1978	13,889
Netherlands	41,160	13,936,000	1978	339
Norway	323,920	4,051,000	1977	13
Poland	312,683	35,032,000	1978	112
Portugal	91,631	9,786,000	1977	107
Romania	237,500	21,953,000	1979	92
San Marino	60.6	20,520	1977	339
Spain (inc. Canary Is.)	504,783	37,109,000	1978	74
Svalbard (Nor.)	62,422	3,500	1976	0.06
Sweden	449,964	8,284,000	1978	18
Switzerland	41,293	6,327,000	1977	153
Turkey – European part	23,764	3,754,000	1977	158
Union of Soviet Socialist Republics	22,274,900	260,040,000	1978	12
U.S.S.R. – European part	5,443,900	191,241,000	1977	35
United Kingdom	244,023	56,010,000	1977	230
Vatican City, State of	0.44	723	1977	1,643
West Berlin	480	1,927,000	1978	4,015
Yugoslavia	255,804	22,014,000	1978	86

15.1% of the world's population live in Europe, **628 million people,** with a population density of 60 persons per sq.km (1978), which makes it is the most densely inhabited continent. The most densely populated countries (leaving aside small countries like Monaco, Gibraltar, West Berlin and the Vatican City) are the Netherlands, Belgium, the Federal Republic of Germany, the United Kingdom and Italy. The Scandinavian countries, however, have a low population density. Europe has a low rate of **population growth – 0.60%** in the years 1970–75, birth rate 16.1 per 1,000, death rate 10.4 per 1,000. In the years 1970–75 the average life expectancy was 71.2 years.

European languages are of three main Indo-European types: Slavonic, Germanic, and Romance. But there are other language groups: Celtic, Semitic, Finno-Ugric, etc. The main European nationalities are the Russians, Germans, Italians, English, French, Ukrainians, Spanish and Poles.

Conurbations are typical features of all industrial areas, in particular in Central England, the Netherlands, the Rhine and Ruhr regions of the Federal Republic of Germany, Saxony in the German Democratic Republic, Upper Silesia in Poland, the Donbas region of the U.S.S.R. and the Po Valley in Italy. Europe has 33 cities with a population over one million and 67.6% of all inhabitants live in towns.

Economically Europe is the world's most advanced continent. It looks back on a tradition of industrial development and intensive plant and livestock farming. The extraction of coal, oil, natural gas, iron ore, bauxite, nickel, mercury, magnesite, phosphates and potassium salts is of world-wide importance. But Europe's main and most important industry is engineering. Europe has a highly developed transport network and is the continent with the greatest volume of foreign trade.

ICELAND
Akureyri
Reykjavík
Seyðisfjördur
Arctic Circle
NORWEGIAN SEA
ATLANTIC OCEAN
Thorshavn
Faeroe Is. (Den.)
Shetland Is. (U. K.)
Orkney Is. (U. K.)
Hebrides
Thurso
UNITED KINGDOM
Aberdeen
GLASGOW
Edinburgh
Londonderry
Belfast
Newcastle u. Tyne
LEEDS
Kingston u. Hull
LIVERPOOL
Manchester
BIRMINGHAM
Cardiff
Bristol
LONDON
Plymouth
Portsmouth
Dover
IRISH SEA
CELTIC SEA
English Ch.
NORTH SEA
IRELAND
DUBLIN
Sligo
Galway
Limerick
Cork
Channel Is. (U. K.)
Brest
Caen
Le Havre
Calais
PARIS
Reims
Le Mans
St.-Nazaire
Nantes
Orléans
La Rochelle
Bay of Biscay
Bordeaux
Limoges
Lyon
Dijon
Toulouse
MARSEILLE
Toulon
Nice
FRANCE
Strasbourg
Genève (Geneva)
Bern
Zürich
SWITZERLAND
LIECHT.
Vaduz
MONACO
Monaco
ANDORRA
Andorra
G. du Lion
SPAIN
La Coruña
Gijón
Vigo
León
Bilbao
Porto
Valladolid
Coimbra
LISBOA (Lisbon)
PORTUGAL
MADRID
ZARAGOZA
BARCELONA
VALENCIA
Duero
Tajo
Guadiana
Ebro
Turia
Guadalquivir
SEVILLA
Córdoba
Faro
Cádiz
Málaga
Almeria
Cartagena
Alicante
Str. of Gibraltar
Gibraltar (U. K.)
Ceuta (Sp.)
Tanger
Islas Baleares
Menorca
Mallorca
Palma
Ibiza
MEDITERRANEAN
Corse (Fr.)
Bastia
Ajaccio
Sassari
Sardegna (It.)
Cagliari
TYRRHENIAN SEA
NORWAY
Trondheim
Molde
Dombås
Hamar
Bergen
Oslo
Stavanger
Skien
Fredrikstad
Vänern
Skagerrak
Kattegat
Frederikshavn
Göteborg
Ålborg
Århus
DENMARK
KØBENHAVN
Esbjerg
Malmö
Kiel
HAMBURG
BREMEN
Rostock
AMSTERDAM
'S-GRAVENHAGE
NETHERLANDS
FEDERAL
REP.
OF GERMANY
GERMAN DEM. REP.
HANNOVER
Magdeburg
ESSEN
KÖLN (Cologne)
Bonn
LEIPZIG
DRESDEN
BRUXELLES (Brussel)
BELGIUM
LUXEMBOURG
Luxembourg
FRANKFURT a. M.
Nürnberg
STUTTGART
MÜNCHEN (Munich)
Plzeň
Linz
Innsbruck
Donau
Rhein
Seine
Loire
Rhône
Garonne
TORINO (Turin)
MILANO
Venezia (Venice)
Udine
Trieste
Pula
GENOVA (Genoa)
Bologna
SAN MARINO
San Marino
Livorno
Firenze
ITALY
VATICAN C.
ROMA (Rome)
NAPOLI (Naples)
PALERMO
Sicilia
MALTA
ALGER
Bejaïa
Annaba
Binzert
Constantine
TUNIS
TUNISIA
ALGERIA
Djelfa
Biskra
Sfax
MOROCCO
Safi
Marrakech
1 : 25 000 000
0 200 400 600 Km
0 100 200 300 400 Mi

BARENTS SEA
7
8
9
10
11
12
30°
40°
50°
60°
70°
60°
50°
40°
20°
Pečenga
Murmansk
Kirovsk
Apatity
Ponoj
Kandalakša
Kemijoki
Muonio
FINLAND
BELOJE MORE
Češskaja G.
Narjan-Mar
Pečora
Mezen'
Uchta
Ber'ozovo
Sergino
Chanty-Mansijsk
Irtyš
Tobol'sk
Tavda
Polunočnoje
Ivdel'
Serov
Archangel'sk
Kemi
Taivalkoski
Kalevala
Kem'
Luleå
Oulu
Onega
Belomorsk
Sev. Dvina
V'yčegda
Syktyvkar
Kotlas
Solikamsk
Berezniki
Kama
T'umen'
Šadrinsk
Iisalmi
Medvežjegorsk
Onežskoje Oz.
Joensuu
Vaasa
Kuopio
Mikkeli
Petrozavodsk
Konoša
Suchona
PERM'
SVERDLOVSK
ČEL'ABINSK
Pori
Tampere
Turku
Kotka
Vyborg
Helsinki
Ladožskoje Oz.
Volchov
LENINGRAD
Belozersk
Vologda
Rybinskoje Vdchr.
Kostroma
Šarja
Kirov
Joškar-Ola
KAZAN'
IŽEVSK
Naberežnyje-Čelny
UFA
Magnitogorsk
Sibaj
Tallinn
Narva
Novgorod
Rybinsk
JAROSLAVL'
Ivanovo
Gor'kovskoje Vdchr.
GOR'KIJ
Volga
Kujbyševskoje Vdchr.
Ul'janovsk
Sterlitamak
Išimbaj
Ural
Pärnu
Pskov
Vladimir
Kalinin
KUJBYŠEV
Buzuluk
Orenburg
Orsk
UNION OF SOVIET SOCIALIST REPUBLICS
RIGA
Daugavpils
MOSKVA (Moscow)
R'azan'
Saransk
Penza
Syzran'
Saratovskoje Vdchr.
Ural'sk
Akt'ubinsk
Šiauliai
Polock
Vitebsk
Smolensk
Zap. Dvina
TULA
Oka
Mičurinsk
Tambov
Vol'sk
SARATOV
Aleksandrov Gaj
Neman
Kaunas
Vilnius
MINSK
Mogil'ov
Br'ansk
Or'ol
VORONEŽ
Volgogradskoje Vdchr.
Kaliningrad
Olsztyn
Grodno
Białystok
Soligorsk
Gomel'
Desna
Kursk
Makat
Emba
WARSZAWA (Warsaw)
Brest
Pinsk
Pripjat'
Don
VOLGOGRAD
Gurjev
Lublin
Korosten'
KIJEV
CHAR'KOV
Poltava
Donec
Cimľanskoje Vdchr.
Volga
CASPIAN SEA
KRAKÓW
L'VOV
Ternopol'
Vinnica
Dnestr
DNEPROPETROVSK
Vorošilovgrad
DONECK
ROSTOV-n.-D.
Astrachan'
Fort-Ševčenko
Ševčenko
Košice
Užgorod
Černovcy
ZAPOROŽJE
Nikolajev
Melitopol'
ŽDANOV
Kuma
Miskolc
Debrecen
KIŠIN'OV
ODESSA
Cherson
AZOVSKOJE M.
KRASNODAR
Stavropol'
Cluj-Napoca
Iaşi
Prut
Dnepr
Kerč'
Armavir
Groznyj
Terek
Machačkala
Szeged
Oradea
ROMANIA
Galaţi
Izmail
Simferopol'
Novorossijsk
Kislovodsk
Nal'čik
Ordžonikidze
Timişoara
Ploieşti
Sevastopol'
Jalta
BLACK SEA
Suchumi
Kutaisi
TBILISI
Kirovabad
BAKU
BEOGRAD (Belgrade)
Craiova
BUCUREŞTI (Bucharest)
Constanţa
Vidin
Dunărea
Kruševac
Pleven
Ruse
Varna
Niš
BULGARIA
Burgas
SOFIJA (Sofia)
Plovdiv
ISTANBUL
Skopje
ALBANIA
GREECE
Alexandroúpolis
TURKEY
Bursa
Thessaloníki
Balıkesir
Trikala
Lárisa
Lamía
AEGEAN SEA
ATHÍNAI (Athens)
Pátrai
İZMİR
Denizli
Kalámai
Ródhos
Iráklion
Kriti
6 a Channel Islands
1 : 2 500 000
C. de la Hague
Alderney
Casquets
St. Annes
English Channel
ATLANTIC OCEAN
3°
2°
49°
St. Sampson
Guernsey
St. Peter Port
Sark
Pas. de la Déroute
Channel Islands (U. K.)
St. Mary
Jersey
Rozel
St. Aubin
St. Helier
Roches Douvres
Barnouic
St.-Pierre-Église
Cherbourg
Barfleur
Diélette
Valognes
Bricquebec
Montebourg
FRANCE
St. Sauveur le V.
Barneville-Carteret
Carentan
La Haye-du-Puits
Lessay
Périers
Coutances
Montmartin-s.-M.
1
2
3
B
C

UNITED KINGDOM

The United Kingdom of Great Britain and Northern Ireland, 244,013 sq.km (with the Channel Islands and the Isle of Man 244,805 sq.km), **population 56,010,000** (mid-year 1978), **constitutional monarchy** (Queen Elizabeth II since 6 February 1952). – **Currency:** £ 1 = = 100 pence.

Geographical position. The United Kingdom occupies the major part of the British Isles off the north-west coast of Europe. It is separated from the continent by the North Sea and the English Channel, which, at its narrowest point, the Strait of Dover, is only 34 km wide. The largest island is Great Britain, 216,325 sq.km; it is separated by the Irish Sea from Ireland, the northern part of which belongs to the United Kingdom. The islands of the Outer and Inner Hebrides stretch along the indented west coast of Scotland. The Orkney Islands and the Shetland Islands form the northernmost part of the British Isles. The Isle of Wight, the Isles of Scilly and the Channel Islands are to be found off the south coast of England, the Isle of Man and Anglesey in the Irish Sea.

Geology. Up to the Pleistocene Era the British Isles formed part of the European continental plate. Its geological structure and topography developed in the same way as that of the continent. Caledonian and Hercynian foldings affected the islands in the Paleozoic Era. The Caledonian folding left its mark upon Scotland, the northern part of Ireland, England and Wales; the southern parts of the islands were shaped in the Hercynian movements with the hills running predominantly in a north-south direction. The old rocks have since been gradually eroded into rolling plains with hills of rounded shapes. The newer Alpine folding caused numerous faults, along which rift valleys and horsts have formed. The south-eastern part of Great Britain is taken up by the London Basin, formed of sediments and fringed by striking ridges of high ground. The Pleistocene ice sheets and ice caps covered the entire British Isles north of the Thames valley. To this day the land bears numerous marks of glacial activity: moraines, corries, glacial lakes, etc. The subsidence of the continent gave rise to many islands and river estuaries forming deep inlets along the coastline.

Hills and mountains. The most mountainous part of the United Kingdom is Scotland. The Scottish Highlands comprise the North West Highlands and the Grampian Mountains with Ben Nevis, 1,343 m, the highest mountain peak in the United Kingdom. The fertile Central Lowlands extend southward and are enclosed by the hilly Southern Uplands. The border between England and Scotland runs along the Cheviot Hills and the lower reaches of the River Tweed.

The main ranges in England – the Pennines (Cross Fell, 893 m) and the Cumbrian Mountains with Scafell Pikes, 978 m, England's highest peak – run in a north-south direction. The Cotswolds rise in the south-west and two rolling uplands, Dartmoor and Exmoor, are in the far south-west. The remaining land consists, for the most part, of either hilly or low-lying country, as in the Midlands. In Wales the highest mountain, at 1,085 m above sea-level, is Snowdon in the Cambrian Mountains. The highest point of Northern Ireland is Slieve Donard, 852 m.

Rivers. The British rivers are short with a strong flow of water throughout the year. The estuaries serve as natural harbours, and some ships have direct access to the hinterland on those rivers with reliable high tides. The sources of the rivers are divided by low watersheds, which facilitated the construction of a wide network of canals. The longest rivers are the Severn (354 km) and the Thames (338 km). Most of the **lakes** are of glacial origin and are to be found in large numbers in Scotland, the Lake District, and in Northern Ireland. Best known are Lough Neagh (396 sq.km), Lough Erne, Loch Lomond and Loch Ness.

Climate. The climate is oceanic with moderate temperatures and abundant precipitation. The temperature is modified in summer and in winter by the Gulf Stream flowing in a north-easterly direction along the west coast. The mean annual temperature in the period 1941–70 in England and Wales was 10.0°C (January 4.0°C, July 16.3°C), Scotland 8.7°C (3.5°C, 14.1°C), Northern Ireland 9.3°C (4.0°C, 14.7°C). The prevailing south-westerly winds bring moisture from the Atlantic Ocean. The long-term annual average precipitation (1941–70) was 912 mm in England and Wales, 1,431 mm in Scotland and 1,095 mm in Northern Ireland. Rainfall is at its highest in November and December and at its lowest in April.

Administrative units. The United Kingdom is composed of four countries, England, Scotland, Wales and Northern Ireland. The main pattern of local government is the division into counties – Scotland has had regions since 1975 – within which there are district authorities. The Channel Islands and the Isle of Man enjoy a degree of autonomy.

System of Government. The monarch (King or Queen) is the head of state, succession following the hereditary principle. The Sovereign is the head of the judiciary, the commander-in-chief of all the armed forces and the temporal 'governor' of the established Church of England. The United Kingdom is governed, in the name of the Queen, by Her Majesty's Government, headed by the Prime Minister. The Government exercises executive power and is responsible to Parliament for its activities.

The supreme organ of state is Parliament composed of the House of Commons and the House of Lords, a law-making body which controls the activities of the Government. All British citizens aged 18 and over have the franchise. General elections are held a minimum of every five years. One Member is returned to the House of Commons for each of the 635 constituencies on the principle of a simple majority.

The political system is based on two major parties, one forming the Government and the other the Opposition. The majority party, at present, is the Conservative Party (with Mrs M. Thatcher as Prime Minister) and the Opposition is formed by the Labour Party.

Population. The estimated population on 1 July 1977 was 55,852,000 persons, of which: 27,184,000 were male and 28,668,000 female. England had a population of 46,351,000, Wales 2,768,000, Scotland 5,196,000 and Northern Ireland 1,537,000. The United Kingdom has a high population density, 229 persons per sq.km, but they are distributed unevenly: England 356 persons per sq.km, Scotland 66, Wales 133 and Northern Ireland 109. More than 80% are English, 9% Scots, 2.6% Irish, 1.6% Welsh and 0.6% Gaels. Coloured immigrants make up roughly 3.5% of the total population. In 1975 89.1% of the population lived in towns. London, **the capital**, is the main administrative, economic, cultural and political centre of country.

There has been a steady decline in the birth rate from 18.8 per 1,000 in 1964 to 11.8 per 1,000 in 1977. Despite the decline in mortality (11.7 per 1,000 in 1977) the natural population increase of 0.1 per thousand is very low and a further decline is expected in coming years. The United Kingdom is a country with a very low rate of infant mortality, 14.1 per thousand (England and Wales 13.8 per thousand, Scotland 16.1 per thousand, Northern Ireland 17.2 per thousand) and a high average life expectancy. In the period 1974–76 men reached an average age of 69.4 years, women 75.6 years. Those aged 0–14 years make up 22.5% of the pópulation, the age group 15–64 years makes up 63.1% and those aged 65 years and over make up 14.4%. There is a considerable migration to and from the United Kingdom. Since the first half of the 20th century the immigration from Commonwealth countries been rising but is now regulated by several Immigration Acts. In 1977 162,000 people came to the U.K., but in the same year 208,700 persons emigrated, mainly to Commonwealth countries.

Major towns and cities in the United Kingdom
(1976, with agglomeration in brackets):

City	Population	Agglomeration	City	Population	Agglomeration
London	7,028,200	(11,175,000)	Sunderland	295,700	
Birmingham	1,058,800		Leicester	289,400	
Glasgow	856,012	(1,890,000)	Doncaster	286,500	
Leeds	744,500	(1,555,000)	Cardiff	281,500	(625,000)
Sheffield	588,000	(725,000)	Kingston upon Hull	276,600	
Liverpool	539,700	(1,575,000)	Wolverhampton	266,400	
Manchester	490,000	(2,835,000)	Plymouth	259,100	
Edinburgh	467,097	(650,000)	Stoke-on-Trent	256,200	
Bradford	458,900		Derby	213,700	
Bristol	416,300	(640,000)	Southampton	213,700	
Belfast	363,000		Aberdeen	209,831	
Coventry	336,800	(650,000)	Portsmouth	198,500	(500,000)
Newcastle upon Tyne	295,800				

The official language is English, together with Welsh in Wales. The Scottish form of Gaelic survives in Scotland, while in Northern Ireland a few people still speak Irish Gaelic.

English is the official language in the Channel Is., with the exception of Jersey, where Norman French (patois) is widely spoken. Manx, the language used on official occasions on the Isle of Man, belongs to the Celtic languages.

Education. School attendance is compulsory from the age of 5 to 16 and is free of charge. Primary schools are coeducational but secondary and independent schools are often single sex. The great majority of schools, attended by over 95% of school children, are publicly maintained. 24.3% of children between the age of 2 and 5 attend nurseries or play groups. There are about 38,000 primary and secondary schools attended by 11 mn. pupils with a teaching staff of 535,000. Over 900,000 students are taking university degree courses. Among the 45 universities are some of the oldest in the world (Oxford, Cambridge, St. Andrews, Aberdeen, Edinburgh and Glasgow). The universities are autonomous institutions with their own forms of government, but receive grants from the University Grants Committee.

Health Services. Medical care is provided free of charge under the National Health Service. In the United Kingdom there are over 56,000 medical practitioners, i.e. 1 doctor to approximately 1,000 inhabitants. 2,700 hospitals with half a million beds and 38,620 doctors provide hospital treatment. It is also possible to obtain private medical care.

Religion. The majority of the population are nominally members of the established Church of England or the Church of Scotland or of the Free Churches. There are at present about 5.5 mn. Roman Catholics in the United Kingdom, mainly in England and Wales.

The economy (data for 1977). The United Kingdom is one of the most highly developed countries in the world. It holds an important position in the fields of international trade, seafaring, banking, and insurance. Apart from industry and trade, important sources of revenue derive from foreign investments, air and sea transport, financial, insurance and banking transactions and tourism. Important industries such as steel and the energy industries (coal, gas, electricity, oil), rail, road haulage and air transport, the post office and the majority of ports are in public ownership. About 7% of the economically active population are employed in the public sector of industry and produce roughly 10% of the gross national product (GNP). The economically active population numbers 26,327,000 persons, i.e. 48% of all inhabitants. 1.6% of these are engaged in agriculture, forestry and fishery, 37.9% are in industry, 52.8% are in services, the rest are made up of the armed forces, employers, the self-employed and the unemployed. 40.7% of all women go out to work. **Industry.** The United Kingdom was the first country to develop factory production, and industry is the most important sector of the British economy. Together with the construction and fuel and power industries it produces 41.6% of the gross national product and employs 37.9% of the economically active population. 81.5% of exports are industrial products. The most important industry is manufacturing, which contributes 28.6% to the GNP. In recent years the most successful industries have been engineering, chemical, coal and oil processing.

Mining (1977). The United Kingdom suffers a shortage of mineral raw materials with the exception of coal. The main coal mining regions are in Central and North-west England (Lancashire, South and West Yorkshire, Derbyshire, Nottinghamshire, Durham, Northumberland), Central Scotland (Clydeside) and South Wales (Bristol Channel). The run-down of the coal mining industry – 120.6 mn. tons – is being reversed with heavy investments. As a consequence of the world energy crisis there has been increasing local production of crude petroleum, 37.9 mn. tons and natural gas, 40.5 billion cub.m, chiefly in the North Sea oil fields (oil: Argyll, Forties, Piper; natural gas: Frigg, Leman Bank, West Sole, Hewett, Indefatigable, Viking). Total crude petroleum reserves in the British Continental Shelf are estimated at 4,500 mn. tons. Ore mining, once of considerable importance, is steadily decreasing. Mining of iron ore produces 973,440 tons (Fe metal content) in Cumbria, Lancashire and Staffordshire, tin ore 2,900 tons (Sn metal content) in Cornwall and lead ore 10,400 tons (Pb metal content) in Clwyd, Durham and Derbyshire. Salt is mined in Cheshire, potash in Cleveland, as well as china and potter's clay.

Industrial production (1977). **Metallurgy.** The United Kingdom is one of the world's largest steel producing nations. Output of crude steel totalled 20.4 mn. tons and pig iron 12.2 mn. tons. The biggest steel works are run by the British Steel Corporation; the main centres of steel production are Sheffield, Swansea, Port Talbot, Teesside, the counties of Lincolnshire, Lancashire, Cheshire and the Greater Glasgow region. The British non-ferrous metal processing and fabricating industry is also very important: virgin aluminium 348,900 tons and 200,800 tons of secondary metal

Shetland Is.
Yell
Unst
Fetlar
Ronas H.
450
St. Magnus B.
Mainland
Sumburgh Hd.
Fair I.
Westray
Rousay
Sanday
Stronsay
Mainland
477
Hoy
Orkney Islands
S. Ronaldsay
Pentland Firth
Dunnet Hd.
Duncansby Hd.
Thurso
North Rona
C. Wrath
Butt of Lewis
The Minch
Lewis
799
Outer Hebrides
Sd. of Harris
Little Minch
North Uist
Benbecula
South Uist
Barra
Barra Hd.
St. Kilda
Skye
1009
Cullin Hs.
Raasay
Carn Eige
1182
Ben More
998
L. Shin
Dornoch Firth
Moray Firth
L. Maree
1045
Ben Wyvis
North West Highlands
Beauly
Inverness
Loch Ness
Glen More
Oich
Spey
Deveron
Don
Ben Macdhui
1311
Dee
Grampian Mts.
1148
Lochnagar
Aberdeen
Kinnairds Hd.
Buchan Ness
Ben Nevis
1343
L. Linnhe
L. Rannoch
Tay
Strathmore
Dundee
F. of Tay
Ben Lawers
1214
L. Tay
Ochil Hills
Firth of Forth
Inner Hebrides
Rhum
Eigg
Coll
Tiree
Passage of Tiree
Mull
966
Ben More
F. of Lorn
Colonsay
Jura
Sd. of Jura
Islay
Gigha I.
874
L. Awe
Ben Lomond
974
Loch Lomond
Glasgow
Clyde
Edinburgh
535
Lammermuir Hs.
Berwick-upon-Tweed
Tweed
Uplands
Hills
OCEAN
NORTH S
1159
1812
143
200
100
100
1
2
3
4
5
6
7
8
9
A
B
C
60°
58°
56°
12°
10°
8°
6°
4°
2°
0°
2°

ATLANTIC
Ireland
Central Plain
IRISH SEA
Great Britain
CELTIC SEA
English Channel
St. George's Channel
Bristol Channel
Strait of Dover
Erris Hd.
Mullet Pen.
Achill I.
Clare I.
Inishbofin
Nephin 807
Mweelrea 819
Donegal B.
Sligo B.
L. Conn
L. Mask
L. Corrib
Galway B.
Aran Is.
Loop Hd.
Shannon
Blasket Is.
953 Brandon Mt.
Slea Hd.
Dingle B.
Valencia I.
Carrauntoohil 1041
Kenmare B.
Caha Mts. 707
Bantry B.
Mizen Hd.
L. Allen
667
Iron Mts.
Lower L. Erne
Upper
Neagh
L. Ree
Suck
Boyne
Liffey
L. Derg
St. Bloom Mts. 529
Limerick
920
Galty Mts.
Blackwater
Suir
792
Waterford
Lee
Cork
Cork Hr.
Barrow
Nore
796
Mt. Leinster
Slaney
Wicklow Mts. 754
Lugnaquillia Mtn. 926
Dublin
Wexford Hr.
Carnsore Pt.
Belfast
Strangford L.
Mourne Mts. 852
St. Donard
Dundrum B.
Dundalk B.
139
Mull of Galloway
St. Bees Hd.
Snaefell 620
I. of Man
Scafell Pikes 978
Cumbrian Mts.
Whernside 737
Morecambe B.
Ribble
Middlesbrough
Tees
Vale of York
North York Moors 454
Flamborough Hd.
Ure
Wharfe
Derwent
Aire
Ouse
Kingston upon Hull
Leeds
Goole
Humber
Spurn Hd.
Formby Pt.
Gt. Ormes Hd.
Liverpool
Mersey
Dee
Cheshire Plain
The Peaks 636
Don
Trent
Sheffield
Lincoln
Witham
The Wash
Carmel Hd.
Anglesey
Menai Str.
Caernarvon B.
Snowdon 1085
Lleyn Pen.
Tremadoc B.
892
Cader Idris
752
Plynlimon
Cardigan Bay
Cambrian Mts.
Severn
Dove
407 The Wrekin
The Fens
Welland
Nene
Gt. Ouse
Breckland
Norwich
Yare
Lowestoft
Birmingham
Teme
Avon
Wye
Teifi
Tywi
578
St. David's Hd.
St. Brides B.
Milford Haven
Brecon Beacons 886
Swansea
Carmarthen B.
Cardiff
Cotswolds 330
Oxford
Thames
Chiltern Hs. 280
Lea
Cambridge
Stour
The Naze
London
North Foreland
Kennet
Bristol
297
Salisbury Plain
Leith Hill 294
North Downs
The Weald
Dover
Exmoor 526 Dunkery Beacon
Parrett
Hartland Pt.
Exe
Southampton
292
South Downs
Brighton
Beachy Hd.
Dungeness
Calais
Boulogne-s.-M.
100
128
High Willhays 621
Tamar
Bodmin Moor
Dartmoor
Plymouth
Lyme Bay
I. of Portland
The Needles
239 I. of Wight
Spithead
Start Pt.
252
Land's End
Falmouth B.
Is. of Scilly
Lizard Pt.
172
Alderney
C. de la Hague
Cherbourg
191
Cotentin
Channel Is.
Guernsey
Jersey
Baie de la Seine
Le Havre
Rouen
Seine
241
1 : 6 000 000
0 50 100 150 Km
0 20 40 60 80 100 Mi
54°
52°
50°
10° 8° 6° 4° 2° 0°
2 3 4 5 6 7 8
E F G

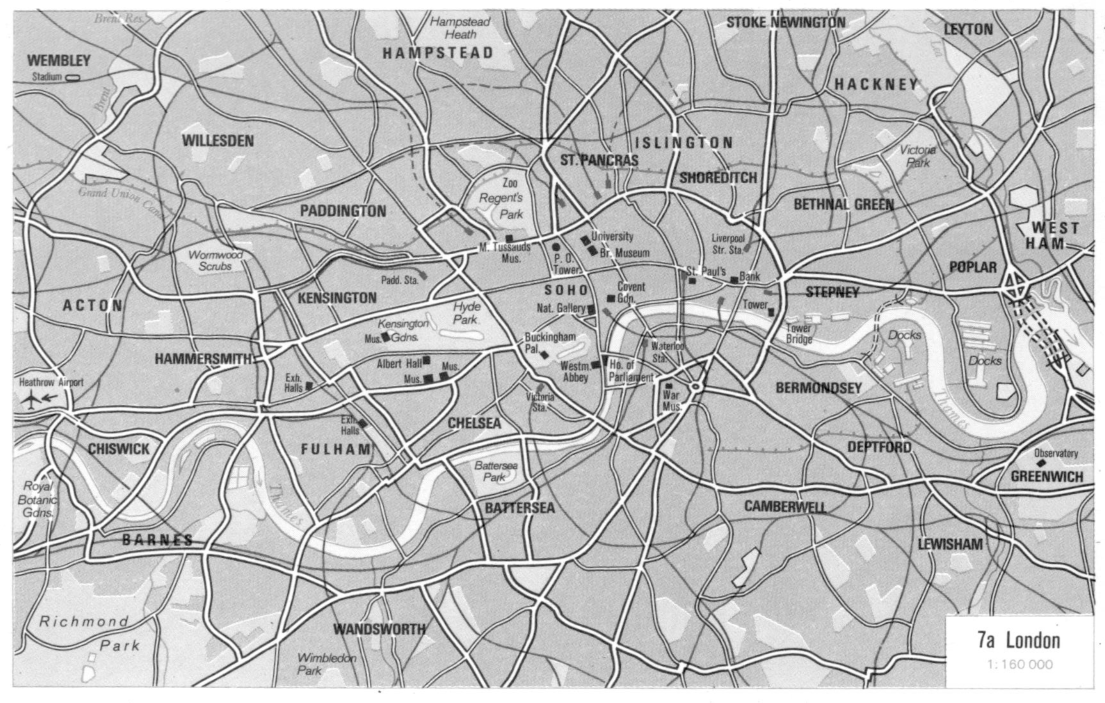
Brent Res.
Hampstead Heath
HAMPSTEAD
STOKE NEWINGTON
LEYTON
WEMBLEY
Stadium
Brent
Lea
HACKNEY
WILLESDEN
ISLINGTON
ST. PANCRAS
Victoria Park
SHOREDITCH
Zoo
Regent's Park
Grand Union Canal
PADDINGTON
BETHNAL GREEN
WEST HAM
University
Br. Museum
Liverpool Str. Sta.
M. Tussauds Mus.
P. O. Tower
Wormwood Scrubs
POPLAR
St. Paul's
Bank
Padd. Sta.
SOHO
Covent Gdn.
ACTON
KENSINGTON
STEPNEY
Tower
Nat. Gallery
Hyde Park
Kensington Gdns.
Mus.
Tower Bridge
Docks
Buckingham Pal.
Waterloo Sta.
Docks
HAMMERSMITH
Albert Hall
Mus.
Mus.
Westm. Abbey
Ho. of Parliament
Exh. Halls
Heathrow Airport
War Mus.
BERMONDSEY
Victoria Sta.
Thames
Exh. Halls
CHELSEA
CHISWICK
FULHAM
DEPTFORD
Observatory
GREENWICH
Battersea Park
Royal Botanic Gdns.
Thames
BATTERSEA
CAMBERWELL
BARNES
LEWISHAM
Richmond Park
WANDSWORTH
Wimbledon Park
7a London
1:160 000

(refineries at Kinlochleven, Foyers, Fort William, Dolgarrog, Resolven); refined virgin copper 44,400 tons, secondary metal 76,100 tons; lead 124,800 tons (Northfleet); zinc 81,500 tons (Avonmouth).

Engineering. The United Kingdom is an important producer of transport equipment (London, Birmingham, Coventry, Manchester, Nottingham, Wolverhampton, Glasgow, Newcastle upon Tyne) 1.32 mn. passenger cars, 398,268 commercial vehicles, 370,635 heavy commercial vehicles, 27,633 buses (of which 1,803 were double-deckers), 1.73 mn. bicycles, 104 ships over 100 GRT (shipyards on Clydeside, Tyneside, Wearside, Merseyside, Teesside and at Barrow-in-Furness), 261 civil and military aircraft (Yeovil, Bristol, Gloucester, Coventry, Luton, Derby, Manchester, Bedford). Engineering industries include the production of textile machinery (Bradford, Halifax, Keighley), industrial plants and electrical and electronic engineering (Manchester, Rugby, Newcastle upon Tyne, Stafford). In the **chemicals industry** (Birmingham, Glasgow, Newcastle upon Tyne, Wilton, Aberdeen) fast growth has been recorded in organic chemicals, the production of plastic materials 2.54 mn. tons and pharmaceutical chemicals. The production of sulphuric acid amounted to 3.4 mn. tons, phosphates 2.77 mn. tons, synthetic rubber 321,600 tons (Hythe).

The textile industry produces traditional wool 30,000 tons, woollen fabrics 150 mn. sq.m (London, Huddersfield, Halifax, Bradford), cotton fibres 97,200 tons and 370 mn. m cotton fabrics (Lancashire) and new man-made fibres 499 mn. m and fabrics (Wilton, Drighlington, Little Heath, Coventry). The linen industry is centred on Northern Ireland and along the east coast of Scotland (Dundee, Arbroath, Montrose). The Dundee area is the sole producer of jute products 43,680 tons. Macclesfield is the centre of the silk industry and Nottingham and Leicester produce hosiery and knitwear. Footwear production is centred on the Rossendale Valley, Leicester and Norwich 160.1 mn. pairs, of which 70.8 mn. are made of leather. The principal manufacturing centres of the clothing industry are London, Leeds and Manchester. The glass industry is centred in Glasgow, London, Birmingham and Sunderland. Cement production 15.46 mn. tons occurs along the Medway and the Humber.

The food, drink and tobacco industries have, in recent years, been subject to mergers amongst leading firms. Exports include 63,437,010 hl beer (London, Burton-upon-Trent, Birmingham, Edinburgh) and alcoholic spirits 4,563,400 hl mainly of Scotch whisky and gin. The tobacco industry (124.0 billion cigarettes – 1976) is centred in Bristol, Nottingham, Liverpool, London and Manchester. The cocoa, chocolate and sugar confectionary industry is located in Yorkshire, Bristol and Nottingham.

Energy. The output capacity of electric power stations is 78,911,000 kW (1975). Electric energy output 262.1 billion kWh (1977); 84% in thermal power stations, 14% in nuclear power stations, 2% in hydro-electric power stations. Of the total power output petroleum accounted for 40.4%, coal 36.3%, natural gas 18.5%, nuclear power 4.2% and hydro-electric power 0.6%. Thermal power stations are situated on the coalfields, in the vicinity of the big cities and along the lower reaches of the Trent, Aire, Calder, Thames. The largest nuclear power stations are Hinkley Point, Hunterston, Hartlepool and Heysham (each of 1,320 MW capacity). The largest refineries are situated at Shell Haven, Stanlow, the Isle of Grain, Coryton, Fawley, Llandarcy, Grangemouth, Milford Haven.

Agriculture (1977). Agriculture, forestry and fisheries provide about 2.8% of the GNP and occupy a mere 1.6% of the economically active population. The United Kingdom continues to be dependent on imported food, although home agriculture output is gradually increasing. 78.2% of the land is used for agricultural purposes, 63.6% of this is under crops and 40.6% of this area is arable land, i.e. 20.2% of the total area of the United Kingdom. Agriculture specializes in **animal husbandry,** cattle breeding 13,854,000 head (of this 4,366,000 dairy cows), sheep 28,104,000 head, pigs 7,736,000 head, poultry 133,886,000. The United Kingdom is one of the world's leading milk producers (14,594 mn. l with an average of 3,342 l per dairy cow per year). Nearly all the eggs consumed in the United Kingdom are home produced, a total of 13,908 mn. Livestock is raised in Scotland, Northern Ireland, in the South and South-west of England. The main corn growing region is the eastern half of England and the east coast of Scotland. **Production (yields):** barley 10.74 mn. tons (4.39 tons per hectare), wheat 5.24 mn. tons (4.9 tons per hectare), oats 0.78 mn. tons (4.06 tons per hectare); other crops: potatoes 6.62 mn. tons (28.5 tons per hectare); vegetables: cabbages, carrots, onions, green peas, etc.; sugar beet 6.38 mn. tons (31.7 tons per hectare). British agriculture is highly mechanized; there is 1 tractor to every 14 hectares of arable land. The fish catch is relatively small 904,300 tons and the high demand is covered by imports. The main fishing ports are Grimsby, Kingston upon Hull, Fleetwood, North Shields, Milford Haven, Lowestoft, Great Yarmouth, Aberdeen, Stornoway.

Transport (data for 1977). Transport is an important modern sector which together with communications contributes 8.3% to the GNP and employs 5.9% of the economicallly active population. Freight trafffic is carried mainly by road 65.5%, railway 15.2% and coastal shipping 13.4%. Pipeline transport is growing at 5.9%. There are 335,186 km of roads, of which 2,286 km are motorways, 13,161 km are trunk roads and 33,534 km are main roads. There were 14.02 mn. licensed passenger cars in the United Kingdom, 112,000 public road transport vehicles, and 1.71 mn. commercial vehicles. Road transport carried 92.2% of all travellers. The railway network measures 17,973 km, of which 3,767 km were electrified. Only 7.3% of travellers used rail transport. Britain's railways link up with the European railway system via the cross-Channel ferries. The construction of a railway tunnel under the Channel is still under consideration. Inland transport uses 4,828 km of navigable waterways and canals. The British merchant fleet is one of the largest in the world. It has 1,545 ships over 500 GRT with a total tonnage of 30.1 mn. GRT (of this 413 tankers are of 15.8 mn. GRT). In 1976 the British merchant fleet transported 46% of the U.K.'s exports and 31% of imports. The biggest port is the Port of London handling roughly 44 mn. tons of goods a year. Other ports include Liverpool, Manchester, the oil port of Milford Haven (43 mn. tons of crude petroleum), Southampton and Grangemouth; there is also a transoceanic passenger terminal at Southampton; most Channel traffic is handled at Dover; there is a coal port at Newcastle upon Tyne, iron ore is handled at Teesside and the main fishing port is at Kingston upon Hull. 34.7 mn. travellers used air transport on domestic and overseas routes. The biggest of the 128 civil airports are London-Heathrow and Gatwick, then Luton, Manchester and Glasgow. British Airways is the largest U.K. air transport operator.

Foreign trade (1977). Foreign trade continues to be an important factor in the British economy, even if the importance of the United Kingdom as an economic power has declined. London remains one of the most important centres of international trade and finance. The pattern of foreign trade is such that imports usually exceed exports; exports were valued at £ 33,331 million, imports £ 36,978 million. The main items of foreign trade consist of over 81.5% of manufactured goods, mainly machinery, transport equipment and chemical products. Other items include food, beverages

1
2
3
4
5
6
7
8
9
A
B
C
12°
10°
8°
6°
4°
2°
0°
2°
60°
58°
56°
Shetland Is.
Yell
Unst
Mainland
Lerwick
Orkney Is.
Mainland
Kirkwall
Hoy
Pentland Firth
Duncansby Hd.
Thurso
Wick
UNITED KI
NORTH S
OCEAN
IC
Outer Hebrides
Lewis
Stornoway
N. Uist
S. Uist
The Minch
Little Minch
Inner Hebrides
Skye
Uig
Rhum
Eigg
Coll
Tiree
Mull
Jura
Islay
Pt. Ellen
Arran
Scotland
L. Shin
Lairg
Moray Firth
Dingwall
Inverness
Kyle of Lochalsh
Mallaig
L. Ness
Fort William
Elgin
Spey
Huntly
Don
Fraserburgh
Peterhead
Aberdeen
Dee
Aviemore
Pitlochry
Montrose
Tay
Dundee
Perth
St. Andrews
Crianlarich
Oban
Firth of Lorn
L. Lomond
Dunfermline
Kirkcaldy
Firth of Forth
Stirling
Falkirk
Edinburgh
GLASGOW
Clydebank
Dumbarton
Greenock
Coatbridge
Motherwell
Paisley
Hamilton
Galashiels
Tweed
Berwick-upon-Tweed
Clyde
Irvine
Kilmarnock
Carstairs
Ayr
Alnwick

IRISH SEA
CELTIC SEA
English Channel
Bristol Channel
St. George's Channel
Cardigan Bay
Strait of Dover
Lyme Bay
The Wash
Ireland
IRELAND
Wales
England
UNITED KINGDOM
FRANCE
DUBLIN (Baile Átha Cliath)
Belfast
Lisburn
Lurgan
Portadown
Armagh
Newry
Omagh
Dundalk
Drogheda
Cavan
Sligo
Ballina
Castlebar
Westport
Claremorris
Longford
Mullingar
Roscommon
Athlone
Athenry
Galway
Loughrea
Tullamore
Birr
Pt. Laoise
Dún Laoghaire
Bray
Wicklow
Arklow
Carlow
Enniscorthy
Ennis
Nenagh
Thurles
Kilkenny
Limerick (Luimneach)
Tipperary
Clonmel
Wexford
Rosslare
Waterford
Dungarvan
Tralee
Killarney
Mallow
Cork (Corcaigh)
Cobh
Youghal
Bantry
Clifden
Achill I.
Clare I.
Aran Is.
Mullet Pen.
Loop Hd.
Slea Hd.
Valencia I.
Mizen Hd.
Carnsore Pt.
Donegal B.
Galway B.
Dingle B.
Mull of Galloway
Isle of Man
Douglas
Holyhead
Anglesey
Bangor
Caernarvon
Colwyn Bay
Pwllheli
Aberystwyth
Newtown
Llandrindod Wells
Fishguard
Milford Haven
Pembroke
Carmarthen
Swansea
Port Talbot
Merthyr Tydfil
Rhondda
Cardiff
Newport
Wrexham
Workington
Barrow-in-Furness
Kendal
Heysham
Lancaster
Blackpool
Preston
Blackburn
Burnley
Bolton
St. Helens
LIVERPOOL
Birkenhead
Widnes
Chester
Salford
Manchester
Oldham
Stockport
Halifax
Huddersfield
Bradford
LEEDS
Harrogate
York
Darlington
Stockton
Middlesbrough
Whitby
Northallerton
Scarborough
Bridlington
Kingston upon Hull
Scunthorpe
Grimsby
Barnsley
Doncaster
Rotherham
SHEFFIELD
Lincoln
Skegness
Mansfield
Newark-u.-T.
Boston
Stoke-on-T.
Crewe
Derby
Nottingham
Stafford
Shrewsbury
Walsall
West Bromwich
Wolverhampton
Dudley
BIRMINGHAM
Solihull
Leicester
Nuneaton
Coventry
Rugby
Corby
Spalding
Peterborough
Ely
King's Lynn
Norwich
Great Yarmouth
Lowestoft
Craven Arms
Worcester
Hereford
Stratford-u.-Av.
Banbury
Northampton
Bedford
Cambridge
Ipswich
Harwich
Colchester
Luton
Cheltenham
Gloucester
Oxford
Swindon
Kingswood
Bath
Bristol
Harlow
St. Albans
Watford
Slough
Reading
LONDON
Basildon
Southend-on-Sea
Chatham
Margate
Canterbury
Dover
Folkestone
Ashford
Tonbridge
Reigate
Guildford
Woking
Basingstoke
Salisbury
Winchester
Southampton
Havant
Portsmouth
Brighton
Worthing
Newhaven
Eastbourne
Hastings
Newport
Isle of Wight
Bournemouth
Poole
Weymouth
Dorchester
Yeovil
Bridgwater
Taunton
Barnstaple
Exeter
Torquay
Newton Abbot
Plymouth
Bodmin
St. Austell
Truro
Camborne
Penzance
Land's End
Is. of Scilly
Lizard Pt.
Alderney
Channel Is. (U.K.)
Guernsey
Jersey
Cherbourg
Bayeux
Caen
Baie de la Seine
Le Havre
Fécamp
Dieppe
Rouen
Le Tréport
Abbeville
Boulogne-s.-M.
Calais
Thames
Severn
Trent
Avon
Wye
Dee
Exe
Ouse
Nene
Welland
Shannon
Blackwater
Suir
Barrow
Boyne
L. Corrib
L. Mask
L. Ree
L. Derg
Lower L. Erne
Upper
Morecambe B.
1 : 6 000 000
0 50 100 150 Km
0 20 40 60 80 100 Mi
54°
52°
50°
10° 8° 6° 4° 2° 0°
2 3 4 5 6 7 8
E F G

ADMINISTRATIVE UNITS

Local Government Area	Area in sq.km	Population in thousands 1977	Density of population inh. per sq.km
ENGLAND AND WALES	151,130	49,119.5	325
ENGLAND	130,367	46,351.3	356
1 Greater London	1,580	6,970.1	4,411
Metropolitan counties (admin. centre):			
2 Greater Manchester (Manchester)	1,287	2,674.8	2,078
3 Merseyside (Liverpool)	652	1,561.8	2,395
4 South Yorkshire (Barnsley)	1,560	1,304.0	836
5 Tyne and Wear (Newcastle upon Tyne)	540	1,174.0	2,174
6 West Midlands (Birmingham)	900	2,729.9	3,033
7 West Yorkshire (Wakefield)	2,039	2,072.5	1,016
Non-metropolitan counties (admin. centre):			
8 Avon (Bristol)	1,338	916.9	685
9 Bedfordshire (Bedford)	1,235	489.6	396
10 Berkshire (Reading)	1,256	667.6	532
11 Buckinghamshire (Aylesbury)	1,883	516.3	274
12 Cambridgeshire (Cambridge)	3,409	564.3	166
13 Cheshire (Chester)	2,322	913.9	394
14 Cleveland (Middlesbrough)	583	569.1	976
15 Cornwall (Truro)	3,546	413.6	117
16 Cumbria (Carlisle)	6,809	474.3	70
17 Derbyshire (Matlock)	2,631	896.5	341
18 Devon (Exeter)	6,715	947.2	141
19 Dorset (Dorchester)	2,654	583.1	220
20 Durham (Durham)	2,436	607.9	250
21 East Sussex (Lewes)	1,795	652.9	364
22 Essex (Chelmsford)	3,674	1,425.9	388
23 Gloucestershire (Gloucester)	2,638	492.7	187
24 Hampshire (Winchester)	3,772	1,450.5	385
25 Hereford and Worcester (Worcester)	3,927	602.7	153
26 Hertfordshire (Hertford)	1,634	941.6	576
27 Humberside (Kingston upon Hull)	3,512	845.4	241
28 Isle of Wight (Newport)	381	113.6	298
29 Kent (Maidstone)	3,732	1,445.1	387
30 Lancashire (Preston)	3,043	1,368.1	450
31 Leicestershire (Leicester)	2,553	832.7	326
32 Lincolnshire (Lincoln)	5,885	529.8	90
33 Norfolk (Norwich)	5,355	674.7	126
34 Northamptonshire (Northampton)	2,367	513.8	217
35 Northumberland (South Shields)	5,033	290.7	58
36 North Yorkshire (Northallerton)	8,317	654.0	79
37 Nottinghamshire (Nottingham)	2,164	974.1	450
38 Oxfordshire (Oxford)	2,612	539.4	207
39 Shropshire (Shrewsbury)	3,490	360.7	103
40 Somerset (Taunton)	3,458	411.4	119
41 Staffordshire (Stafford)	2,716	994.0	366
42 Suffolk (Ipswich)	3,801	588.4	155
43 Surrey (Kingston)	1,655	995.8	602
44 Warwickshire (Warwick)	1,981	467.0	236
45 West Sussex (Chichester)	2,016	625.1	310
46 Wiltshire (Trowbridge)	3,481	513.8	148
WALES	20,763	2,768.2	133
47 Clwyd (Mold)	2,425	379.8	157
48 Dyfed (Carmarthen)	5,766	324.8	56
49 Gwent (Cwmbran)	1,376	439.3	319
50 Gwynedd (Caernarvon)	3,868	226.8	59
51 Mid Glamorgan (Cardiff)	1,019	539.2	529
52 Powys (Llandrindod Wells)	5,078	105.5	21
53 South Glamorgan (Cardiff)	416	385.2	926
54 West Glamorgan (Swansea)	815	367.6	451
SCOTLAND – regions (admin. centre):	78,773+	5,195.6	66
55 Borders (Newtown St. Boswells)	4,670	100.4	21
56 Central (Stirling)	2,621	271.5	104
57 Dumfries and Galloway (Dumfries)	6,369	143.4	23
58 Fife (Glenrothes)	1,305	339.2	260

59 Grampian (Aberdeen)	8,702	458.7	53
60 Highland (Inverness)	25,141	189.8	8
61 Lothian (Edinburgh)	1,753	756.2	431
62 Strathclyde (Glasgow)	13,849	2,466.3	178
63 Tayside (Dundee)	7,501	401.9	54
64 Orkney (Kirkwall)	975	18.0	19
65 Shetland (Lerwick)	1,429	20.4	14
66 Western Isles (Stornoway)	2,898	29.7	10
NORTHERN IRELAND – districts (admin. centre):	14,120	1,537.3	109
1 Antrim (Antrim)	563	38.9	69
2 Ards (Newtownards)	361	52.5	145
3 Armagh (Armagh)	675	47.8	71
4 Ballymena (Ballymena)	637	53.1	83
5 Ballymoney (Ballymoney)	418	22.2	53
6 Banbridge (Banbridge)	445	28.6	64
7 Belfast (Belfast)	115	357.6	3,110
8 Carrickfergus (Carrickfergus)	77	27.8	361
9 Castlereagh (Castlereagh)	85	63.7	749
10 Coleraine (Coleraine)	485	45.4	94
11 Cookstown (Cookstown)	611	27.8	45
12 Craigavon (Portadown)	388	72.5	187
13 Down (Downpatrick)	646	41.9	65
14 Dungannon (Dungannon)	780	43.1	55
15 Fermanagh (Enniskillen)	1,876	50.9	27
16 Larne (Larne)	340	28.7	84
17 Limavady (Limavady)	590	25.6	43
18 Lisburn (Hillsborough)	447	81.9	183
19 Londonderry (Londonderry)	387	87.9	227
20 Magherafelt (Magherafelt)	562	32.0	57
21 Moyle (Ballycastle)	495	13.0	26
22 Newry and Mourne (Newry)	910	75.8	83
23 Newtownabbey (Newtownabbey)	139	73.2	527
24 North Down (Bangor)	74	60.9	823
25 Omagh (Omagh)	1,125	41.9	37
26 Strabane (Strabane)	862	36.0	42
*including water area			

and tobacco (6.7%), fuels (6.3%), raw materials (2.7%). Raw materials for industry and semi-manufactured goods form 36.5% of imports and the share of finished manufactures is growing at 31.9%. The United Kingdom is now less dependent on food imports, which have dropped to 16.1%. Fuels 14.2% form a big portion of imports. British trade with Commonwealth countries has greatly declined in recent years. In the last decade the main trading partners of the United Kingdom have been the EEC countries taking 36.5% of exports, and providing 38.3% of imports, while other countries in Western Europe handled 16.8% of exports and 15.1% of imports. The United States and Canada continue to play an important role, 11.5% and 13.4%, as do other industrially developed countries 6.3% and 7.3%. Trade with the developing countries represents 25.9% of exports and 22.1% of imports.

Membership of international organizations. The United Kingdom stands at the head of the Commonwealth, which is a free association of sovereign independent countries. The original dominions and most of the former colonies in the British Empire have retained membership in the Commonwealth after declaring independence. Individual member countries are linked by treaties and agreements on mutual cooperation in political life, economic affairs, finance, education, science, and culture. The British Queen stands at the head of the Commonwealth. The following were members of the Commonwealth in 1979: America: the Bahamas, Barbados, Canada, Dominica, Grenada, Guyana, Jamaica, Saint Lucia, Saint Vincent, Trinidad and Tobago; Europe: United Kingdom, Malta; Africa: Botswana, Gambia, Ghana, Kenya, Lesotho, Malawi, Mauritius, Nigeria, Seychelles, Sierra Leone, Swaziland, Tanzania, Uganda, Zambia, Zimbabwe; Asia: Bangladesh, Cyprus, India, Malaysia, Singapore, Sri Lanka; Australia and Oceania: Australia, Fiji, Kiribati, Nauru, New Zealand, Papua New Guinea, Samoa, Solomon Islands, Tonga, Tuvalu.

The United Kingdom is a founder-member of the United Nations and a permanent member of the Security Council with the right of veto. It takes an active part in the NATO and SEATO pacts. It is a member of the following international organizations: The International Monetary Fund (IMF), the International Bank for Reconstruction and Development (IBRD), the General Agreement on Tariffs and Trade (GATT), the Organization for Economic Co-operation and Development (OECD), the United Nations Conference on Trade and Development (UNCTAD), the European Atomic Energy Community (EAEC or Euratom), the European Coal and Steel Community (ECSC), and others. Since 1973 the United Kingdom has been a member of the European Economic Community (EEC).

The Isle of Man

588 sq.km, 61,000 inhabitants (1977), **autonomous region of the United Kingdom. Currency:** £ 1 – 100 pence. **Position:** Island in the Irish Sea; **capital:** Douglas 19,897 inhabitants (1976). **Official language:** English; Manx sometimes used.

Economy: Agriculture (1975): barley 7,000 tons, oats 4,000 tons, wheat 2,000 tons, potatoes 13,000 tons. Fish catch 17,852 tons; **animal husbandry**: sheep 101,000 head, cattle 39,000 head, pigs 4,000 head, poultry 115,000. Tourism; airport. The island is the setting of the annual Tourist Trophy motorcycle race.

NORTH SEA
IRISH
North Channel
Scotland
England
Northern Ireland
Mull
Colonsay
Oban
Inveraray
Crianlarich
Perth
Cupar
St. Andrews
Kinross
Glenrothes
Loch Lomond
Stirling
Alloa
Kirkcaldy
Firth of Forth
Dumbarton
Falkirk
Dunfermline
Greenock
Clydebank
GLASGOW
Grangemouth
Musselburgh
Haddington
Coatbridge
Edinburgh
Paisley
Motherwell
Berwick-upon-Tweed
Largs
East Kilbride
Hamilton
Carstairs
Duns
Saltcoats
Irvine
Kilmarnock
Lanark
Peebles
Galashiels
Newtown St. Boswells
Troon
Selkirk
Ayr
Cumnock
Hawick
Alnwick
Moffat
Tweed
Teviot
Coquet
Clyde
Jura
Islay
Port Ellen
Sd. of Jura
Firth of Lorn
Kintyre
Tarbert
Arran
Brodick
Firth of Clyde
Campbeltown
Mull of Kintyre
Rathlin Island
Fair Hd.
Ballycastle
Girvan
Ballymoney
Nith
Annan
Dumfries
Gretna Green
Newton Stewart
Stranraer
Wigtown
Kirkcudbright
Luce Bay
Solway Firth
Mull of Galloway
Morpeth
Ashington
Bedlington
Blyth
N. Tyne
S. Tyne
Newcastle upon Tyne
Tynemouth
South Shields
Sunderland
Gateshead
Hexham
Consett
Durham
Stanhope
Wear
Carlisle
Penrith
Maryport
Workington
Keswick
Appleby
Eden
Whitehaven
St. Bees Hd.
Windermere
Kendal
Hartlepool
Redcar
Stockton
Darlington
Middlesbrough
Whitby
Tees
Swale
Northallerton
Ure
Thirsk
Ripon
Norton
Scarborough
Flamborough Hd.
Bridlington
Bridlington Bay
Ballymena
Larne
Antrim
Carrickfergus
Belfast L.
Newtownabbey
Bangor
Belfast
Newtownards
Lough Neagh
Lurgan
Lisburn
Hillsborough
Strangford L.
Portadown
Banbridge
Downpatrick
Bann
Newry
Dundrum Bay
Dundalk
Dundalk Bay
Louth
Drogheda
Meath
Balbriggan
DUBLIN
(Baile Átha Cliath)
Dún Laoghaire
Ramsey
Isle of Man
Peel
Douglas
Spanish Hd.
Barrow-in-Furness
Walney I.
Morecambe
Morecambe Bay
Lancaster
Heysham
Fleetwood
Blackpool
Preston
Skipton
Harrogate
Wharfe
Ouse
York
Derwent
Keighley
Bradford
LEEDS
Aire
Burnley
Blackburn
Halifax
Dewsbury
Wakefield
Selby
Beverley
Kingston upon Hull
Humber
Spurn Hd.
Darwen
Bolton
Bury
Rochdale
Huddersfield
Barnsley
Scunthorpe
Grimsby
Cleethorpes
Southport
Wigan
St. Helens
Salford
Oldham
Ashton-u.-L.
Doncaster
Trent
Liverpool Bay
LIVERPOOL
Wallasey
Birkenhead
Stretford
Manchester
Rotherham
Gainsborough
Louth
Carmel Hd.
Holyhead
Menai Str.
Rhyl
Colwyn Bay
Dee
Widnes
Warrington
Stockport
SHEFFIELD
Worksop
Chesterfield
East Retford
Lincoln
Horncastle
Skegness

Bardsey I.
Dolgellau
Welshpool
Shrewsbury
Wolverhampton
Walsall
West Bromwich
Nuneaton
Leicester
Loughborough
Stamford
Peterborough
Wisbech
East Dereham
Norwich
Great Yarmouth
Beccles
Lowestoft
Tywyn
Machynlleth
Newtown
Craven Arms
Dudley
Smethwick
BIRMINGHAM
Corby
Kettering
March
Huntingdon
Ely
Thetford
Cardigan Bay
Aberystwyth
Kidderminster
Solihull
Coventry
Rugby
Northampton
Newmarket
Cambridge
Bury St. Edmunds
Stowmarket
Ipswich
Wexford Harbour
Rosslare
Carnsore Pt.
Llandrindod Wells
Bromsgrove
Redditch
Royal Leamington Spa
Warwick
Stratford -u.-Av.
Leominster
Worcester
Lampeter
Builth Wells
Hereford
Banbury
Bedford
Sudbury
Felixstowe
Harwich
The Naze
Cardigan
St. George's Channel
Fishguard
Brecon
Cheltenham
Hitchin
Luton
Stevenage
Colchester
Clacton-on-Sea
St. David's Hd.
Gloucester
Aylesbury
Hemel Hempstead
Hertford
St. Albans
Harlow
Chelmsford
St. Brides Bay
Haverfordwest
Carmarthen
Ebbw Vale
Pontypool
Monmouth
Cirencester
Oxford
Watford
Brentwood
Basildon
Milford Haven
Llanelli
Merthyr Tydfil
Rhondda
Newport
High Wycombe
Southend-on-Sea
North Foreland
Pembroke
Carmarthen Bay
Swansea
Neath
Port Talbot
Pontypridd
Caerphilly
Slough
LONDON
Thames
Porthcawl
Cardiff
Kingswood
Swindon
Reading
Windsor
Staines
Gravesend
Chatham
Gillingham
Margate
Ramsgate
Barry
Bristol
Bath
Chippenham
Kennet
Newbury
Whitstable
Canterbury
Maidstone
Bristol Channel
Weston-s.-M.
Trowbridge
Basingstoke
Woking
Epsom
Reigate
Tonbridge
Ashford
Dover
Aldershot
Guildford
Tunbridge Wells
Folkestone
Ilfracombe
Lundy
Wells
Andover
Wilton
Winchester
Crawley
Barnstaple
Bridgwater
Salisbury
Eastleigh
Hartland Pt.
Bideford
Taunton
Yeovil
Southampton
Havant
Chichester
Hove
Lewes
Hastings
Bexhill
Boulogne -s.-M.
Fareham
Gosport
Portsmouth
Worthing
Brighton
Newhaven
Eastbourne
Honiton
Okehampton
Exeter
Dorchester
Poole
Cowes
Newport
Ventnor
Bognor Regis
Strait of Dover
Étaples
Exmouth
Weymouth
Bournemouth
The Needles
Isle of Wight
Launceston
Newton Abbot
Lyme Bay
Berck-s.-M.
Wadebridge
Bodmin
Torquay
Paignton
Newquay
St. Austell
Plymouth
Dartmouth
St.-Valéry-s.-S.
Truro
St. Ives
Camborne
Falmouth
English Channel
Le Tréport
Penzance
Falmouth Bay
Mount's Bay
Isles of Scilly
Land's End
Lizard Pt.
1 : 3 500 000
0 20 40 60 80 Km
0 10 20 30 40 50 Mi
Dieppe
Neufchâtel-en-B.
Alderney (U. K.)
C. de la Hague
Cherbourg
Barfleur
Étretat
Fécamp
Yvetot

The Channel Islands

194 sq.km, 126,000 inhabitants (1977), **autonomous region of the United Kingdom. Currency:** £1 = 100 pence. **Position:** Islands in the English Channel off the coast of France. **Jersey** 116 sq.km, 74,470 inhabitants (1976), administrative centre Saint Helier (28,135 inhab.) and **Guernsey** 63 sq.km, 53,637 inhabitants (1976), administrative centre Saint Peter Port; dependencies of Guernsey: Alderney 8 sq.km, 1,785 inhabitants (1975); Sark 5 sq.km, 604 inhabitants (1976); Brechou 0.3 sq.km; Herm 1.29 sq.km; Jethou 0.18 sq.km; Lihou 0.15 sq.km. **Capital:** Saint Peter Port 16,303 inhabitants (1976). **Official language:** English, French on the island of Jersey; Norman French is spoken on the smaller islands.
Economy: Agriculture (1975): potatoes 60,000 tons, tomatoes 60,000 tons, spring vegetables, flowers; cattle breeding 12,000 head, pigs 3,000 head, poultry 118,000. Fishery. Tourism, 3 airports.

ALBANIA

Republika Popullore Socialiste e Shqipërisë, area 28,748 sq.km, population 2,616,000 (1977), **socialist republic** (Chairman of the Presidium of the People's Assembly Haxhi Lleshi since 1979).
Administrative units: 26 districts (Rrethët). **Capital:** Tiranë 192,000 inhab. (1975); **other towns** (1973, in 1,000 inhab.): Shkodër 59, Durrës 57, Vlorë 53, Korçë 49, Elbasan 46, Berat 28, Fier 26, Lushnjë 21. **Population:** Albanians. **Density** 91 persons per sq.km; average annual rate of population increase 2.7% (1970–75); urban population 36.6% (1975). 62.1% of inhabitants employed in agriculture (1977). – **Currency:** lek = 100 quindarkas.
Economy: agricultural and industrial country. **Mining** (1976, in 1,000 tons, metal content): brown coal 950, crude petroleum 2,500 (Qytet Stalin, Patos), chromium 340 (Kam, Bulqizë), copper 10, nickel 7, asphalt. Electricity 2 billion kWh (1976). **Industries:** textiles and foodstuffs. **Agriculture** (1977, in 1,000 tons): maize (corn) 300, wheat 400, rice, potatoes, sugar beet, cotton, olives, grapes, tobacco; livestock (1977, in 1,000 head): cattle 472, sheep 1,163, goats 674; fish catch 4,000 tons (1976); roundwood 2.5 mn. cub.m (1976).- **Communications** (1975): railways 302 km, roads 4,827 km. Merchant shipping 57,368 GRT (1976).- **Exports:** petroleum and petroleum products, ores and metals, cigarettes, fresh and canned fruit and vegetables. **Imports:** machines and equipment, transport equipment.

ANDORRA

Principat d'Andorra – Principauté d'Andorre, area 453 sq.km, population 30,584 (1977), **republic under the joint suzerainty of France and the Bishop of Seo de Urgel** (Spain), headed by two Syndics.
Administrative units: 6 villages. **Capital:** Andorra 5,500 inhab. **Official languages:** Catalan, Spanish, French. – **Currency:** French franc, Spanish peseta. **Economy:** mountain agriculture and grazing, cultivation of cereals, vines, tobacco in the valleys; raising of sheep and goats. Tourism is the chief source of inhabitants' incomes.

AUSTRIA

Republik Österreich, area 83,853 sq.km, population 7,512,000 inhab. (1978), **neutral federal republic** (President Dr Rudolf Kirchschläger since 1974).

Administrative units: 9 federal countries (Wien, Niederösterreich, Burgenland, Oberösterreich, Salzburg, Steiermark, Kärnten, Tirol, Vorarlberg). **Capital:** Wien (Vienna) 1,590,000 inhab. (1977); **other towns** (1977, in 1,000 inhab.): Graz 249, Linz 208, Salzburg 139, Innsbruck 123, Klagenfurt 85. **Population:** German speaking Austrians 98%, small Croatian, Czech and Slovenian minorities. **Density** 90 persons per sq.km; average annual rate of population increase 0.24%; urban population 52.8% (1975). 21.4% of inhabitants employed in industry. – **Currency:** schilling = 100 groschen.
Economy: highly developed industrial country with considerable mineral resources. **Principal industries:** metallurgy (Steiermark, Linz), engineering (Steiermark, Wien, W. Neustadt), petroleum refining, energy production, chemicals and electronics. **Mining** (1977, in 1,000 tons, metal content): brown coal 3,127, crude petroleum 1,788 (Marchfeld – Matzen, Weinviertel – Mühlberg etc.), natural gas 2.4 billion cub.m, iron ore 1,068 (Steiermark: Erzberg, Radmer), copper 1.2 (1976), lead 5.4 (1976), zinc 22.7 (1976), tungsten 682 tons (1976), antimony 510 tons, magnesite 1,003 (Tirol – Hochfilzen, Kärnten – Radenthein, Steiermark – Trieben, Breitenau), bauxite, salt 665 (1976, Salzkammergut), graphite 29.8. Electricity 37.7 billion kWh (1977), of which 66% hydro-electric power stations. **Production** (1977, in 1,000 tons): pig iron 2,964, crude steel 4,608, lead 10, zinc 16.4, refined copper 35, aluminium (primary) 91.8, nitrogenous fertilizers 239 (1976), plastics 404 (1976), rayon and acetate fibres 15 (1976), rayon and acetate staple fibres 89 (1976), cotton yarn 20, woven cotton fabrics 73 mn. m, cement 5,988, chemical wood pulp 852 (1976), paper 1,396 (1976), musical instruments (Wien), sugar 495, meat 591, milk 3,318, butter 43, cheese 83, eggs 90, wine 259, beer 7.8 mn. hl (1976); 14 billion cigarettes (1976).
Agriculture: principal branch is livestock raising. Land use: arable land 19.6%, meadows and pastures 25%, forests 39.5% (with important hunting grounds in the Alps). **Crops** (1977, in 1,000 tons): wheat 1,072, rye 351, barley 1,212, oats, maize (corn) 1,159, potatoes 1,352, sugar beet 2,721, grapes 369, fruit 898, vegetables 587; **livestock** (1977, in 1,000 head): cattle 2,502, pigs 3,878, sheep 174, goats, horses, poultry 13,359; roundwood 13.1 mn. cub.m – **Communications:** railways 5,858 km (1976), of which 2,728 km electrified, roads 33,143 km, motorways 722 km (1976), passenger cars 2,050,000 (1977). Navigable waterways 1,733 km. Civil aviation (1976): 16.8 mn. km flown, 1,061,000 passengers carried. Important tourism 11.7 mn. visitors (1977). – **Exports:** finished products, especially machines, steel, chemicals, textiles, metals, wood, paper, salt, electricity. **Imports:** coal, metals, chemicals, transport equipment, machines, finished electronic products, textile raw materials, foodstuffs. Chief trading partners: Fed. Rep. of Germany, Italy, Switzerland, United Kingdom.

BELGIUM

Royaume de Belgique – Koninkrijk België, area 30,521 sq.km, population 9,837,113 (1978), **kingdom** (King Baudouin I since 1951).

Administrative units: 9 provinces. **Capital:** Bruxelles – Brussel 103,712 inhab., with agglomeration 1,042,052 (1976); **other towns** (1976, in 1,000 inhab.): Antwerpen 223 (with agglom. 659), Gent 148, Liège 145 (with agglom. 427), Brugge 120, Schaerbeek 113, Anderlecht 101, Deurne 81, Ixelles 81, Oostende 72, Mechelen 65. **Population:** Walloons, Flemings. **Density** 322 persons per sq.km; average annual rate of population increase 0.43% (1970–75); urban population 71.7%. 32.4% of inhabitants employed in industry. – **Currency:** Belgian franc = 100 centimes.
Economy: highly developed industrial country with large concentration of industry and intensive agriculture. **Principal industries:** metallurgy, engineering, chemicals, textiles. **Mining:** coal 7,068,000 tons (1977, reg. Borinage, Liège, reg. Campine, Charleroi-Namur), natural gas, iron ore. Electricity 44.2 billion kWh (1977), of which 21% are nuclear power stations (Mol). **Production** (1977, in 1,000 tons): pig iron 8,916, crude steel 11,268 (prov. Liège, Hainaut and Brabant), coke oven coke 6,200 (1976), lead 122, zinc 259 (Flône, Balen), refined copper 553, tin 5,064 tons (Hoboken), merchant vessels 169,000 GRT (Hoboken, Temse), assembly of passenger cars 986,000 units (Antwerpen, Bruxelles), radio receivers 1,726,000 units (1976), television receivers 558,000 (1976), sulphuric acid 2,016, nitric acid 711 (1976), synthetic nitrogen 570, nitrogenous fertilizers 652⁺, phosphorus fertilizers 589⁺, plastics 802 (1974), pharmaceuticals (Bruxelles), capacity of petroleum refineries 56 mn. tons (1976, Antwerpen, Gent, Bruxelles), motor spirit 5,000, mineral oils 19,800 (1976), cement 7,800, paper products 762, woven cotton fabrics 48 (Gent and surroundings), woollen – yarn 77, woven fabrics 27 (Verviers), flax industry (Vlaanderen) – sugar 721 (Tienen), meat 1,010⁺, butter 97.4⁺, milk 3,830⁺, cheese 43.7⁺, eggs 216⁺, beer 14.5 mn. hl (1976); 26.5 billion cigarettes (1976). (⁺Figures for Belgium and Luxembourg combined.)
Agriculture: highly productive with animal production predominating. Land use: arable land 26%, meadows and pastures 23.9%, forests 20.2%. **Crops** (1977, in 1,000 tons): wheat 742, barley 676, oats, potatoes 1,488, sugar beet 4,220, fruit; **livestock** (1977, in 1,000 head): cattle 2,987, pigs 4,886; fish catch 44,400 tons (1976); roundwood 2.6 mn. cub.m (1976).
Communications (1976): railways 3,998 km, of which 1,296 km electrified, roads 94,000 km, motorways 1,051 km, passenger cars 2,738,000. Navigable rivers and canals 1,569 km. Merchant shipping 1,595,000 GRT (1977). Largest ports: Antwerpen and Gent. Civil aviation: 47.8 mn. km flown, 1,682,000 passengers carried. Tourism 7.9 mn. visitors.
Exports: metals, machines and equipment, cars, chemical and pharmaceutical products, textiles etc. **Imports:** raw materials, especially for the power industry, foodstuffs, machines and equipment. Chief trading partners (1977): EEC countries, Fed. Rep. of Germany 22%, France 17.5%, Netherlands 17%.

BULGARIA

Narodna Republika Bălgarija, area 110,912 sq.km, population 8,822,000 (1977), **socialist republic** (Chairman of the State Council Todor Zhivkov since 1972).

Administrative units: 27 regions (okrăg) and the capital. **Capital:** Sofija (Sofia) 965,355 inhab. (1975); **other towns** (1975, in 1,000 inhab.): Plovdiv 303, Varna 262, Ruse 166, Burgas 146, Stara Zagora 122, Pleven 112, Sliven 90, Gabrovo 90, Pernik 87, Tolbuhin 86, Šumen 84, Jambol 76, Haskovo 75, Pazardžik 68. **Population:** Bulgarians 91%, Gipsies, Macedonians etc. **Density** 80 persons per sq.km; average annual rate of population increase 0.7% (1970–75); urban population 57.7% (1975). 38.3% of inhabitants employed in agriculture, 28% in industry (1976). – **Currency:** lev = 100 stotinki.
Economy: industrial and agricultural country. **Principal industries:** metallurgy (Kremikovci, Pernik), engineering (Sofija, Plovdiv, Ruse), chemicals (Sofija, Burgas, Plovdiv, Dimitrovgrad, Stara Zagora, Devnja), textiles (Gabrovo, Sofija), foodstuffs.
Mining (1977, in 1,000 tons, metal content): brown coal 24,864 (Marica-East, Dimitrovgrad, Pernik, Bobovdol), crude petroleum 120, natural gas, iron ore 748 (Kremikovci), lead 115, zinc 87, copper 57 (1976, Sredna Gora Mts.), manganese 11.2 (1976), silver 23 tons (1976), mercury 191 tons (1976), uranium, salt 75 (1976). Electricity 29.7 billion kWh (1977), of which 10% are hydro-electric and 18% nuclear power stations (near Kozloduj). **Production** (1977, in 1,000 tons): pig iron 1,620, crude steel 2,592, copper smelted 60 (1976), lead 120, zinc 90, vessels 172,000 GRT (1976, Varna, Ruse), sulphuric acid 853, nitric acid 783 (1976), nitrogenous fertilizers 663, phosphorus fertilizers 242, soda ash 1,218, plastics and resins 169, capacity of petroleum refineries 12 mn. tons (1976), motor spirit 1,700 (1976), oils 8,200 (1976), cement 4,668, cotton yarn 86, woven fabrics – cotton 363 mn.m, woollen 34 mn.m, rayon and acetate 32 mn. m (1976) – synthetic staple fibre 29 (1976), leather shoes 20.8 mn. pairs, meat 607, milk 1,933, cheese 148, eggs 103.6, wine 533, canned vegetables 222, beer 4.7 mn. hl (1976); 73.1 billion cigarettes (1976).
Agriculture: vegetable production predominates. Land use: arable land 39.1%, meadows and pastures 16.9%, forests 34.4%. **Crops** (1977, in 1,000 tons): wheat 3,012, barley 1,750, maize (corn) 2,555, rice, potatoes 350, cotton-seed 20, sugar beet 2,434, sunflower seeds 370, soybeans 99, fruit 2,316, grapes 1,232, strawberries 19, vegetables 2,052, tomatoes 800, legumes, roses (Kazanlâk), tobacco 167, walnuts 28.8; **livestock** (1977, in 1,000 head): cattle 1,722, buffaloes 65, horses 128, asses 320, pigs 3,456, sheep 9,723, goats 308; poultry 37,329, silkworms 240 tons, wool 17,600 tons, honey 6,500 tons; fish catch 167,100 tons (1976); roundwood 4.4 mn. cub.m (1976).
Communications: railways 4,307 km, of which 1,425 km electrified (1976), roads 31,434 km (1975), passenger cars 197,500 (1975). Navigable waterways 471 km. Merchant shipping 964,000 GRT (1977). Chief ports: Varna and Burgas. Civil aviation (1976): 10.3 mn. km flown, 1,370,000 passengers carried. Tourism 4,033,404 visitors (1976).
Exports: machines and equipment, foodstuffs (fresh and canned fruit and vegetables), rose oil, wine, cigarettes, tobacco, chemical products and textiles, furs. **Imports:** petroleum, fuels and raw materials, machines, consumer goods. Chief trading partners (1977): the socialist countries 80%, of which the U.S.S.R. 55.6%, German Dem. Rep., Poland, Czechoslovakia and Fed. Rep. of Germany.

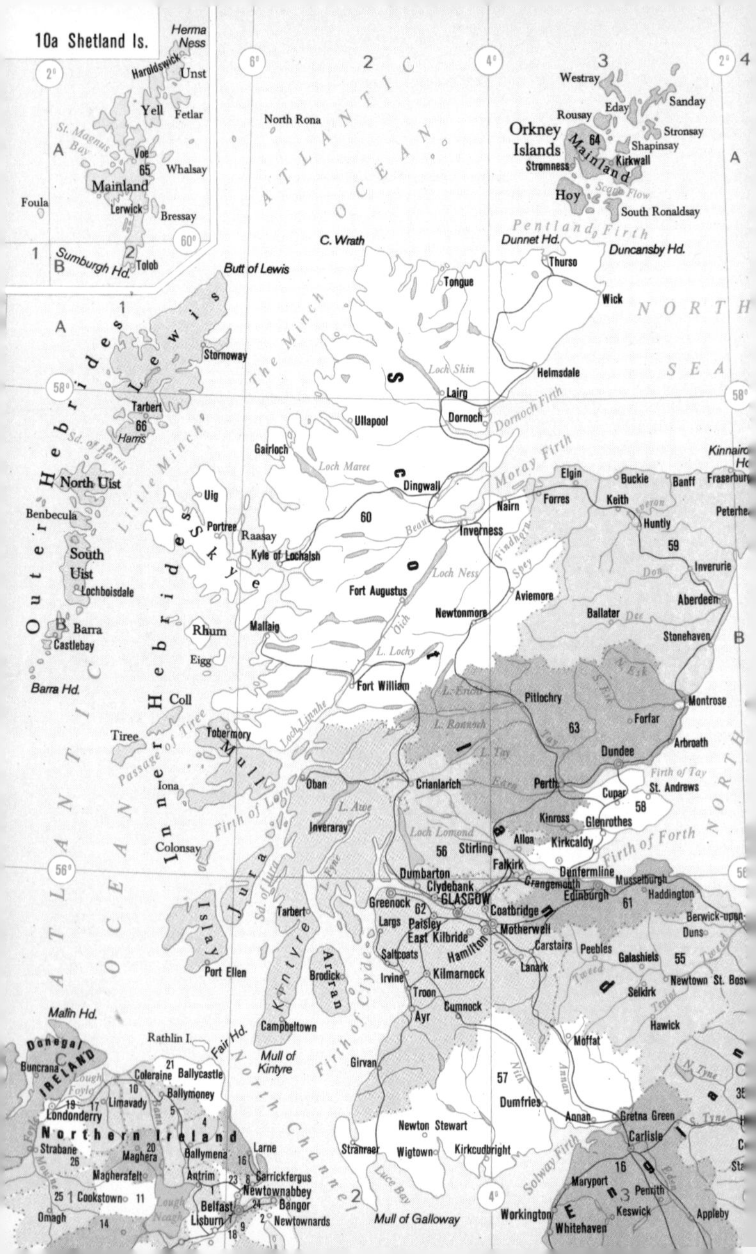

10a Shetland Is.
Herma Ness
Haroldswick
Unst
Yell
Fetlar
St. Magnus Bay
Voe
65
Whalsay
Mainland
Foula
Lerwick
Bressay
Sumburgh Hd.
Tolob
North Rona
ATLANTIC OCEAN
Westray
Eday
Sanday
Rousay
Orkney Islands
Mainland
64
Stronsay
Shapinsay
Kirkwall
Stromness
Hoy
Scapa Flow
South Ronaldsay
Pentland Firth
Dunnet Hd.
Duncansby Hd.
C. Wrath
Thurso
Tongue
Wick
NORTH SEA
Butt of Lewis
Lewis
Stornoway
The Minch
Loch Shin
Helmsdale
Lairg
Dornoch
Dornoch Firth
Outer Hebrides
Tarbert
66
Harris
Sd. of Harris
Ullapool
Gairloch
Loch Maree
Little Minch
North Uist
Benbecula
Uig
Dingwall
Moray Firth
Elgin
Buckie
Banff
Fraserburgh
Kinnaird Hd
Nairn
Forres
Keith
Deveron
Peterhead
60
Portree
Raasay
Skye
Inverness
Huntly
59
South Uist
Kyle of Lochalsh
Beauly
Findhorn
Spey
Loch Ness
Don
Inverurie
Lochboisdale
Fort Augustus
Aviemore
Aberdeen
Barra
Castlebay
Rhum
Mallaig
Newtonmore
Ballater
Dee
Stonehaven
Eigg
Oich
L. Lochy
N. Esk
S. Esk
Barra Hd.
Fort William
L. Ericht
Pitlochry
Montrose
Coll
Loch Linnhe
L. Rannoch
Forfar
Inner Hebrides
63
Tiree
Tobermory
Mull
L. Tay
Tay
Arbroath
Dundee
Passage of Tiree
Firth of Tay
Iona
Oban
Crianlarich
Earn
Perth
St. Andrews
Cupar
Firth of Lorn
L. Awe
58
Kinross
Glenrothes
Inveraray
Loch Lomond
Colonsay
Alloa
Kirkcaldy
56
Stirling
Firth of Forth
Jura
Sd. of Jura
L. Fyne
Falkirk
Dunfermline
Dumbarton
Grangemouth
Musselburgh
Clydebank
Edinburgh
Haddington
Greenock
62
GLASGOW
Coatbridge
61
Islay
Tarbert
Largs
Paisley
Motherwell
Berwick-upon-
East Kilbride
Duns
Hamilton
Clyde
Carstairs
Peebles
Galashiels
Saltcoats
55
Port Ellen
Kintyre
Arran
Brodick
Firth of Clyde
Irvine
Kilmarnock
Lanark
Tweed
Newtown St. Bos
Troon
Selkirk
Teviot
Ayr
Cumnock
Hawick
Malin Hd.
Campbeltown
Moffat
Rathlin I.
Fair Hd.
Donegal
IRELAND
Buncrana
Lough Foyle
21
Coleraine
Ballycastle
Mull of Kintyre
Girvan
North Channel
Nith
Annan
N. Tyne
10
Ballymoney
57
19
17
Limavady
Bann
5
Dumfries
Londonderry
4
Annan
Gretna Green
S. Tyne
Northern Ireland
Newton Stewart
Carlisle
Strabane
20
Larne
Stranraer
Wigtown
Kirkcudbright
26
Maghera
Ballymena
16
Solway Firth
Magherafelt
Antrim
23
8
Carrickfergus
Luce Bay
Eden
Maryport
Penrith
16
Newtownabbey
25
1
Cookstown
11
Lough Neagh
Belfast
24
Bangor
Workington
Keswick
Omagh
14
Lisburn
7
2
Newtownards
Mull of Galloway
Whitehaven
Appleby
18
9
England
Scotland

Colonsay
Jura
Islay
Scotland
Kintyre
Port Ellen
Campbeltown
Mull of Kintyre
Rathlin I.
Fair Hd.
North Channel
Malin Hd.
Lough Swilly
Tory I.
Aran Island
Gweedore
Buncrana
Donegal
Coleraine
Ballycastle
Limavady
Ballymoney
Londonderry
Letterkenny
Northern Ireland
Lifford
Strabane
Maghera
Ballymena
Larne
Carrickfergus
Killybegs
Donegal
Magherafelt
Antrim
Newtownabbey
Belfast L.
Bangor
Cookstown
Omagh
Belfast
Newtownards
Ballyshannon
Donegal Bay
Lough Neagh
Lower L. Erne
Dungannon
Lurgan
Lisburn
Erris Hd.
Sligo Bay
Portadown
Hillsborough
Strangford L.
Belmullet
Easky
Sligo
Enniskillen
Banbridge
Mullet Peninsula
Leitrim
Armagh
Downpatrick
Ballina
Upper L. Erne
Monaghan
Newry
Dundrum Bay
Achill Island
Mayo
L. Conn
L. Allen
Cavan
Monaghan
Louth
Boyle
Dundalk
Clare Island
Clew Bay
Castlebar
Carrick-o.-Sh.
Cavan
Kingscourt
Dundalk Bay
Westport
Claremorris
Castlerea
Roscommon
Inishbofin
Longford
Meath
Roscommon
Longford
Drogheda
L. Mask
Navan
Clifden
Tuam
Mullingar
Trim
Balbriggan
Lough Corrib
L. Ree
Westmeath
Galway
Athenry
Ballinasloe
Athlone
DUBLIN
(Baile Átha Cliath)
Galway Bay
Dún Laoghaire
Aran Islands
Loughrea
Offaly
Tullamore
Naas
Droichead Nua
Bray
Kildare
Dublin
Gort
Birr
Port Laoise
Kildare
Wicklow
Ennistymon
Lough Derg
Roscrea
Laois
Wicklow
Athy
Clare
Ennis
Nenagh
Carlow
IRISH SEA
Kilkee
Kilrush
Tipperary
Carlow
Arklow
Loop Hd.
Thurles
Kilkenny
Kilkenny
Limerick
(Luimneach)
Gorey
Shannon
Listowel
Limerick
Wexford
Newcastle West
Tipperary
Cashel
Enniscorthy
Clonmel
Tralee
Kerry
Suir
Wexford
Wexford Harbour
Dingle
Dingle Bay
Mallow
Fermoy
Waterford
Waterford
Rosslare
Killarney
Blackwater
Tramore
Carnsore Pt.
Dungarvan
Hook Hd.
Cahirciveen
Kenmare
Macroom
Lee
Cork (Corcaigh)
Youghal
Cobh
St. David's Hd.
Bandon
Cork
Cork Harbour
St. George's Channel
Bear I.
Bantry Bay
Bantry
Clonakilty
Skibbereen
Mizen Hd.
Clear I.
1 : 3 500 000
0 20 40 60 80 Km
0 10 20 30 40 50 Mi

CZECHOSLOVAKIA

Československá socialistická republika, area 127,876 sq.km, population 15,184,323 (1979), **socialist federal republic** (President Dr Gustáv Husák since 1975).

Administrative units: 2 socialist federal republics – the Czech Socialist Republic (7 regions and the capital of the ČSSR Praha with the status of a region), the Slovak Socialist Republic (3 regions and the capital of the SSR Bratislava with the status of a region). **Capital:** Praha (Prague) 1,191,125 inhab. (1979); **other towns** (1979, in 1,000 inhab.): Bratislava 372, Brno 371, Ostrava 324, Košice 199, Plzeň 169, Olomouc 102, Havířov 94, Hradec Králové 93, Pardubice 92, České Budějovice 89, Liberec 85, Gottwaldov 83, Ústí n. Labem 80, Karviná 80, Nitra 72, Prešov 68, Kladno 66, Žilina 66, Banská Bystrica 65, Karlovy Vary 62, Most 61, Trnava 61. **Population:** Czechs 64%, Slovaks 30.3%, Hungarians 4%, Poles, Germans. **Density** 119 persons per sq.km; average annual rate of population increase 0.62% (1970–75); urban population 57.9% (1975). 35.5% of inhabitants employed in industry (1975). – **Currency:** koruna (Kčs) = 100 halers.

Economy: advanced industrial and agricultural country. **Principal industries:** mineral mining, machinery, metallurgy, foodstuffs, textiles and wood-working. **Mining** (1977, in 1,000 tons, metal content): coal 27,450 (Ostrava – Karviná), brown coal 90,696 (Chomutov, Most, Sokolov), crude petroleum 123, natural gas 936 mn. cub.m (1976), iron ore 518, copper 10 (1976), lead 15.6 (1976), tin 180 tons (1976), mercury 183 tons, antimony 285 tons (1976), silver 40 tons (1976), magnesite 2,885 (1975, leading world producer, Slovenské rudohorie), uranium (Českomoravská vrchovina, Příbram), salt 48 (1976), asbestos 43 (1976), graphite, kaolin and glass sands. Electricity 66.5 billion kWh (1977), of which 5% are hydro-electric and 1% nuclear power stations.

Production (1977, in 1,000 tons): pig iron 9,715, crude steel 15,064, coke oven coke 10,861 (Ostrava, Třinec, Kladno, Košice), aluminium 36.5 (Žiar nad Hronom), smelted copper 23 (Krompachy), lead 19; engineering – Praha, Plzeň ("Škoda" Works), Brno, Váh Valley (1977, in 1,000 units): metal cutting machines 37.6, passenger cars 159 (Mladá Boleslav), lorries 38.6 (Kopřivnice), motorcycles 109, tractors 35, electric locomotives 114 units, television receivers 461; chemical industry – Most and Ostrava vicinity, Elbeland, south-western Slovakia (1977, in 1,000 tons): sulphuric acid 1,276, nitrogenous fertilizers 605, phosphorus fertilizers 389, plastics and resins 738, synthetic rubber 58.8 (Kralupy nad Vltavou), capacity of petroleum refineries 17 mn. tons (1976) – Litvínov-Záluží, Bratislava, pipeline for the import of Soviet petroleum; motor spirit 1,600 (1976), oils 12,800 (1976), cement 9,749, paper 857 (Štětí, Ružomberok, Větřní); textile industry uses mainly imported raw materials (northern and north-eastern Bohemia, northern Moravia, Žilina, Ružomberok), cotton yarn 125, woven fabrics – cotton 533.4 mn. m, linen 77 mn. m, woollen 63.5 mn. m, silk 136,000 m, rayon 60.4 mn. m – synthetic fibres 50.2, shoes 128 mn. pairs (Gottwaldov), glass (northern Bohemia), porcelain (Karlovy Vary), ceramics (Horní Bříza near Plzeň), costume jewellery (Jablonec n. N.), musical instruments (Kraslice, Hradec Králové, Krnov), sugar 939 (Elbeland), meat 1,560, milk 5,761, condensed milk 135, butter 121, cheese 151, eggs 233, malt, beer 22.4 mn. hl (Plzeň, Praha, České Budějovice), wine 140; 24 billion cigarettes.

Agriculture: arable land 41.9%, meadows and pastures 13.8%, forests 35.9%. **Crops** (1977, in 1,000 tons): wheat 5,214, rye 641, barley 3,207, maize (corn) 792, oats 600, potatoes 3,760, flax 10, hemp 2, rapeseed 120, sunflower 9, sugar beet 8,229, hops 11 (third greatest producer in the world, Žatec), fruit 667, grapes 218 (Malé Karpaty, southern Moravia), vegetables 850; **livestock** (1977, in 1,000 head): cattle 4,758, pigs 7,510, sheep 841, goats, horses; poultry 44,774; fish catch 17,200 tons (1976); roundwood 17.5 mn. cub.m (1977).

Communications (1977): railways 13,190 km, of which 2,830 km electrified, roads 73,719 km, passenger cars 1,827,688. Navigable waterways 475 km. Merchant shipping 149,000 GRT. Civil aviation (1977): 31.7 mn. km flown, 1,760,000 passengers carried. Tourism 16.8 mn. visitors (1977).

Exports: machines, equipment, machine tools and transport equipment (cutting machines, cars, power stations and other plant equipment, electromotors), fuels (coal, coke), raw materials, hops, malt, beer, glass, porcelain, wood, shoes, textiles. **Imports:** fuels (petroleum, natural gas, coal) and raw materials (iron ores, non-ferrous metals, cotton), machines and equipment, consumer goods, foodstuffs (cereals, fruit, vegetables, beer). Chief trading partners: U.S.S.R., German Dem. Rep., Poland, Hungary, Fed. Rep. of Germany.

DENMARK

Kongeriget Danmark, area 43,075 sq.km, population 5,105,423 (1978), **kingdom** (Queen Margrethe II since 1972).

Administrative units: 14 districts (Ämter) and 2 cities (København and Frederiksberg). The autonomous Faeroe Islands (Færøerne) and Greenland belong to Denmark. **Capital:** København (Copenhagen) 529,154 inhab. (1977), with agglomeration 1,327,940 (1974); **other towns** (1977, in 1,000 inhab.): Århus 246, Odense 168, Ålborg 155, Frederiksberg 91, Esbjerg 79, Gentofte 69, Randers 64, Helsingør 56, Horsens 54. **Population:** Danes. **Density** 119 persons per sq.km; average annual rate of population increase 0.39% (1970–75); urban population 81.5% (1975). 15.8% of inhabitants employed in industry (1975). – **Currency:** Danish krone = 100 øre.

Economy: advanced industrial and agricultural country. **Principal industries:** metal-working and engineering (especially shipbuilding, electrical and radio engineering), foodstuffs, chemicals, paper and textiles. København is the chief industrial centre. **Mining:** petroleum mining under the sea near the west coast (0.5 mn. tons in 1977), lignite, salt 349,000 tons (1976), sulphur. Electricity 20.6 billion kWh (1977), 100% thermal power stations. **Production** (1977, in 1,000 tons): crude steel 684, vessels 633,000 GRT (København and others), cement 2,316, nitrogenous fertilizers 109.[illegible] (1976), meat 1,094, milk 5,139, butter 131, cheese 177, eggs 68.2, fish flour 320, sugar 566, beer 8.3 mn. hl (1976). **Agriculture** is highly developed with intensive animal production. Arable land 63.1%, meadows and pastures 6.2%, forests 11.7%. **Crops** (1977, in 1,000 tons): barley 6,084, wheat 605 (leading world yield per ha 5,261 kg), oats, rye, potatoes 800, sugar beet 3,521; **livestock** (1977, in 1,000 head): cattle 3,095, pigs 7,811, poultry 15,417; fish catch 1,911,600 tons (1976); roundwood 1.6 mn. cub.m (1976).

Communications (1976): railways 2,487 km, roads 66,137 km, motorways 397 km, passenger cars 1,344,200. Merchant shipping 5,331,000 GRT (1977). Chief port København. Civil aviation (1976): 37.8 mn. km flown, 2,796,000 passengers carried.
Exports: industrial products 70% (machinery, transport equipment, vessels, chemical products), agricultural products 30% (meat, live animals, butter, milk products, eggs, fish products). **Imports:** fuels and raw materials, machines, equipment and transport equipment. Chief trading partners: Fed. Rep. of Germany, Sweden, United Kingdom.

FAEROE ISLANDS

Færøerne, area 1,399 sq.km, population 41,575 (1977), **autonomous region of Denmark. Capital:** Thorshavn 11,474 inhab. (1976). **Economy:** potatoes, sheep, fish 341,962 tons (1976), whaling.

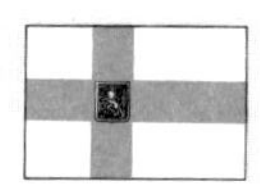

FINLAND

Suomen Tasavalta – Republiken Finland, area 337,032 sq.km, population 4,760,000 (1979), **republic** (President Mauno Koivisto since 1982).

Administrative units: 12 regions (Laäni), of which islands Ahvenanmaa (Åland) are an autonomous province. **Capital:** Helsinki 490,205 inhab. (1977); **other towns** (1977, in 1,000 inhab.): Tampere 165, Turku 165, Espoo 126, Vantaa 125, Lahti 95, Oulu 93, Pori 80, Kuopio 73. **Population:** Finns 93.3%, Swedes 6.5%. **Density** 14 persons per sq.km; average annual rate of population increase 0.2% (1970–75); urban population 54.9% (1975). 22.1% of inhabitants employed in industry (1976). – **Currency:** markka = 100 penni.
Economy: developed industrial and agricultural country with modern industry, intensive agriculture and forestry. Large resources of water power. **Chief industries:** wood working, chemical wood pulp and paper manufacturing (Lahti, Kuopio, Kotka, Kemi, Oulu and others), metallurgy, shipbuilding, engineering (Helsinki, Turku, Tampere), foodstuffs and textiles. **Mining** (1976, in 1,000 tons, metal content): iron ore 768, chromium 162.6, copper 46.8 (1977, Ylöjärvi), lead 1.1, zinc 61.3 (1977, Kisko), nickel 6,433 tons (Leppävirta), cobalt, vanadium 1,534 tons, mercury 13 tons, gold 818 kg (Haveri), silver 24 tons, sulphur 84 (1975), pyrites, asbestos. Electricity 31.6 billion kWh (1977), of which hydro-electric power stations 33%. **Production** (1977, in 1,000 tons): pig iron 1,764, crude steel 2,304 (Tampere, Imatra, Raahe), copper – smelted 60.8, refined 38.1 (Harjavalta, 1976) – zinc 138, vessels 324,000 GRT (Raahe, Vaasa, Pori), sulphuric acid 985 (Harjavalta), nitrogenous fertilizers 190, phosphorus fertilizers 148.6, capacity of petroleum refineries 15 mn. tons (1976, Naantali), motor spirit 1.7 mn. tons (1976), oils 7.4 mn. tons (1976), cement 1,712, wood working and paper industry – Lahti, Kuopio, Kemi, Kotka, Oulu and others (1977, in 1,000 tons): chemical wood pulp 3,228, paper products 4,180, plywood 360,000 cub.m, sawnwood 6,269,000 cub.m; textile industry (Tampere, Turku, Pori): rayon and acetate staple fibre 31.2 (1976), glass industry (Riihimäki, Lahti), porcelain (Helsinki), meat 267, milk 3,269, cheese 62.6, butter 74.3, eggs 84; 6.4 billion cigarettes (1976).
Agriculture: arable land 8.6%, meadows and pastures 0.5%, forests 74.2% – the chief natural wealth of the country. **Crops** (1977, in 1,000 tons): cereals 2,888 (barley 1,447, oats 1,022, wheat), hay 1,879, potatoes 737, sugar beet 560; **livestock** (1977, in 1,000 head): cattle 1,762, pigs 1,145, sheep; poultry 8,708, raising of reindeer 234 (1976) and fur animals; fish catch 120,500 tons (1976); roundwood 33 mn. cub.m (1976).– **Communications** (1976): railways 6,034 km, roads 73,341 km (1975), motorways 230 km, passenger cars 1,032,900. Navigable waterways 6,675 km. Merchant shipping 2,262,000 GRT (1977). Chief port Helsinki. Civil aviation (1976): 30 mn. km flown, 2,040,000 passengers carried. Tourism 4.9 mn. visitors (1974).
Exports: wood working and paper industry products, machines and equipment, iron and steel, chemical and food products. **Imports:** raw materials, semi-finished products, fuels. Chief trading partners: U.S.S.R., Sweden, United Kingdom.

FRANCE

République Française, area 543,965 sq.km, population 53,241,000 (1978), **republic** (President François Mitterrand since 1981).

Administrative units: The Republic of France comprises 96 metropolitan departments which make up 22 administrative regions. **Overseas departments:** Guadeloupe, Martinique, French Guiana, Réunion, St. Pierre et Miquelon, Mayotte; overseas territories New Caledonia, the Wallis and Futuna islands, and French Polynesia.
Capital: Paris 2,317,227 inhab. (1975, with agglomeration 9,878,524 inhab.); **other towns** (1975, in 1,000 inhab.): Marseille 914, Lyon 463, Toulouse 383, Nice 347, Nantes 264, Strasbourg 257, Bordeaux 226, Saint-Étienne 222, Le Havre 220, Rennes 206, Montpellier 196, Toulon 185, Reims 184, Lille 177, Brest 172, Grenoble 170, Clermont-Ferrand 161, Dijon 157, Le Mans 155. **Population:** French 90%, Italians, Spaniards, Algerians, Portuguese and others. **Density** 98 persons per sq.km; average annual rate of population increase 0.87% (1970–75); urban population 75.9% (1975). 25.7% of inhabitants employed in industry (1976). – **Currency:** French franc = 100 centimes.
Economy: highly developed industrial and agricultural country. With its high concentration of industrial plants France belongs to the leading group of industrial countries in the world. Industry has an insufficient fuel and power base and has to import coal, coke and petroleum. **Chief industries:** mineral mining, metallurgy, engineering, energy production, chemicals, electrical and radio engineering, ship and aircraft building, textile and food industries.
Mining (1977, in 1,000 tons, metal content): coal 23,304 (Nord, Pas-de-Calais, reg. Lorraine), lignite 3,132, crude petroleum 1,044 (reg. Aquitaine, Paris Basin, reg. Alsace), natural gas 7.7 billion cub.m (Lacq, Saint Marcet), iron ore 10,991 (Metz-Thionville and Briey-Longwy areas), bauxite 2,028 (Brignoles, Bédarieux, Les Baux), manganese, lead 31.5, zinc 41.8, tungsten 1,015 tons (1976), vanadium 90 tons (1974), antimony, gold 1,586 kg (1976), silver 114 tons (1976), uranium 2,063 tons (1976, Massif Central), salt 5,575 (1976), potassium salt 1,718 (at Mulhouse), phosphates,

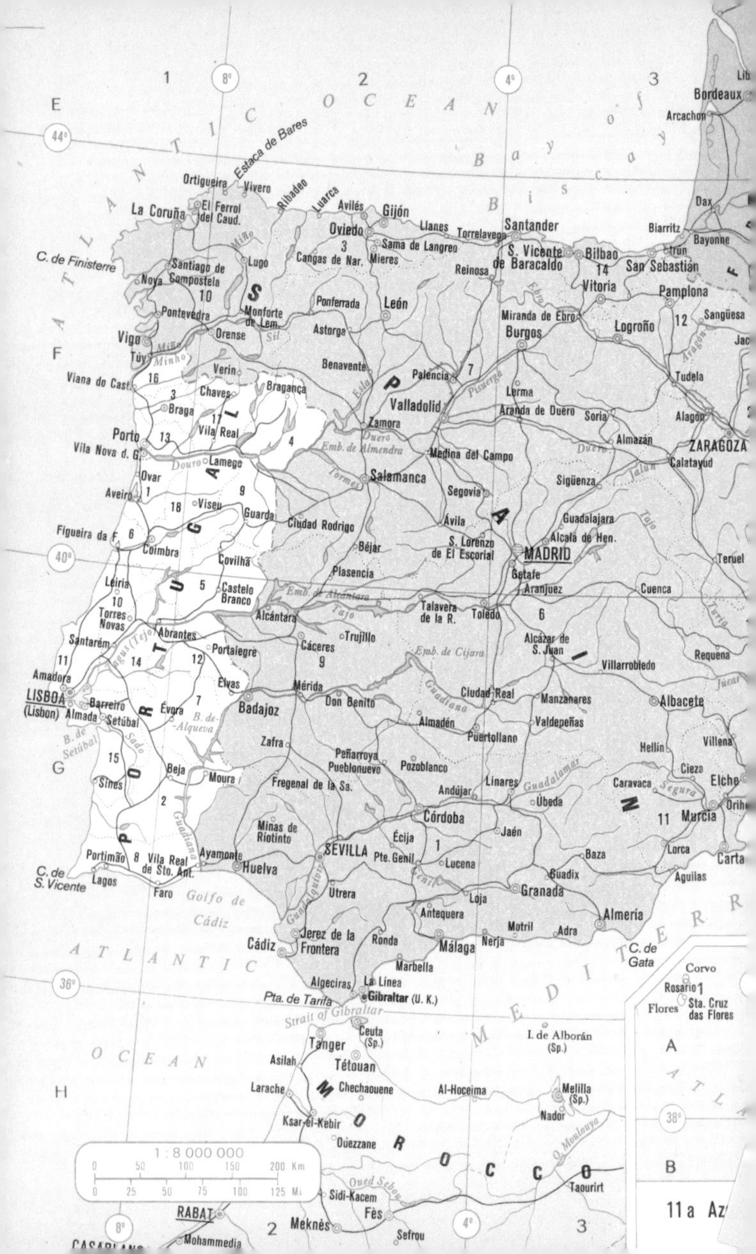
ATLANTIC OCEAN
Bay of Biscay
SPAIN
PORTUGAL
MOROCCO
MEDITERR
Estaca de Bares
Ortigueira
Vivero
Ribadeo
Luarca
Avilés
Gijón
La Coruña
El Ferrol del Caud.
Oviedo
Llanes
Torrelavega
Santander
Sama de Langreo
Cangas de Nar.
Mieres
C. de Finisterre
Santiago de Compostela
Lugo
Noya
Reinosa
S. Vicente de Baracaldo
Bilbao
San Sebastián
Vitoria
Pamplona
Sangüesa
Bordeaux
Arcachon
Dax
Biarritz
Bayonne
Irun
Pontevedra
Monforte de Lem.
Ponferrada
León
Miranda de Ebro
Logroño
Vigo
Túy
Orense
Astorga
Burgos
Viana do Cast.
Verin
Chaves
Bragança
Benavente
Palencia
Lerma
Tudela
Braga
Valladolid
Aranda de Duero
Soria
Alagón
Porto
Vila Real
Zamora
Almazán
Zaragoza
Vila Nova d. G.
Lamego
Medina del Campo
Calatayud
Ovar
Salamanca
Sigüenza
Aveiro
Viseu
Segovia
Guarda
Ciudad Rodrigo
Ávila
Guadalajara
Figueira da F.
Coimbra
Béjar
S. Lorenzo de El Escorial
Alcalá de Hen.
MADRID
Covilhã
Teruel
Plasencia
Getafe
Leiria
Castelo Branco
Aranjuez
Cuenca
Torres Novas
Alcántara
Talavera de la R.
Toledo
Santarém
Abrantes
Trujillo
Alcázar de S. Juan
Portalegre
Cáceres
Requena
Villarrobledo
Amadora
Elvas
Mérida
LISBOA (Lisbon)
Barreiro
Almada
Setúbal
Évora
Badajoz
Don Benito
Ciudad Real
Manzanares
Albacete
Almadén
Valdepeñas
Puertollano
Zafra
Peñarroya Pueblonuevo
Pozoblanco
Hellín
Villena
Sines
Beja
Moura
Fregenal de la Sa.
Linares
Andújar
Úbeda
Caravaca
Cieza
Elche
Córdoba
Murcia
Minas de Riotinto
Écija
Jaén
Portimão
Vila Real de Sto. Ant.
Ayamonte
SEVILLA
Pte. Genil
Lucena
Baza
Lorca
C. de S. Vicente
Lagos
Huelva
Guadix
Aguilas
Faro
Golfo de Cádiz
Utrera
Loja
Granada
Antequera
Almería
Jerez de la Frontera
Ronda
Motril
Adra
Cádiz
Málaga
Nerja
C. de Gata
Marbella
Algeciras
La Línea
Gibraltar (U. K.)
Pta. de Tarifa
Strait of Gibraltar
Ceuta (Sp.)
Tanger
I. de Alborán (Sp.)
Asilah
Tétouan
Larache
Chechaouene
Al-Hoceima
Melilla (Sp.)
Ksar-el-Kebir
Nador
Ouezzane
Taourirt
Sidi-Kacem
Oued Sebou
O. Moulouya
RABAT
Meknès
Fès
Sefrou
Mohammedia
Corvo
Rosario
Flores
Sta. Cruz das Flores
11a Az
1:8 000 000
0 50 100 150 200 Km
0 25 50 75 100 125 Mi
Miño
Minho
Sil
Esla
Duero
Douro
Emb. de Almendra
Tormes
Pisuerga
Ebro
Aragón
Jalón
Tajo
Tagus (Tejo)
Emb. de Alcántara
Emb. de Cijara
Guadiana
B. de Alqueva
Sado
B. de Setúbal
Guadalquivir
Genil
Guadalimar
Segura
Júcar
Turia

sulphur 1,900. Electricity 201.7 billion kWh (1977), of which 38% hydro-electric power stations (Alps, Massif Central and Pyrenees) and 13.5% nuclear power stations (Chinon, Marcoule, Fessenheim).

Industry: manufacturing is concentrated in the Paris Basin and the territory of the lower Seine, the North, East, Lyon district, Atlantic ports and Marseille. A major part of metallurgy is situated near the resources: reg. Lorraine (from Longwy to Nancy), Nord (Dunkerque, Valenciennes and others), centre (Le Creusot, Saint-Étienne). The main concentration of engineering is to be found around Paris, Lille and its surroundings and Lyon. Chemical industry: in the coal and metallurgical regions, Paris region, petroleum processing in the ports. Textile industry: reg. Alsace, Lille and vicinity, Lyon.

Production (1977, in 1,000 tons): pig iron 18,312, crude steel 22,116, coke oven coke 11,311 (1976), aluminium 399.7 (Saint-Jean-de-Maurienne, Noguères), zinc 238, lead 184, refined copper 44.4, magnesium 8,006 tons (1976), synthetic nitrogen 1,781, nitric acid 3,069 (1976), sulphuric acid 4,504, hydrochloric acid 210.9 (1976), nitrogenous fertilizers 1,462, caustic soda 1,274 (1976), soda ash 1,320, phosphorus fertilizers 1,490, synthetic rubber 480, plastics and resins 2,639; capacity of petroleum refineries 169.5 mn. tons (1976), oil pipelines network 103,351 km (1976), naphtha 4,977 (1976), motor spirit 17,783, oils 79,847 (1976), cosmetics and pharmaceuticals (Paris, Lyon), tyres 46.6 mn. units, motor vehicles – passenger 3,564,000 (Paris, Le Mans, Rennes, Flins, Sochaux), – commercial 540,000 (Lyon-Vénissieux) – ships 1,148,000 GRT (shipyards Saint-Nazaire, La Ciotat, Dunkerque), aircraft (Paris, Toulouse, Bordeaux), locomotives, carriages, railway equipment (Le Creusot, Lille, Belfort), agricultural machines (Vierzon, Beauvais, Saint-Dizier), radio receivers 3,458,000 units (1976), television receivers 1,777,000 (1976); cement 28,700, chemical wood pulp 1,402 (1976), paper 4,610 (1976), yarn – cotton 229.2, woollen 141.5, jute 24.7 – woven fabrics – cotton 176, woollen 62.1, silk 47.9, jute 13.6 – rayon and acetate – fibres 20.8, staple fibre 62.2 – synthetic – fibres 101.3, staple fibre 149.1 – sugar 4,248, meat 4,712, milk 31,381 (third world producer), condensed milk 151, butter 545, cheese 1,004, eggs 744, canned fish 100 (1976), wine 5,420 (second world producer, reg. Languedoc, Bordeaux, reg. Bourgogne, reg. Champagne), alcoholic spirits (Cognac, Fécamp, reg. Armagnac, Isère), beer 24.6 mn. hl (1976); 86.6 billion cigarettes (1976), tobacco 8,299 tons (1976).

Agriculture: France has many large agricultural establishments, although small and medium-sized farms predominate. Animal production exceeds vegetable production. Land use: arable land 34.3%, meadows and pastures 24.4%, forests 26.7%; tractors 1,380,000 (1976), combine harvesters 153,000 (1976). **Crops** (1977, in 1,000 tons): wheat 17,450 (Paris Basin, reg. Picardie), barley 10,290, oats 1,928, rye, maize (corn) 8,614 (reg. Aquitaine, Paris Basin), rice (Camargue), sorghum, potatoes 8,190, flax 36, sugar beet 24,500 (second world producer, Nord, Paris Basin), rapeseed 400, olives, sunflower seed 70, hops, fruit 11,234, apples 2,190 (second world producer), grapes 8,100 (second world producer, principal vine-growing region from the lower Rhône to the Pyrenees and Gironde), strawberries 73.9, vegetables 6,589, tomatoes 625, tobacco 53; **livestock** (1977, in 1,000 head): cattle 23,898, sheep 10,915, goats 1,012, pigs 11,638, horses 375, asses, mules; poultry 207,543; wool 10,900 tons, honey 6,000 tons; fish catch 805,900 tons (1976); roundwood 30.4 mn. cub.m (1976).

Communications: railways 34,299 km, of which 9,341 km are electrified (1976), roads 806,903 km, of which 73,306 km are national and 289,295 km departmental roads; motorways 4,500 km (1978), private cars 15.9 mn. (1976). Navigable waterways 6,931 km, of which canals take up 4,228 km (1976). Merchant shipping 11,614,000 GRT (1977). Chief ports: Marseille (second largest European port, 104 mn. tons of freight in 1976), Le Havre, Dunkerque, Rouen, Nantes, Bordeaux. Civil aviation (1976): 265.6 mn. km flown, 14,302,000 passengers carried. Tourism 13.5 mn. visitors (1976).

Exports: machines, cars, aircraft, raw materials and semi-finished products (iron, ores), textile and chemical products, agricultural products, wine and others. **Imports:** fuels (petroleum, coal) 21%, finished products and equipment (Boeing aircraft and miscellaneous goods) 16.5%, agricultural products (fruit, early vegetables) 16%, raw materials and semi-finished products (cotton, wool, rubber) 13.5%, consumer goods 12%. Chief trading partners: Fed. Rep. of Germany, Italy, Belgium-Luxembourg and other EEC countries, U.S.A.

GERMAN DEMOCRATIC REPUBLIC

Deutsche Demokratische Republik, area 108,178 sq.km, population 16,757,857 (1978), **socialist republic** (Chairman of the Council of State Erich Honecker since 1976).

Administrative units: 14 regions (Bezirke), **the capital** Berlin 1,118,142 inhab. (1977) also has the status of a region; **other towns** (1977, in 1,000 inhab.): Leipzig 564, Dresden 512, Karl-Marx-Stadt 311, Magdeburg 282, Halle 231, Rostock 221, Erfurt 207, Potsdam 125, Zwickau 123, Gera 119, Schwerin 113, Cottbus 105, Dessau 101, Jena 101, Stralsund 73, Weimar 63, Gotha 59. **Population:** Germans (over 99%). **Density** 155 persons per sq.km; average annual rate of population increase only 0.08%; urban population 74.8% (1975). 36% of inhabitants employed in industry. – **Currency:** Deutsche mark = 100 pfennigs.

Economy: highly developed industrial country with intensive agriculture. **Principal industries:** engineering (Saxony, Thuringia, Magdeburg, Berlin with surroundings), electrotechnical (Berlin), chemicals (region of brown coal deposits, Leuna), electronics, precision mechanics and optics (Jena), textiles. **Mining** (1977, in 1,000 tons, metal content): brown coal 253,704 (leading world producer, Thuringian-Saxon Basin – Leipzig, Halle, Merseburg; Lower-Lusatian Basin – Senftenberg, Spremberg), natural gas 9 billion cub.m, copper 16 (1976, Harz), nickel 2,500 tons (1976), silver 50 tons (1976), uranium (Aue), potassium salt 3,161 (1976) and salt 2,560 (1976, Stassfurt, Halberstadt, Bleicherode, valleys of the Werra and Unstrut). Limited mining of coal, iron ore, lead, zinc, tungsten. Electricity 92 billion kWh (1977), of which 93% are thermal and 6% nuclear power stations (at Rheinsberg).

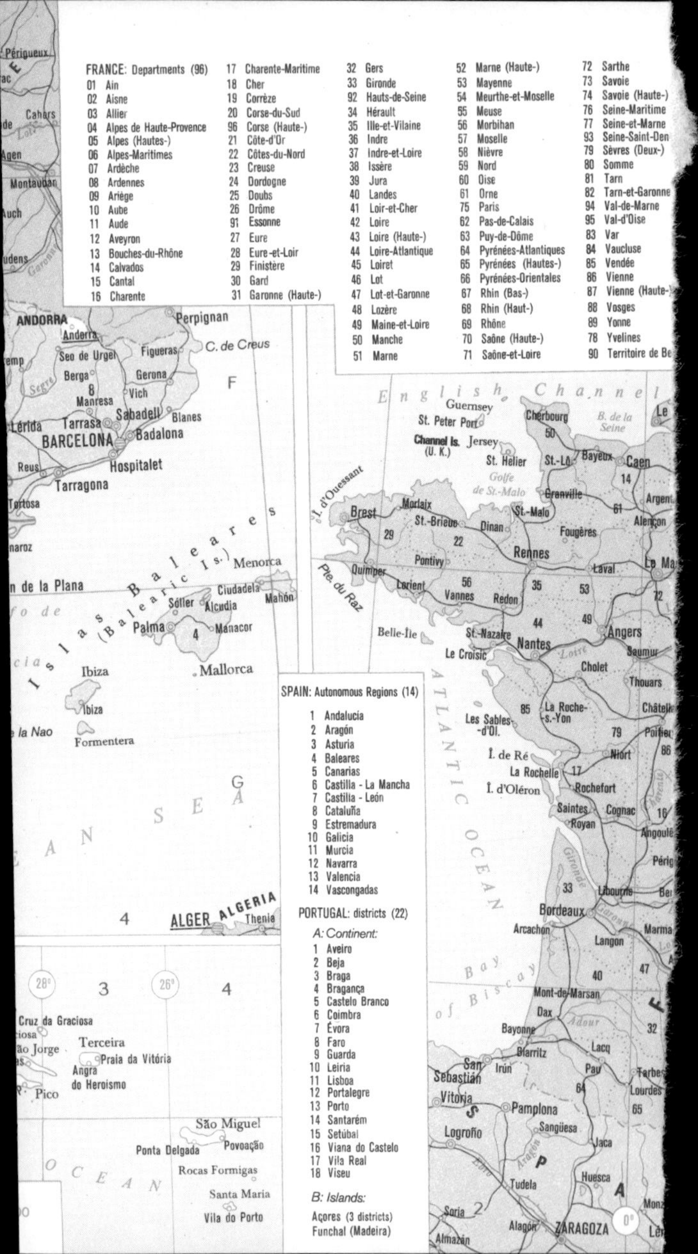

FRANCE: Departments (96)

01 Ain
02 Aisne
03 Allier
04 Alpes de Haute-Provence
05 Alpes (Hautes-)
06 Alpes-Maritimes
07 Ardèche
08 Ardennes
09 Ariège
10 Aube
11 Aude
12 Aveyron
13 Bouches-du-Rhône
14 Calvados
15 Cantal
16 Charente
17 Charente-Maritime
18 Cher
19 Corrèze
20 Corse-du-Sud
96 Corse (Haute-)
21 Côte-d'Or
22 Côtes-du-Nord
23 Creuse
24 Dordogne
25 Doubs
26 Drôme
91 Essonne
27 Eure
28 Eure-et-Loir
29 Finistère
30 Gard
31 Garonne (Haute-)
32 Gers
33 Gironde
92 Hauts-de-Seine
34 Hérault
35 Ille-et-Vilaine
36 Indre
37 Indre-et-Loire
38 Issère
39 Jura
40 Landes
41 Loir-et-Cher
42 Loire
43 Loire (Haute-)
44 Loire-Atlantique
45 Loiret
46 Lot
47 Lot-et-Garonne
48 Lozère
49 Maine-et-Loire
50 Manche
51 Marne
52 Marne (Haute-)
53 Mayenne
54 Meurthe-et-Moselle
55 Meuse
56 Morbihan
57 Moselle
58 Nièvre
59 Nord
60 Oise
61 Orne
75 Paris
62 Pas-de-Calais
63 Puy-de-Dôme
64 Pyrénées-Atlantiques
65 Pyrénées (Hautes-)
66 Pyrénées-Orientales
67 Rhin (Bas-)
68 Rhin (Haut-)
69 Rhône
70 Saône (Haute-)
71 Saône-et-Loire
72 Sarthe
73 Savoie
74 Savoie (Haute-)
76 Seine-Maritime
77 Seine-et-Marne
93 Seine-Saint-Den
79 Sèvres (Deux-)
80 Somme
81 Tarn
82 Tarn-et-Garonne
94 Val-de-Marne
95 Val-d'Oise
83 Var
84 Vaucluse
85 Vendée
86 Vienne
87 Vienne (Haute-)
88 Vosges
89 Yonne
78 Yvelines
90 Territoire de Be

SPAIN: Autonomous Regions (14)

1 Andalucía
2 Aragón
3 Asturia
4 Baleares
5 Canarias
6 Castilla - La Mancha
7 Castilla - León
8 Cataluña
9 Estremadura
10 Galicia
11 Murcia
12 Navarra
13 Valencia
14 Vascongadas

PORTUGAL: districts (22)

A: Continent:

1 Aveiro
2 Beja
3 Braga
4 Bragança
5 Castelo Branco
6 Coimbra
7 Évora
8 Faro
9 Guarda
10 Leiria
11 Lisboa
12 Portalegre
13 Porto
14 Santarém
15 Setúbal
16 Viana do Castelo
17 Vila Real
18 Viseu

B: Islands:

Açores (3 districts)
Funchal (Madeira)

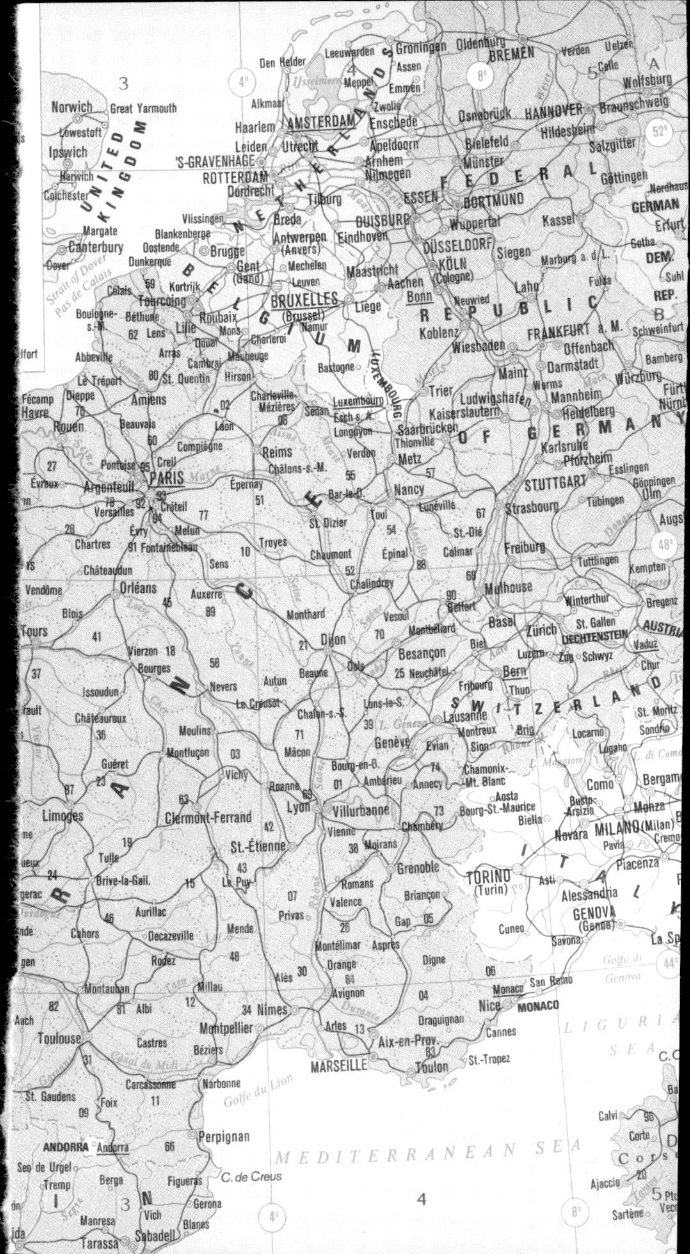

UNITED KINGDOM
NETHERLANDS
BELGIUM
LUXEMBOURG
FEDERAL REPUBLIC OF GERMANY
GERMAN DEM. REP.
FRANCE
SWITZERLAND
LIECHTENSTEIN
AUSTRIA
ITALY
MONACO
ANDORRA
SPAIN
MEDITERRANEAN SEA
LIGURIAN SEA
Golfe du Lion
Golfo di Genova
Strait of Dover
Pas de Calais
IJsselmeer
C. de Creus
Norwich
Great Yarmouth
Lowestoft
Ipswich
Harwich
Colchester
Margate
Canterbury
Dover
Den Helder
Leeuwarden
Groningen
Assen
Meppel
Emmen
Alkmaar
Zwolle
Haarlem
AMSTERDAM
Enschede
Leiden
Utrecht
Apeldoorn
'S-GRAVENHAGE
Arnhem
ROTTERDAM
Nijmegen
Dordrecht
Tilburg
Vlissingen
Breda
Blankenberge
Antwerpen
(Anvers)
Eindhoven
Oostende
Brugge
Gent
(Gand)
Mechelen
Leuven
Maastricht
Dunkerque
Kortrijk
Calais
Tourcoing
BRUXELLES
(Brussel)
Liège
Boulogne-s.-M.
Béthune
Roubaix
Lille
Lens
Mons
Namur
Douai
Charleroi
Arras
Cambrai
Maubeuge
Bastogne
Luxembourg
Esch s. A.
Abbeville
Le Tréport
St. Quentin
Hirson
Charleville-Mézières
Sedan
Longuyon
Fécamp
Dieppe
Havre
Amiens
Rouen
Beauvais
Laon
Compiègne
Reims
Verdun
Pontoise
Creil
Châlons-s.-M.
Évreux
Argenteuil
PARIS
Épernay
Bar-le-D.
Versailles
Créteil
St. Dizier
Toul
Évry
Melun
Chartres
Fontainebleau
Troyes
Chaumont
Épinal
Sens
Châteaudun
Chalindrey
Vendôme
Orléans
Auxerre
Blois
Monthard
Tours
Dijon
Vierzon
Bourges
Beaune
Dole
Nevers
Autun
Issoudun
Le Creusot
Chalon-s.-S.
Lons-le-S.
Châteauroux
Moulins
Montluçon
Mâcon
Bourg-en-B.
Guéret
Vichy
Ambérieu
Roanne
Limoges
Lyon
Villurbanne
Clermont-Ferrand
Vienne
St.-Étienne
Tulle
Brive-la-Gail.
Le Puy
Aurillac
Privas
Valence
Romans
Cahors
Decazeville
Mende
Montélimar
Rodez
Orange
Alès
Avignon
Montauban
Millau
Albi
Nîmes
Arles
Aix-en-Prov.
Toulouse
Montpellier
Castres
Béziers
MARSEILLE
Toulon
Carcassonne
Narbonne
St. Gaudens
Foix
Perpignan
Andorra
Seo de Urgel
Tremp
Berga
Figueras
Gerona
Manresa
Vich
Blanes
Sabadell
Tarassa
Vesoul
Montbéliard
Besançon
Neuchâtel
Belfort
Mulhouse
Basel
Colmar
St.-Dié
Lunéville
Nancy
Metz
Thionville
Saarbrücken
Strasbourg
Freiburg
Genève
Evian
Annecy
Chambéry
Moirans
Grenoble
Briançon
Gap
Aspres
Digne
Draguignan
Cannes
St.-Tropez
Nice
Monaco
San Remo
Chamonix-Mt. Blanc
Bourg-St.-Maurice
Aosta
Biella
Lausanne
Montreux
Sion
Brig
Fribourg
Bern
Thun
Biel
Luzern
Zug
Schwyz
Zürich
Winterthur
St. Gallen
Vaduz
Chur
Bregenz
Locarno
Lugano
St. Moritz
Sondrio
Como
Bergamo
Busto-Arsizio
Monza
MILANO (Milan)
Novara
Pavia
Cremona
Piacenza
TORINO (Turin)
Asti
Alessandria
GENOVA (Genoa)
Cuneo
Savona
La Spezia
Calvi
Corte
Ajaccio
Sartène
Corse
Oldenburg
BREMEN
Verden
Uelzen
Celle
Wolfsburg
Osnabrück
HANNOVER
Braunschweig
Hildesheim
Bielefeld
Salzgitter
Münster
Göttingen
Nordhausen
ESSEN
DORTMUND
DUISBURG
Wuppertal
Kassel
Erfurt
Gotha
DÜSSELDORF
Siegen
Marburg a. d. L.
KÖLN
(Cologne)
Aachen
Bonn
Lahn
Fulda
Suhl
Neuwied
Koblenz
FRANKFURT a. M.
Schweinfurt
Wiesbaden
Offenbach
Mainz
Darmstadt
Bamberg
Würzburg
Trier
Worms
Ludwigshafen
Mannheim
Kaiserslautern
Heidelberg
Karlsruhe
Pforzheim
Esslingen
STUTTGART
Göppingen
Ulm
Tübingen
Tuttlingen
Kempten
Rhein
Maas
Mosel
Main
Donau
Seine
Marne
Oise
Aisne
Meuse
Loire
Yonne
Saône
Rhône
Isère
Durance
Lot
Tarn
Garonne
Dordogne
Allier
Cher
Somme
Doubs
Aare
L. Genève
L. Maggiore
L. di Como
Po
Canal du Midi
Weser
Segre

Production (1977, in 1,000 tons): pig iron 2,628, crude steel 6,852, aluminium 60, copper – smelted 16, refined 50 (1976) – vessels 372,000 GRT (Rostock, Wismar), passenger cars 167,000 units (Eisenach, Zwickau), radio receivers 1.1 mn. units (1976), television receivers 526,000 units, sulphuric acid 928, soda ash 903, caustic soda 424, fertilizers: nitrogenous 776, phosphorus 384; synthetic rubber 146 (Schkopau), plastics and resins 729, photographic materials (Wolfen), electrochemistry (Bitterfeld); capacity of petroleum refineries 18.8 mn. tons (1976), pipeline for the import of Soviet petroleum, motor spirit 3 mn. tons (1976), oils 14.3 mn. tons (1976), cement 12.1 mn. tons, chemical wood pulp 454 (1976, Premnitz), paper 1,215 (1976). Textile industry mainly in Saxony, Thuringia (1976, in 1,000 tons): yarn – cotton 138.1, woollen 83.6 – woven fabrics – cotton 274 mn. sq.m, woollen 39 mn. sq.m, rayon and acetate 58.3 mn. sq.m – fibres – rayon and acetate 34, synthetic 54 – staple fibre – rayon and acetate 137, synthetic 68 – shoes 80 mn. pairs. Food industry (1977, in 1,000 tons): meat 1,671, milk 8,026, butter 270, cheese 180.5, eggs 293, sugar 780, beer 21.2 mn. hl (1976); 19.8 billion cigarettes (1976). Glass industry (Jena), porcelain (Meissen), printing (Leipzig, Gotha).
Agriculture: mainly cultivation of cereals and potatoes. Land use: arable land 47.1%, meadows and pastures 12.2%, forests 27.8%. **Crops** (1977, in 1,000 tons): wheat 3,100, barley 3,400, rye 1,500, oats 650, potatoes 9,976, flax, sugar beet 5,264, rapeseed 315, vegetables 926, tobacco; **livestock** (1977, in 1,000 head): cattle 5,471, pigs 11,291, sheep 1,870; poultry 48,445, honey 8,000 tons; fish catch 279,200 tons (1976); roundwood 8.7 mn. cub.m (1976).– **Communications:** railways 14,306 km (1976), of which 1,508 km are electrified, roads 47,500 km (1976), passenger cars 2,052,200 (1976). Navigable waterways 2,538 km (1976). Merchant shipping 1,487,000 GRT (1977). Chief ports: Wismar, Rostock, Stralsund. Civil aviation (1976): 1,088,000 passengers carried. Tourism 1.1 mn. visitors (1976).
Exports: machines and equipment, transport equipment, precision instruments, optics and electronics, chemical products, dyes, photo materials, brown coal, potassium salts and consumer goods. **Imports:** petroleum, ores and other raw materials, foodstuffs. Chief trading partners (1977): socialist countries 68%, of which U.S.S.R. 34%, Fed. Rep. of Germany, Netherlands and United Kingdom.

GERMANY, FEDERAL REPUBLIC OF

Bundesrepublik Deutschland, area 248,097 sq.km, population 59,398,000 (1978), **federal republic** (President Dr Karl Carstens since 1979).

Administrative units: 10 federal countries (Schleswig-Holstein, Hamburg, Niedersachsen, Bremen, Nordrhein-Westfalen, Hessen, Rheinland-Pfalz, Baden-Württemberg, Bayern /Bavaria/, Saarland). **Capital:** Bonn 284,300 inhab. (1977); **other towns** (1977, in 1,000 inhab.): Hamburg 1,688, München (Munich) 1,315, Köln (Cologne) 978, Essen 667, Frankfurt am Main 635, Dortmund 621, Düsseldorf 612, Stuttgart 588, Duisburg 578, Bremen 565, Hannover 544, Nürnberg 491, Bochum 411, Wuppertal 400, Gelsenkirchen 316, Bielefeld 314, Mannheim 307, Karlsruhe 275, Wiesbaden 270, Münster 266, Braunschweig 266, Mönchengladbach 259, Kiel 257, Augsburg 245, Aachen 242, Oberhausen 234, Saarbrücken 202.
Population: Germans. **Density** 239 persons per sq.km; average annual rate of population increase 0.32% (1970–75); urban population 83.3% (1975). 26,4% of inhabitants employed in industry (1976). There were 1.9 mn. foreign workers in 1977. – **Currency:** Deutsche mark = 100 pfennigs.
Economy: highly developed industrial country with advanced agriculture. The Fed. Rep. of Germany belongs economically among the most advanced countries of the world, ranking fourth in value of production, after the U.S.A., the U.S.S.R. and Japan, and it holds a decisive position in the EEC.
Industry: the 50 largest firms produce more than 50% of the total industrial output. **Principal industries:** mining, metallurgy, engineering (shipbuilding, manufacture of motor vehicles, electrotechnical), chemicals, building, textiles and food processing. The production of optical instruments, watches, toys, musical instruments and jewellery is important, too. The principal economic region is the Ruhr agglomeration with more than one third of total industrial production in the country, followed by the Saarland, Siegerland, Peine-Salzgitter and metropolitan agglomerations. **Mining** (1977, in 1,000 tons, metal content): coal 85,500 (the Ruhr Basin, Aachen, Saarland), brown coal 122,928 (third world producer, reg. Ville near Köln), crude petroleum 5,412 (Emsland, Hannover region, smaller resources near Hamburg and in Bayern, extensive pipeline network from abroad; Ingolstadt is a major centre of petroleum industry, natural gas 18.9 billion cub.m (Emsland, Niedersachsen, Rehden, Hengstlage; gas pipeline from Netherlands resources in Groningen to Hamburg), iron ore 791 (Peine, Salzgitter, Siegen and Amberg), lead 31, zinc 114 (Harz, Sauerland), copper 1.2, gold 10,789 kg (1976), silver 33 tons (1976), uranium 38 tons (1976, Schwarzwald), salt 11,602 (1976, Schwäbisch Hall, Berchtesgaden), potassium salt 2,340 (valleys of the Leine and Werra rivers), pyrites 554, sulphur. Electricity 324 billion kWh (1977), of which 4% are hydro-electric and 7% nuclear power stations (Biblis, Neckarwestheim, Brunsbüttel, Würgassen).
Production: metallurgy (1977, in 1,000 tons): pig iron 29,148, crude steel 39,988 (most plants in the Ruhr Basin – Duisburg, Oberhausen, Bochum, Gelsenkirchen, in Niedersachsen – Peine, Salzgitter, in the Saarland and elsewhere), lead 310 (Braubach, Nordenham), zinc – primary 210, secondary 278.4 (Datteln, Harlingerode), aluminium – primary 741.6, secondary 390 (Töging, Rheinfelden), copper – smelted 190, refined 440 (Hamburg, Lünen), magnesium 545 tons (1976), tin 6,800 tons (Essen), coke oven coke 28 mn. tons.
Engineering: the Ruhr Basin (Düsseldorf, Wuppertal, Köln), Hamburg, Bremen, Solingen, Stuttgart and others (1977, in 1,000 units): motor vehicles – passenger 3,796 (third world producer, Wolfsburg – "Volkswagen", Rüsselsheim, Bochum – "Opel", Köln – "Ford", Stuttgart – "Mercedes", München – "BMW"), commercial 276 – locomotives (München, Essen, Düsseldorf), carriages (Köln, Braunschweig), tractors 129 (Hannover, Kassel), vessels 1,373,000 GRT (third world producer, Hamburg, Bremen, Emden, Kiel), printing

add page 58

machines (Augsburg, Offenbach), textile machines (Mönchengladbach, Esslingen), agricultural machines (Mannheim, Hannover), precision engineering (München, Kassel, Göttingen), watches (Schwarzwald), radio receivers 5,400 (1976), television receivers 3,700 (1976).
Chemical industry in the lower Rhine zone (Köln–Leverkusen–Ruhr Basin), southern zone (from Mannheim–Ludwigshafen to Frankfurt am Main, 1977, in 1,000 tons): acids: sulphuric 4,584, hydrochloric 870 (1976), nitrogenous 2,792 (1976); soda ash 1,340, caustic soda 3,108 (second world producer), synthetic nitrogen 1,923, chlorine 2,820; fertilizers: nitrogenous 1,290, phosphorus 733, potash 2,217; plastics and resins 6,260, synthetic rubber 410, dyes (Frankfurt am Main), pharmaceuticals and photo materials (Leverkusen), synthetic fibres 837; capacity of petroleum refineries 154 mn. tons (1976, Ruhr Basin, Karlsruhe, Ingolstadt, Hamburg), naphtha 2,500 (1976), motor spirit 14,500, oils 69,700 (1976), tyres 38.2 mn. units. Production of cement 32.2 mn. tons (1977). **Chemical wood pulp and paper industry** (1976, in 1,000 tons): chemical wood pulp 813, newsprint 501, paper and paper products 5,919. **Textile industry** in the Rhineland (from Aachen, Krefeld to Bielefeld), Münster, Osnabrück, south-west zone (Esslingen, Reutlingen), Augsburg, Kempten, Hof (1977, in 1,000 tons): yarn – cotton 177.6, woollen 53.9, flax and hemp 4.6, jute 11, woven fabrics (in mn. sq.m): – cotton 169.2, woollen 36.7, rayon and acetate 448.4 (1976), footwear (Pirmasens, Stuttgart), sale of furs (Frankfurt am Main). Glass industry (Ruhr Basin, Saarland), optics, photographic apparatus (München, Stuttgart), ceramics (München), musical instruments (Trossingen, Mittenwald), jewellery (Pforzheim), toys.
Food industry: mainly in large cities and surroundings (1977, in 1,000 tons): meat 4,269, milk 22,546, condensed milk 478, butter 536, cheese 692, margarine 532 (1976, third world producer), eggs 894, honey 16, sugar 2,940 (Braunschweig), wine 956, beer 91.4 mn. hl (1976, second world producer, Bayern – München, Nürnberg); 149 billion cigarettes (1976), 2.4 billion cigars (1976).
Agriculture is very intensive. 80% of the food supply derives from domestic agricultural resources. Animal production predominates (3/4 of agricultural production). Land use: arable land 33%, meadows and pastures 21.4%, forests 29.4%. High average hectare yields, extensive use of synthetic fertilizers. Tractors 1,452,661 (1976), combine harvesters 170,900 (1976). **Crops** (1977, in 1,000 tons): wheat 7,181 (Rhineland and Danubeland), barley 7,497, rye 2,538 (North German Lowlands), oats 2,723, maize (corn) 571, potatoes 11,251, sugar beet 20,294 (surroundigs of Braunschweig and Köln), rapeseed 279, fruit 3,496, grapes 1,330 (cultivation of fruit and viniculture – middle Rhineland, valleys of the Mosel, Main, Neckar and others), vegetables 1,851, hops 37 (leading world producer, Danubeland), flax, tobacco; **livestock** (1977, in 1,000 head): cattle 14,496, pigs 20,589, horses 355, sheep 1,091, goats; poultry 88,085; fish catch 454,400 tons (1976); roundwood 30 mn. cub.m (1976).
Communications (1976): railways 28,576 km, of which 10,349 km are electrified, roads 135,465 km, motorways 6,207 km, passenger cars 18,919,700. Navigable waterways 4,283 km, (river transport – the Rhein and the Ruhr and North German canal system), the largest river port Duisburg. Merchant shipping 9,592,000 GRT (1977). Chief ports: Hamburg (turnover 51.5 mn. tons in 1976), Bremen, Wilhelmshaven, Emden, Lübeck. Civil aviation (1976): 180.3 mn. km flown, 10,419,000 passengers carried. Tourism 7.9 mn. visitors (1976).
Foreign trade; The Fed. Rep. of Germany is the second most important trading country in the world.-**Exports:** machines of all kinds, cars, chemical and electrotechnical products, iron and steel, textiles, products of precision mechanics, coal, metals. **Imports:** finished products (machines, motor vehicles, electrotechnical products, textile and clothes, paper and paper products, semi-finished products (non-ferrous metals, fuel and lubricating oils), raw materials (petroleum, iron ore, cotton, wool), foodstuffs (fruit, vegetables, meat, coffee, tobacco). Chief trading partners: EEC countries, Austria, U.S.A., Switzerland.

GIBRALTAR

Dominion of Gibraltar, area 6 sq.km, population 28,275 (1977), **British territory** since 1704, **with extended internal autonomy** according to the 1969 Constitution. Security, foreign affairs and defence fall within the competence of the British Governor (Governor Sir William Jackson). – **Currency:** Gibraltar pound = 100 pence. **Importance:** British naval and air base of great strategic importance, also a merchant port. Transit trade, fishing, food processing. Merchant shipping 28,850 GRT (1975). Tourism 125,219 visitors (1976).

GREECE

Elliniki Dimokratia, area 131,944 sq.km, population 9,284,000 (1977), **republic** (President Konstantinos Karamanlis since 1980).

Administrative units: 9 provinces, 51 prefectures (Nomói). One of these is the monastic state of Mount Áthos (Áyion Óros), area 336 sq.km, 1,732 inhab. in 1971 on the Khalkidhiki Peninsula. **Capital:** Athínai (Athens) 867,023 inhab. (1971), Greater Athínai with agglomeration 2,540,241 inhab.; **other towns** (1971, in 1,000 inhab.): Thessaloniki (Salonica) 346, Pátrai 112, Iráklion 78, Lárisa 72, Vólos 51, Kavála 47, Ioánnina 40. **Population:** Greeks 95%, Macedonians, Turks, Albanians. **Density** 70 persons per sq.km; average annual rate of population increase 0.31% (1970–75); urban population 56.7% (1975). 39.6% of inhabitants employed in agriculture (1977), 15.2% in industry (1974). – **Currency:** drachma = 100 lepta.
Economy: industrial and agricultural country with heavy foreign investments (Philips, Pirelli, Benz, Péchiney and others). **Principal industries:** textiles, food processing, chemicals and mining. Heavy industry is only developing. **Mining** (1976, in 1,000 tons, metal content): brown coal 23,400 (1977, Ptolemais, Alivérion, Megalópolis), iron ore 882 (1977, Khalkidhiki), manganese 3.7, chromium 14.7 (reg. Thessalia, Kozáni), lead 30.3, zinc 28.3, nickel 9,600 tons (1977, Larimna), silver 15 tons, bauxite 2,880 (1977, Elevsis, Distomon), magnesite 1,251 (island Évvoia), salt 149, pyrite 110 (1975), emery (island Náxos), marble. Electricity 17.2 billion kWh (1977). **Production** (1977, in 1,000 tons): crude steel 780, aluminium 130.8 (Distomon); capacity of petroleum refineries 12.9 mn. tons (1976 – Thessaloniki, Athínai), petroleum products 8.5 mn. tons; sulphuric acid 1,098, phosphorus fertilizers 169, cement 10,600, cotton yarn 87, woollen yarn 14, carpets (Thessaloniki, Athínai), sugar 330, meat 437, milk 1,732, cheese 165, eggs 118, canned food, raisins 131, wine 452, olive oil 254 (third world producer); 22.9 billion cigarettes (1976).

1
4°
2
8°
12°
A
B
C
D
56°
52°
48°
NORTH SEA
Grenen
Hjørring
Frederikshavn
Göteborg
Borås
Thisted
Læsø
Varberg
Ålborg
Kattegat
Holstebro
Viborg
Randers
Halmstad
Herning
Århus
Hässleholm
Helsingør
Helsingborg
Horsens
Vejle
KØBENHAVN
(Copenhagen)
Lund
The Sound
Malmö
Esbjerg
Kolding
Kalundborg
Odense
Nyborg
Roskilde
Ribe
Fyn
Korsør
Sjælland
Trelleborg
Åbenrå
Svendborg
Næstved
Møn
DENMARK
Sylt
North Frisian Is.
Nakskov
Falster
Flensburg
Lolland
Nykøbing
Kap Arkona
Kieler B.
Rødby
Schleswig
Gedser
Zingst
Fehmarn
Mecklenburger B.
Kiel
Neumünster
Deutsche B.
Stralsund
Rostock
Greifswald
Ostfriesische In.
Lübeck
Wismar
Waddeneilanden
Wilhelmshaven
Cuxhaven
HAMBURG
Waddenzee
Groningen
Emden
Bremerhaven
Schwerin
Neubrandenburg
UNITED KINGDOM
Gr. Yarmouth
Lowestoft
Den Helder
Leeuwarden
Oldenburg
Lüneburg
Neustrelitz
Assen
BREMEN
IJsselmeer
Meppel
Emmen
Verden
Wittenberge
Alkmaar
Uelzen
Zwolle
Neuruppin
Eberswalde
Haarlem
AMSTERDAM
Celle
Leiden
Utrecht
Enschede
Mittellandkanal
HANNOVER
Braunschweig
Stendal
'S-GRAVENHAGE
Wolfsburg
WESTBERLIN
Apeldoorn
ROTTERDAM
Rijn
Osnabrück
Hildesheim
Brandenburg
Potsdam
Arnhem
NETHERLANDS
Maas
Bielefeld
Vlissingen
Münster
DEMOCRATIC
Blankenberge
Dordrecht
Nijmegen
Salzgitter
Magdeburg
Oostende
Breda
Tilburg
ESSEN
Wittenberg
Brugge
DORTMUND
Bernburg
Dunkerque
Eindhoven
Göttingen
Dessau
Schelde
DUISBURG
Antwerpen (Anvers)
Halle
Kortrijk
Gent
(Gand)
Mechelen
Wuppertal
Kassel
Nordhausen
Merseburg
LEIPZIG
Riesa
Roubaix
Leuven
DÜSSELDORF
Lille
BRUXELLES
(Brussel)
Maastricht
KÖLN
(Cologne)
Siegen
REPUBLIC
Erfurt
Valenciennes
Mons
Aachen
Marburg a. d. L.
Jena
Gotha
Weimar
Gera
Karl-Marx-
Arras
Maubeuge
Charleroi
Namur
Liège
Bonn
BELGIUM
Neuwied
Lahn
Zwickau
Teplice
Koblenz
Limburg
Fulda
Suhl
Plauen
Most
St. Quentin
Bastogne
Wiesbaden
FRANKFURT a. M.
Amiens
LUXEMBOURG
Hanau
Schweinfurt
Hof
Cheb
Karlovy Vary
Charleville-Mézières
Offenbach
Laon
Sedan
Luxembourg
Trier
Mainz
Main
Bayreuth
Compiègne
Esch-s.-A.
Worms
Darmstadt
Bamberg
Mariánské Lázně
Reims
Ludwigshafen
Würzburg
Marne
Thionville
Kaiserslautern
Mannheim
Châlons-s.-Marne
Verdun
Heidelberg
Fürth
Amberg
Plzeň
Épernay
Metz
Saarbrücken
Meuse
Heilbronn
Nürnberg
Cham
Klatovy
Bar-le-Duc
OF GERMANY
Karlsruhe
Nancy
St.-Dizier
Pforzheim
STUTTGART
Ingolstadt
Regensburg
Troyes
Strasbourg
Baden-Baden
Göppingen
Sens
Esslingen
Straubing
Seine
St.-Dié
Passau
Chaumont
Tübingen
Landshut
Rhein
Donau
Ulm
Augsburg
Colmar
Épinal
Auxerre
Freiburg
Isar
Montbard
Mulhouse
Memmingen
MÜNCHEN
(Munich)
Inn
Tuttlingen
Lech
FRANCE
Belfort
Saône
Vesoul
Bodensee
Rosenheim
Yonne
Dijon
Kempten
Salzburg
Besançon
Doubs
Basel
Winterthur
Bad Ischl
Nevers
Zürich
Aare
St. Gallen
Bregenz
Kufstein
Autun
LIECHTENSTEIN
Le Creusot
Dole
Neuchâtel
Biel
Luzern
Zug
Vaduz
AUSTRIA
Inn
Innsbruck
Bischofshofen
Loire
Chalon-s.-Saône
Fribourg
Bern
SWITZERLAND
Schwyz
Chur
Landeck
Moulins
Lienz
Merano
Drau
1 : 8 000 000
0 50 100 150 200 Km
0 25 50 75 100 125 Mi
Roanne
St. Moritz
Adige
Bolzano
Piave
Villach
Thiers
Rhône
Brig
ITALY
Jesenice
Lyon
Annecy
Chamonix-Mt. Blanc
Lugano
Sondrio
Trento
Belluno
Udine
Vienne
Aosta

12 a The Ruhr Basin
1 : 2 000 000
Wesel
Marl
Recklinghausen
Hamm
Kamen
Unna
Dinslaken
Castrop-Rauxel
Bottrop
Herne
DORTMUND
Gelsenkirchen
Oberhausen
ESSEN
Bochum
Moers
Mülheim a.d.R.
Witten
Schwerte
DUISBURG
Hattingen
Kempen
Iserlohn
Velbert
Gevelsberg
Hagen
Krefeld
Ratingen
Altena
Viersen
Wuppertal
DÜSSELDORF
Mönchen-gladbach
Remscheid
Neuss
Solingen
Grevenbroich
Leichlingen
Västervik
Visby
Slite
Gotland
Ventspils
Oskarshamn
Burgsvik
Borgholm
Kalmar
Öland
Liepāja
Latvian S.S.R.
Karlskrona
Palanga
Klaipėda
Lithuanian S.S.R.
Svetlogorsk
Sovetsk
Kaunas
Vilnius
Smorgon'
Vilejka
Przyl. Rozewie
Kaliningrad
Kapsukas
Ustka
Lębork
Gdynia
Baltijsk
R. S. F. S. R.
Černachovsk
Lithuanian S.S.R.
Alytus
Molodečno
Słupsk
Sopot
Suwałki
Druskininkai
Lida
MINSK
Gdańsk
Elbląg
Koszalin
Tczew
Kętrzyn
Ełk
Augustów
Grodno
Novogrudok
Dzeržinsk
Białogard
Malbork
Olsztyn
Baranoviči
Sluck
Szczecinek
Chojnice
Iława
Szczytno
Volkovysk
Slonim
Soligorsk
Grudziądz
Łomża
Bydgoszcz
Toruń
Mława
Ostrołęka
Białystok
Piła
Ciechanów
Ostrów Maz.
Hajnówka
Bereza
Byelorussian S.S.R.
Krzyż
Inowrocław
Płock
Luninec
Włocławek
Kobrin
Pinsk
Gniezno
WARSZAWA (Warsaw)
David-Gorodok
POZNAŃ
Konin
Kutno
Brest
Łowicz
Siedlce
Żyrardów
Biała Podl.
POLAND
Kościan
ŁÓDŹ
Skierniewice
Łuków
Kamen'-Kaširskij
Kalisz
Leszno
Tomaszów Maz.
Dęblin
Sarny
Ostrów Wlkp.
Sieradz
Pabianice
Radom
Puławy
Chełm
Kovel'
Ukrainian S.S.R.
Głogów
Piotrków Tryb.
Skarżysko-Kam.
Lublin
Vladimir-Volynskij
Oleśnica
Radomsko
Ostrowiec Świętokrzyski
Hrubieszów
Legnica
WROCŁAW
Kielce
Luck
Rovno
Dubno
Jelenia Góra
Zamość
Brzeg
Sandomierz
Częstochowa
Wałbrzych
Opole
Bytom
Jędrzejów
Tarnobrzeg
Chorzów
Mielec
Nysa
Gliwice
Sosnowiec
Králové
Kłodzko
Rzeszów
Jarosław
Racibórz
Zabrze
Katowice
KRAKÓW
Wieliczka
Tarnów
Przemyśl
Šumperk
Rybnik
Pardubice
Opava
Krosno
Sambor
Čes. Třebová
Ostrava
Bielsko-Biała
Sanok
Olomouc
Havířov
Nowy Sącz
Prostějov
Přerov
Zakopane
Gottwaldov
Žilina
Poprad
Prešov
Turka
CZECHOSLOVAKIA
Brno
Martin
Znojmo
Břeclav
Trenčín
Košice
Michalovce
Užgorod
Mukačevo
Banská Bystrica
Piešťany
Chust
Zvolen
Trnava
Nitra
Lučenec
Levice
WIEN (Vienna)
Bratislava
Miskolc
Mátészalka
Salgótarján
Nyíregyháza
Satu Mare
Neustadt
Eisenstadt
Eger
Komárno
Vác
Sopron
Tatabánya
Győr
BUDAPEST
Székefehérvár
Dunaújváros
Cegléd
Szombathely
Veszprém
Szolnok
Balaton
Siófok
Kecskemét
HUNGARY
Zalaegerszeg
Maribor
Nagykanizsa
Kiskunhalas
Kaposvár
Szekszárd
Baja
Varaždin
Pécs
Subotica
Mohács
GERMAN DEMOCRATIC REPUBLIC
1 Berlin
2 Cottbus
3 Dresden
4 Erfurt
5 Frankfurt
6 Gera
7 Halle
8 Karl-Marx-Stadt
9 Leipzig
10 Magdeburg
11 Neubrandenburg
12 Potsdam
13 Rostock
14 Schwerin
15 Suhl
FEDERAL REPUBLIC OF GERMANY
1 Baden-Württemberg
2 Bayern
3 Bremen
4 Hamburg
5 Hessen
6 Niedersachsen
7 Nordrhein-Westfalen
8 Rheinland-Pfalz
9 Saarland
10 Schleswig-Holstein
POLAND
49 Districts (województw)
The seat of the districts is underlined
AUSTRIA
1 Burgenland
2 Kärnten
3 Niederösterreich
4 Oberösterreich
5 Salzburg
6 Steiermark
7 Tirol
8 Vorarlberg
9 Wien
CZECHOSLOVAKIA
1 Středočeský
2 Jihočeský
3 Západočeský
4 Severočeský
5 Východočeský
6 Jihomoravský
7 Severomoravský
8 Západoslovenský
9 Stredoslovenský
10 Východoslovenský

Agriculture: arable land 29.7%, meadows and pastures 40.2%, forests 20%. **Crops** (1977, in 1,000 tons): wheat 1,716, barley 702, maize (corn) 556, rice, potatoes 936, cotton – seed 267, – lint 155, sugar beet 2,900, olives 1,370, groundnuts, sesame, fruit 3,379–watermelons 600, oranges 582, lemons 202, grapes 1,585, figs 130 (1975)– vegetables 3,196, tomatoes 1,457, tobacco 112, walnuts 25.4; **livestock** (1977, in 1,000 head): cattle 1,116, sheep 8,135, goats 4,524, pigs 830, horses 149, asses 280, mules, buffaloes; poultry 29,000; silkworms (87 tons of raw silk in 1976), honey 8,000 tons, wool 6,370 tons; fish catch 70,700 tons (1976); roundwood 2.9 mn. cub.m (1976), resins 14,157 tons (1976).
Communications (1976): railways 2,479 km, roads 36,721 km, passenger cars 510,000. Merchant shipping 29,517,000 GRT (1977). Chief port is Athínai-Piraiévs. Corinth canal: length 6,345 m, depth 7 m, width at level 24.6 m, at bottom 21 m, opened in 1893. Civil aviation (1976): 43.5 mn. km flown, 3,541,000 passengers carried. Tourism 4.5 mn. visitors.
Exports: tobacco, iron and steel, fresh and dried fruit, aluminium, chemical products, cotton. **Imports:** machines and equipment, consumer goods, fuels. Chief trading partners: Fed. Rep. of Germany, U.S.A., Italy, France.

HUNGARY

Magyar Népköztársaság, area 93,032 sq.km, population 10,698,000 (1978), **socialist republic** (Chairman of the Presidential Council Pál Losonczi since 1967).

Administrative units: 19 counties (megye), and the **capital** – Budapest 2,081,696 inhab. (1977); **other towns** (1977, in 1,000 inhab.): Miskolc 203, Debrecen 192, Szeged 173, Pécs 165, Győr 122, Székesfehérvár 99, Nyíregyháza 95, Kecskemét 93, Szombathely 79, Szolnok 74, Tatabánya 73, Kaposvár 71, Békéscsaba 64, Eger 58, Dunaújváros 57, Sopron 54, Hódmezővásárhely 54. **Population:** Hungarians 96%, Serbs, Croats, Slovenians, Germans, Slovaks. **Density** 115 persons per sq.km; average annual rate of population increase 0.38% (1970–75); urban population 47.5% (1975). 19% of inhabitants employed in agriculture, 29.9% in industry (1976). – **Currency:** forint = 100 fillers.
Economy: industrial and agricultural country. **Principal industries:** engineering, metallurgy, chemicals, textiles and food processing. Half the industrial production occurs in Budapest. Shortage of energy resources. **Mining** (1977, in 1,000 tons, metal content): coal 2,928 (Pécs, Komló), brown coal 22,524 (Salgótarján, Tatabánya, Gyöngyös), crude petroleum 2,196 (Nagylengyel, Algyő, Demjén, Szolnok), natural gas 6.6 billion cub.m (Karcag, Szolnok), bauxite 2,952 (Iszkaszentgyörgy, Gúttamási, Halimba), iron ore 126, manganese 25.9 (1976), uranium (Pécs), copper, lead, zinc. Electricity 23.4 billion kWh (1977), 99% thermal power stations. **Production** (1977, in 1,000 tons): pig iron 2,292 and crude steel 3,720 (Dunaújváros, Miskolc-Diósgyőr, Ózd), aluminium 71.3 (Ajka, Várpalota, Tatabánya), refined copper 13.9, engineering (Győr, Eger, Debrecen, Pécs, Miskolc) – lorries 14,300 units, buses 11,890 units (Budapest), railway carriages (Győr)–television receivers 423,000 units, cutting machines, sulphuric acid 632, nitric acid 902 (1976), nitrogenous fertilizers 492 (Leninváros); capacity of petroleum refineries 10.2 mn. tons (1976), oil pipeline for the import of Soviet petroleum, photochemical (Vác) and pharmaceutical products (Debrecen, Tiszavasvári), cement 4,620 (1976), textile industry (Szeged, Vác, Sopron) – woven fabrics – cotton 333 mn. sq.m, – woollen 24 mn. sq.m, – rayon and acetate 54 mn. sq.m, yarn – cotton 60.1, – woollen 10.6, synthetic fibres 13.7, artificial fibres 9.1, leather shoes 45.7 mn. pairs, food industry (Debrecen, Szeged) – meat 1,324, milk 2,407, butter 26.3, cheese 50.6, eggs 222, Hungarian salami for export, sugar 438, wine 510 (Tokaj), beer 6.8 mn. hl (1976); 24.9 billion cigarettes.
Agriculture: vegetable production and cereals predominate; important viniculture, cultivation of fruit and vegetables. Land use: arable land 59.2%, meadows and pastures 13.9%, forests 16.8%. **Crops** (1977, in 1,000 tons): wheat 5,312, maize (corn) 6,150, barley 750, rye, rice, potatoes 1,416, flax 19, hemp 14, sugar beet 3,888, sunflower seed 160, soybeans, hops, fruit 2,509, grapes 860, vegetables 2,291, walnuts 19.1; **livestock** (1977, in 1,000 head): cattle 1,887, pigs 7,854, sheep 2,350, horses 147; poultry 60,498, honey 10,500 tons; fish catch 31,900 tons (1976); roundwood 5.5 mn. cub.m (1976).
Communications: railways 8,336 km, of which 1,303 km electrified (1976), roads 29,895 km (1976), passenger cars 654,800 (1976). Navigable waterways 1,688 km (1977). Civil aviation (1976): 10.7 mn. km flown, 427,000 passengers carried. Tourism 5.6 mn. visitors (1976).
Exports: machines and industrial equipment, transport equipment, bauxite, aluminium, chemical products, foodstuffs (meat, smoked meat products, canned products, wine, fruit, vegetables). **Imports:** fuels (coal, petroleum), raw materials (iron ore, non-ferrous metals), semi-finished products, machines and equipment, cars, industrial consumer goods. Chief trading partners: U.S.S.R. 29%, socialist countries (mainly German Dem. Rep., Czechoslovakia), Fed. Rep. of Germany.

ICELAND

Lýđveldiđ Ísland, area 102,829 sq.km, population 222,000 (1977), **republic** (President Mrs Vigdis Finnbogadottir since 1980).

Administrative units: 7 districts. **Capital:** Reykjavík 84,493 inhab. (1976); **other towns** (1975, in 1,000 inhab.): Kópavogur 13, Akureyri 12, Hafnarfjörđur 12. **Population:** Icelandic. **Density** 2 persons per sq.km; average annual rate of population increase 1.1% (1970–75); urban population 86.5% (1975). – **Currency:** Icelandic króna = 100 aurars.
Economy: agricultural country without raw material resources. The economy is based on fishing – catch 1,365,000 tons (1977) – whaling, raising of sheep 871,000 head (1977), cattle 61,000 (1977). Electricity 2.4 billion kWh (1976), of which 97% are hydro-electric power stations; hot springs. **Industry:** fish processing, canning, freezing plants, production of aluminium and textiles. – **Communications:** roads 11,533 km (1975). Merchant shipping 167,000 GRT (1977). – **Exports:** fresh and canned fish 75%, aluminium, diatomite, woollen products. **Imports:** fuels, industrial products and foodstuffs. Chief trading partners: U.S.A., United Kingdom, Fed. Rep. of Germany, U.S.S.R.

IRELAND

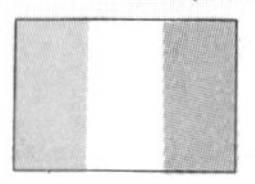

Éire, area 70,282 sq.km, population 3,221,000 (1978), **republic** (President Dr Padraig Ohlrighile /Dr Patrick J. Hillery since 1976).

Administrative units: 4 provinces (26 counties). **Capital:** Dublin (Baile Átha Cliath) 567,866 inhab. (1971); **other towns** (1971, in 1,000 inhab.): Cork 129, Limerick 57, Dún Laoghaire 53. **Population:** Irish. **Density** 46 persons per sq.km; average annual rate of population increase 1.16% (1975); urban population 55.1% (1975). 23% of inhabitants employed in agriculture, 31.5% in industry. – **Currency:** Irish pound = 100 pence.
Economy: industrial and agricultural country. **Principal industries:** mining of minerals, metallurgy, engineering, chemicals, textiles and food processing. Centres of industry: Dublin, Cork, Cobh. **Mining** (1976, in 1,000 tons, metal content): coal, peat (Timahoe), zinc 116 (1977), lead 32.6 (Tynagh), copper 4.1, silver 26 tons. Electricity 8.4 billion kWh (1977). **Production** (1977, in 1,000 tons): crude steel 44, cement 1,600, woven fabrics – cotton 19 mn. sq.m (1976), woollen 3 mn. sq.m (1977), rayon and acetate 16.7 mn. sq.m (1976), sugar 355 (1976), meat 602, milk 4,600, butter 101.6, cheese 53.4, eggs 39.7, beer 5 mn. hl (1976, "Guinness"), alcoholic spirits; 6.9 billion cigarettes (1976).
Agriculture – animal production predominates. Land use: arable land 14.4%, meadows and pastures 68.8%, forests 4.4%. **Crops** (1977, in 1,000 tons): wheat 239, oats 127, barley 1,359, potatoes 1,200, sugar beet 1,483; **livestock** (1977, in 1,000 head): cattle 7,155, sheep 3,526, pigs 947; poultry 10,500; fish catch 94,300 tons (1976).
Communications: railways 2,010 km (1976), roads 88,490 km, of which 15,915 km are main roads (1975), passenger cars 556,400 (1976). Navigable waterways 1,040 km. Merchant shipping 212,000 GRT (1977). Chief ports: Dublin, Cobh. Civil aviation (1976): 19.4 mn. km flown, 1,578,000 passengers carried. Tourism 1.3 mn. visitors (1976).
Exports: meat, live cattle, machines, textiles, chemicals, pharmaceuticals, beverages. Chief trading partners: United Kingdom (47% of export and 48% of import in 1977), Fed. Rep. of Germany, U.S.A., France.

ITALY

Repubblica Italiana, area 301,260 sq.km, population 56,696,000 (1978), **republic** (President Dr Sandro Pertini since 1978).

Administrative units: 20 regions (95 provinces). **Capital:** Roma (Rome) 2,897,505 inhab. (1977); **other towns** (1977, in 1,000 inhab.): Milano (Milan) 1,706, Napoli (Naples) 1,225, Torino (Turin) 1,182, Genova (Genoa) 795, Palermo 679, Bologna 481, Firenze (Florence) 464, Catania 400, Bari 387, Venezia (Venice) 360, Verona 271, Messina 267, Trieste 265, Taranto 245, Padova 242, Cagliari 242, Brescia 215, Modena 180, Reggio di Calabria 179, Parma 178, Livorno 178, Salerno 162, Prato 156, Foggia 155, Ferrara 154. **Population:** Italians 98%. **Density** 188 persons per sq.km; average annual rate of population increase 0.54% (1970–75); urban population 61.4% (1975). 13.8% of inhabitants employed in agriculture, 17.6% in industry (1976). – **Currency:** Italian lira.
Economy: highly developed industrial and agricultural country, economically the most advanced in southern Europe. There is considerable difference between the advanced industrial North with its large modern plants and the under-developed agricultural South. **Principal industries:** engineering, hydroenergetics, electrometallurgy, electrotechnics, electronics, chemistry, textiles and food processing. **Mining** (1977, in 1,000 tons, metal content): lignite 1,844, crude petroleum 1,083 (Gela, Ragusa), natural gas 13.7 billion cub.m (The Po Plain and others), iron ore 478 (island I. d'Elba, Cogne in Valle d'Aosta), manganese 9,314 tons, lead 33, zinc 121 (Iglesias), mercury 768 tons (1976, Monte Amiata, Grosseto), antimony 1,009 tons (1976, Sardegna), copper, bauxite 34.8 (1976), gold (Pestarena in Valle Anzasca), silver 48 tons (1976), uranium (Piemonte Alps, Novazza), salt 3,600 (Sicilia), potassium salts 1,879 (Sicilia), pyrites 864 (Grosseto), sulphur 628 (Sicilia), marble 1,308, magnesite, asbestos 165, fluorite 185, barytes 152, graphite, clay. Electricity 159.8 billion kWh (1977), of which 25% is hydro-electric, nuclear power stations 3% (Caorso, Trino, Garigliano, Latina). **Production** (1977, in 1,000 tons): **metallurgy:** pig iron 11,410 (Trieste, Napoli, Piombino, Aosta), crude steel 23,334 (reg. Lombardia – Milano, reg. Liguria – Genova, reg. Piemonte – Torino; Taranto and others), lead 45.6 (Sardegna, La Spezia), zinc 181 (Monteponi), cadmium 449 tons, aluminium 262.4 (Marghera near Venezia), magnesium 8,836 tons (1975). **Engineering** (1977, in 1,000 units): vessels 580,000 GRT (Genova, La Spezia, Livorno, Napoli, Palermo), electric locomotives 203 units, railway carriages 1,884 units (Torino, Pinerolo, Vado Ligure), aircraft (Torino, Varese), passenger cars 1,440 ("Fiat" – Torino, 75% of production, "Alfa Romeo" – Milano, "Lancia" – Torino, "Ferrari" – Maranello), bicycles 2,394 (Milano), motorcycles 305, tractors 138 (Torino), precision mechanics (microtechnics, photo and cinema apparatus – Milano, Torino), spectacles, electrotechnical apparatus (Milano, Roma, Torino), radio receivers 791 (1975), television receivers 1,595, calculating machines 203, typewriters 695 ("Olivetti" – Ivrea).
Chemical industry (Lombardia, 1977, in 1,000 tons): acid – sulphuric 2,946, hydrochloric 462 (1976), nitric 929 – caustic soda 1,115, synthetic nitrogen 1,378, nitrogenous fertilizers 1,166 (Novara, Merano), phosphorus fertilizers 1,094, potash fertilizers 299, coke oven coke 7,676, pharmaceuticals, dyes (Milano and surroundings), plastics and resins 2,053 (1975, Ferrara, Castellanza), synthetic fibres 410, synthetic rubber 250 (1976); capacity of petroleum refineries 207.5 mn. tons (1976), oil pipeline network 2,505 km, gas pipelines 14,270 km, motor spirit 15.9 mn. tons, oils 72.8 mn. tons; production of cement 38.2 mn. tons, sheet and crystal glass (Marghera near Venezia, Milano), chandeliers (Murano), chemical wood pulp 239, paper and paper products 4,785, furniture (Cantù, Lissone), musical instruments (Emilia-Romagna, Marche, Milano).
Textile industry – cotton processing (The Po Plain, Lombardia), wool processing (reg. Piemonte - Biella), silk processing (reg. Lombardia - Como), flax processing (Lombardia and Piemonte). Production (1977, in 1,000 tons): cotton yarn 149, woven fabrics – cotton 111, woollen 159 (1976), silk 16 (1976), jute 5.2 – rayon and acetate – fibres 46.7 (1976), staple fibre 62 (1976), shoes 199 mn. pairs (Vigevano).
Food industry (1977, in 1,000 tons): sugar 1,230, meat 3,143, milk 10,488, butter 72, cheese 525, eggs 633, production of spaghetti etc., sweets, canned fish and foodstuffs 57 (1976), olive oil 692 (leading world producer), wine 6,363 (leading world producer, "Chianti", "Barbero", "Cinzano", "Martini" and others), 73.6 billion cigarettes.

Fribourg
Thun
Chur
Lausanne
Montreux
Sion
Brig
L. Geneva
Rhône
SWITZERLAND
Rhein
8°
12°
16°
AUSTRIA
Lienz
Graz
Zalaeg
Merano
Villach
Klagenfurt
Drau
Mur
St. Moritz
Sondrio
Adige
17
Bolzano
Maribor
Drava
Lugano
L. Maggiore
Belluno
6
Jesenice
Kranj
Varaždin
Chamonix-Mt. Blanc
Aosta
L. di Como
Trento
Udine
5
Sava
Celje
Como
Bergamo
L. di Garda
Pordenone
19
Biella
Busto Arsizio
20
Gorizia
Ljubljana
ZAGREB
Monza
Vicenza
Piave
Postojna
3
Novara
MILANO (Milan)
9
Brescia
Verona
Treviso
G. di Trieste
Trieste
Sisak
Kupa
Cremona
Padova
Venezia (Venice)
Koper
Rijeka
Karlovac
TORINO (Turin)
12
Asti
Pavia
Mantova
Opatija
Briançon
Po
Adige
G. di Venezia
O. Krk
Crikvenica
Piacenza
Reggio n. Emilia
Istrie
Alessandria
Parma
Ferrara
Pula
Bormida
GENOVA (Genoa)
8
Modena
V. di Comacchio
O. Cres
Bihać
Cuneo
Kvarner
O. Pag
Gospić
Savona
La Spezia
5
Bologna
Ravenna
44°
San Remo
Golfo di Genova
Carrara
Forlì
Cesena
Zadar
Nice
Monaco
MONACO
Imperia
Viareggio
Lucca
Pistoia
Prato
San Marino
Rimini
Dugi O.
Knin
Cannes
Pisa
SAN MARINO
Pesaro
Fano
Šibenik
Firenze (Florence)
LIGURIAN SEA
Livorno
Arezzo
10
Ancona
Trogir
Macerata
C. Corse
Siena
18
Portoferraio
Piombino
16
Foligno
O. Hvar
Perugia
Ascoli Piceno
O. Vis
Calvi
Bastia
I. d'Elba
Grosseto
Spoleto
Corte
Terni
Teramo
Pescara
Orbetello
L'Aquila
B
Ajaccio
Viterbo
Rieti
Chieti
Tevere
Civitavecchia
7
1
Vasto
Taravo
Corse
Tivoli
VATICAN C.
Avezzano
ROMA (Rome)
Sartène
Porto-Vecchio
Frosinone
San Severo
11
Campobasso
St. of Bonifacio
La Maddalena
Latina
Foggia
Manfredoni
Bar
Pto. Torres
Sassari
Gaeta
Caserta
Cerignola
4
Benevento
Ofanto
Olbia
Andria
Alghero
NAPOLI (Naples)
Avellino
13
Nuoro
Torre del Greco
Salerno
Potenza
2
Tirso
I. di Capri
Castellammare di St.
G. di Salerno
40°
Oristano
14
Agri
Sapri
Tortoli
Lauria
TYRRHENIAN SEA
Castrovillar
Iglesias
Sardegna
Cagliari
Carbonia
Paola
G. di Cagliari
C. Teulada
I. Stromboli
Lamezia Terme
I. di Ustica
Ie. Eolie
I. Lipari
C
Messina
Palmi
Trapani
PALERMO
Stretto di Messina
Reggio di
Ie. Egadi
Alcamo
Cefalù
Barcellona
Marsala
Sicilia
C. Spa
R. Ben Sekka
Acireale
Sciacca
Enna
Catania
Menzel Bourguiba
Binzert (C. Blanc)
Caltanissetta
Annaba
Golfe de Tunis
R. Addar
Agrigento
15
El Kala
Tabarka
Mateur
TUNIS
Gela
Salso
Licata
Siracusa
Béja
Kelibia
Strait of Sicily
Ragusa
TUNISIA
Oued Medjerda
Hammam Lif
I. di Pantelleria (It.)
C. Passero
MEDITERRANEAN SEA
ALGERIA
1 : 8 000 000
0 50 100 150 200 Km
0 25 50 75 100 125 Mi
36°
Sousse
El Kairouan
Monastir
12°
3
16°
MALTA
Valletta
D
Ksar Hellal
ITALY
FRANCE
YUGOSLAVIA

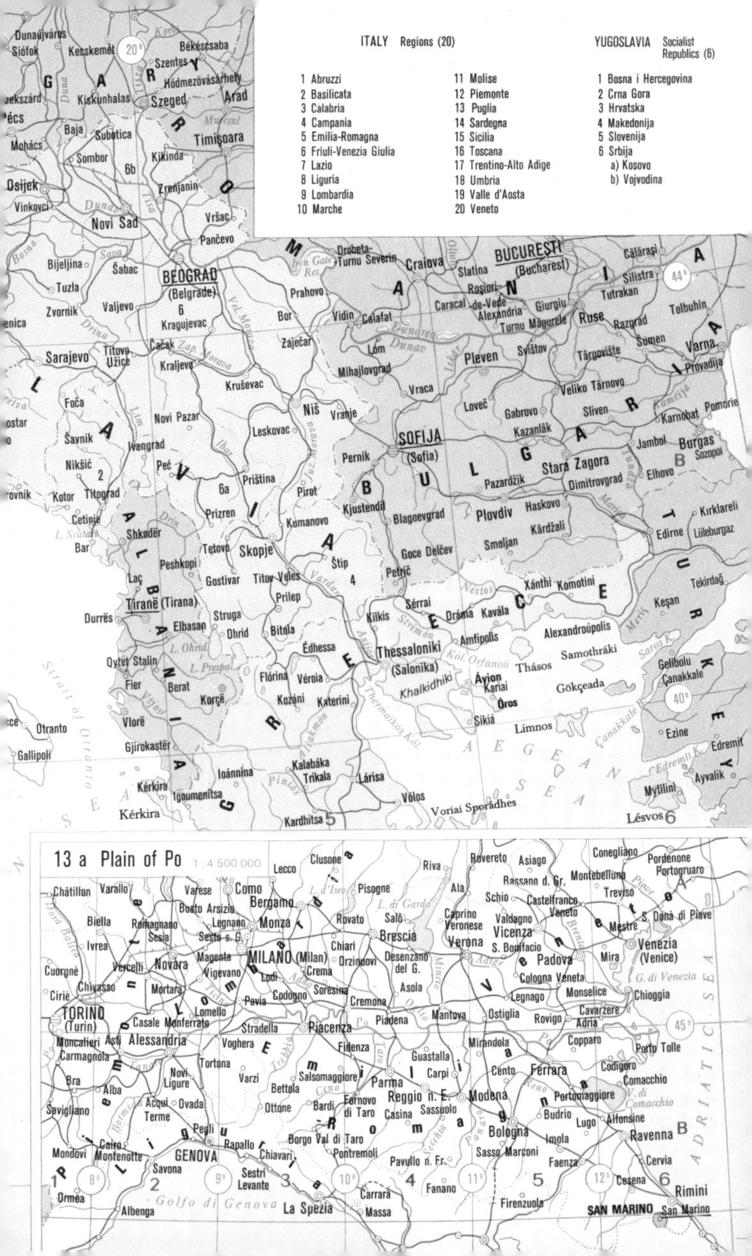
ITALY Regions (20)
1 Abruzzi
2 Basilicata
3 Calabria
4 Campania
5 Emilia-Romagna
6 Friuli-Venezia Giulia
7 Lazio
8 Liguria
9 Lombardia
10 Marche
11 Molise
12 Piemonte
13 Puglia
14 Sardegna
15 Sicilia
16 Toscana
17 Trentino-Alto Adige
18 Umbria
19 Valle d'Aosta
20 Veneto
YUGOSLAVIA Socialist Republics (6)
1 Bosna i Hercegovina
2 Crna Gora
3 Hrvatska
4 Makedonija
5 Slovenija
6 Srbija
a) Kosovo
b) Vojvodina
HUNGARY
ROMANIA
YUGOSLAVIA
BULGARIA
ALBANIA
GREECE
TURKEY
BEOGRAD (Belgrade)
BUCUREŞTI (Bucharest)
SOFIJA (Sofia)
Tiranë (Tirana)
Thessaloniki (Salonika)
Szeged
Timişoara
Novi Sad
Sarajevo
Skopje
Niš
Plovdiv
Varna
Burgas
Edirne
AEGEAN SEA
13 a Plain of Po 1 : 4 500 000
MILANO (Milan)
TORINO (Turin)
GENOVA
Venezia (Venice)
Bologna
Verona
Padova
Parma
Modena
Ferrara
Ravenna
Golfo di Genova
SAN MARINO
ADRIATIC SEA

Agriculture: arable land 31.8%, meadows and pastures 17.6%, forests 21.5%. Number of tractors 900,000 (1977). **Crops** (1977, in 1,000 tons): wheat 6,329 (The Po Plain), barley 677, oats 355, rye, maize (corn) 6,456 (Veneto and Lombardia), rice 721 (Piemonte and Lombardia), potatoes 3,310, flax 5.9, hemp, cotton, sugar beet 11,557 (Po delta, Emilia-Romagna), sunflower seed 50.7, olives 3,626 (leading world producer, Puglia, Calabria), fruit 18,895 - oranges 1,650 (Sicilia), tangerines 243, lemons 867 (second world producer), almonds 181, figs 98.6, apples 1,810, pears 1,260 (leading world producer), peaches 1,187 (second world producer), grapes 10,900 (leading world producer, Veneto, Puglia, Piemonte, Sicilia), sweet chestnuts 56.3, walnuts 60 - vegetables 12,141, watermelons 769, tomatoes 3,066, strawberries 152, tobacco 110. **Livestock** (1977, in 1,000 head): cattle 8,737, horses 264, asses 163, mules, buffaloes 76, sheep 8,445, goats 948, pigs 9,097; poultry 117,552; wool 5,800 tons, raising of silkworms (75 tons of cocoons); fish catch 420,300 tons (1976); roundwood 6.6 mn. cub.m (1976).
Communications (1976): railways 20,088 km, of which 9,546 km are electrified, roads 292,343 km, motorways 5,529 km, passenger cars 15,925,267. Navigable waterways 2,237 km, of which 849 km are canals. Merchant shipping 11,111,000 GRT (1977). Chief ports: Genova, Trieste, Augusta, Taranto, Venezia. Civil aviation (1976): 134.6 mn. km flown, 8,238,000 passengers carried. Important tourism, 13.9 mn. visitors (1976).
Exports: machines and equipment, cars, tractors, chemical products (plastics, pharmaceuticals, fertilizers), metals, products of precision mechanics (calculating machines and typewriters), metallurgical products, textiles, shoes, rubber products, wine, fresh fruit, vegetables.
Imports: raw materials and semi-finished products, fuels, machines and equipment, metals, cars, agricultural products (cereals, meat, wool, cotton, rubber, cattle, sugar, coffee etc.). Chief trading partners: Fed. Rep. of Germany, France, U.S.A., United Kingdom, Belgium.

LIECHTENSTEIN

Fürstentum Liechtenstein, area 157 sq.km, population 25,000 (1978), **principality** (Prince Franciz Joseph II since 1938).
Administrative units: 11 villages. **Capital:** Vaduz 4,620 inhab. (1976). – **Currency:** Swiss franc, customs union with Switzerland. **Economy:** agriculture (cereals, potatoes), livestock raising. Textiles and other industries. Tourism. Chief sources of revenue are the numerous registered foreign firms and postage stamps.

LUXEMBOURG

Grand-Duché de Luxembourg – Grousherzogdem Lezebuurg, area 2,586 sq.km, population 356,000 (1977), **grand duchy** (Grand Duke Jean since 1964).

Administrative units: 12 cantons. **Capital:** Luxembourg 78,400 inhab. (1975). **Density** 138 persons per sq.km; average annual rate of population increase 0.18% (1970–75); urban population 70.1% (1975). 45% of inhabitants employed in industry (1977). **The official language** is French, but the inhabitants speak mainly Luxemburgish (German dialect). – **Currency:** Luxembourg franc = 100 centimes.
Economy: advanced industrial country. Mining and metallurgy are the principal industries. Financial centre of Western Europe. **Mining:** iron ore 448,920 tons (1977, metal content). Electricity 1.2 billion kWh (1977). **Production** (1977, in 1,000 tons): pig iron 3,576, crude steel 4,332. **Agriculture:** cultivation of cereals, potatoes, fruit and vines; raising of cattle and pigs. – **Communications** (1977): railways 274 km, roads 5,051 km, passenger cars 130,700. – **Exports:** steel, plastics, textiles and others. Chief trading partners: Fed. Rep. of Germany, Belgium, France. Customs union with Belgium.

MALTA

Republika Ta Malta, Republic of Malta, area 316 sq.km, population 336,000 (1978), **republic, member of the Commonwealth** (President Dr Anton Buttigieg since 1976).

Administrative units: 6 regions. **Capital:** Valletta 14,071 inhab. (1976). **Population:** Maltese of Italian-Arabic descent. **Density** 1,063 persons per sq.km; average annual rate of population increase 0.23% (1970–75); urban population 78.6% (1975). – **Currency:** Maltese pound = 100 pence.
Economy: cultivation of wheat, potatoes, vines, tomatoes, citrus fruit; livestock (1977, in 1,000 head): cattle 15, sheep 8, goats 10, pigs 25; poultry 1.2 mn. Electricity 382 mn. kWh (1976). Manual production of lace on Gozo I (Ghawdex). Naval base. – **Communications:** roads 1,267 km (1974), passenger cars 56,400 (1976). Merchant shipping 45,950 GRT (1975). Civil aviation (1976): 3.4 mn. km flown, 221,000 passengers carried. Tourism 339,500 visitors (1976). – **Exports:** domestic manufactures. Chief trading partners: United Kingdom and Fed. Rep. of Germany.

MONACO

Principauté de Monaco, area 1.8 sq.km, population 25,000 (1978), **principality** (Prince Rainier III since 1949).
The state consists of 3 joint urban districts: Monaco, Monte Carlo and La Condamine. **Capital:** Monaco. Tourist centre on the French Riviera. – **Currency:** French franc. Customs union with France. Chief sources of revenue are tourism and gambling.

NETHERLANDS

Koninkrijk der Nederlanden, area 41,160 sq.km, population 13,936,000 (1978), **kingdom** (Queen Beatrix Wilhelmina Armgard since 1980).

Administrative units: 13 provinces; autonomous state Netherlands Antilles. **Capital:** Amsterdam 738,441 inhab. (1977, with agglomeration 975,506 inhab.); **other towns** (1977, in 1,000 inhab.): Rotterdam 601, 's-Gravenhage 471 (the seat of the Royal Court and Government), Utrecht 245, Eindhoven 193, Haarlem 163, Groningen 162, Tilburg 151, Nijmegen 148, Enschede 141, Arnhem 126, Breda 119. **Population:** Dutch, small number of Frisians in the North. **Density** 339 persons per sq.km is the highest in Europe (excluding miniature countries); average annual rate of population increase 0.85% (1970–75); urban population 79.3% (1975). 20% of inhabitants employed in industry (1976). – **Currency:** gulden = 100 cents.
Economy: highly developed industrial and agricultural country. **Principal industries:** engineering (especially shipbuilding), electrotechnics, metallurgy, chemistry, textiles and food processing. **Mining** (1977, in 1,000 tons): crude petroleum 1,380 (Coevorden-Schoonebeek), natural gas 86.5 billion cub.m (third world producer, Groningen surroundings, North Sea shelf), coal 758 (1974, prov. Limburg), salt 3,026 (1976, Hengelo, Delfzijl). Electricity 57.8 billion kWh (1977), of which 93% are thermal and 7% nuclear power stations (Borsselen, Dodewaard). **Production** (1977, in 1,000 tons): pig iron 3,924, crude steel 4,920 (IJmuiden), coke oven coke 2,811 (1976), zinc 109 (Budel), aluminium 241 (Delfzilj), lead 22 (1976), vessels 380,000 GRT (shipyards Amsterdam, Rotterdam), passenger cars 68,000 units (Eindhoven, Born), electrotechnics ("Philips" in Eindhoven, vacuum tubes, radio receivers, telephones), nitrogenous fertilizers 1,253 (IJmuiden), phosphorus fertilizers 307, sulphuric acid 1,572, synthetic rubber 239, soda ash 271 (1976), plastics and resins 1,783; capacity of petroleum refineries 102 mn. tons (near Rotterdam and Amsterdam), naphtha 5.4 mn. tons (1976), motor spirit 6.5 mn. tons, oils 40.2 mn. tons, cement 3,890, paper 1,623, production of porcelain and ceramics (Delft, Maastricht), yarn – cotton 28, woollen 10.3 – woven fabrics – cotton 36.9 (reg. Twente), woollen 24 mn. sq.m (Tilburg) – synthetic fibres 31.2, sugar 904, meat 1,724, milk 10,682, condensed milk 514.5, cheese 405 (Edam, Gouda, Hoorn), butter 181.6, eggs 340, canned fish 32.3 (1976), chocolate and cocoa, beer 13.8 mn. hl (1976), alcoholic spirits (curacao – Amsterdam, gin); 30.5 billion cigarettes (1976), production of quinine (Amsterdam), diamond cutting and polishing (Amsterdam).
Agriculture is highly productive. There is a shortage of land which is partly overcome by the reclamation of polders from the sea. Cereal yields are among the highest in the world. Land use: arable land 24.9%, meadows and pastures 36.4%, forests 9.1%. **Crops** (1977, in 1,000 tons): wheat 661, barley 287, potatoes 5,752, flax, sugar beet 6,017, fruit 484, vegetables 2,463, tomatoes 378, important cultivation of flowers (hyacinths, tulips etc); **livestock** (1977, in 1,000 head): cattle 4,877, pigs 8,288, sheep 800; poultry 69,875; fish catch 284,400 tons (1976); roundwood 900,000 cub.m.
Communications (1976): railways 2,832 km, of which 1,712 km electrified, roads 86,354 km (1975), motorways 1,423 km (1974), passenger cars 3,768,000. Navigable waterways and canals 4,343 km. Highly developed sea transport and trade. Chief ports: Rotterdam (the world's largest port, turnover of cargo 283 mn. tons in 1976), Amsterdam. Merchant shipping 5,290,000 GRT (1977). Civil aviation (1976): 95.2 mn. km flown, 4,081,000 passengers carried. Tourism 2,910,500 visitors (1976).
Exports: foodstuffs (dairy products, eggs, meat), machines and transport equipment, petroleum products, chemicals, electrotechnical goods, manufactured products, flowers. **Imports:** raw materials (petroleum, metals, tropical fruit), machines and equipment, cars, and others. Chief trading partners: Fed. Rep. of Germany, Belgium and Luxembourg, France, United Kingdom, U.S.A.

NORWAY

Kongeriket Norge, area 323,920 sq.km, population 4,051,000 (1977), **kingdom** (King Olav V since 1957).

Administrative units: 19 counties (fylker), overseas territories in Europe: Svalbard, Bjørnøya and Jan Mayen; in the Antarctica: island Bouvetøya. **Capital:** Oslo 461,881 inhab. (1977); **other towns** (1977, in 1,000 inhab.): Bergen 213, Trondheim 136, Stavanger 87, Baerum 81, Kristiansand 60, Drammen 51, Tromsø 44. **Population:** Norwegians 97.5%, Lapps, Finns. **Density** 13 persons per sq.km; average annual rate of population increase 0.66% (1970–75); urban population 45.3% (1975). 24.8% of inhabitants employed in industry (1975). – **Currency:** Norwegian krone = 100 øre.
Economy: developed industrial and agricultural country. Large resources of hydro-electricity and timber. **Principal industries:** mining of minerals, shipbuilding, electrometallurgy, electrochemistry, radioelectronics, wood and paper processing, fishery. **Mining** (1977, in 1,000 tons, metal content): coal 456 (Svalbard), crude petroleum 13.7 mn. tons (North Sea shelf), natural gas, iron ore 2,418 (Fossdalen, Rana), copper 29.5, lead 3,360 tons, zinc 30.6, molybdenum 203 tons (1973), titanium, vanadium 526 tons (1976), pyrites 659 (1974, Løkken, Sulitjelma). Electricity 72.6 billion kWh (1977), almost 100% from hydro-electric power stations. **Production** (1977, in 1,000 tons): pig iron 1,212, crude steel 732 (Stavanger, Arendal, Rana), copper – smelted 26, refined 20 – zinc 69, nickel 38, aluminium – primary 628 (Sunndalsøra, Øvre Årdal, Eydehamn) – magnesium 38.8 (1976, second world producer), vessels 530,000 GRT (shipyards in Oslo, Bergen, Fredrikstad), sulphuric acid 388 (chemical industry – Rjukan, Notodden, Odda), capacity of petroleum refineries 12.8 mn. tons (1976), petroleum products 8.5 mn. tons, cement 2,328, chemical wood pulp 872 (1976), paper and paper products 1,187, rayon and acetate staple fibre 26 (1976 – textile industry in Bergen, Oslo, Sandnes), meat 189, milk 1,898, butter 25, cheese 59, eggs 36,5.
Agriculture: intensive animal production predominates. Land use: arable land 2.6%, meadows and pastures 0.3%, forests 27%. **Crops** (1977, in 1,000 tons): cereals 1,078 (barley, oats), potatoes 605; **livestock** (1977, in 1,000 head): cattle 942, sheep 1,779, pigs 702; raising of fur animals (foxes and minks), reindeer in the North (142,000 head in 1973); fish catch 3,435,300 tons (1976); whaling; roundwood 9 mn. cub.m (1976).
Communications (1976): railways 4,241 km, of which 2,440 km electrified, roads 77,117 km, passenger cars 1,022,918.

HUNGARY:
19 counties and capital district
1 Bács-Kiskun
2 Baranya
3 Békés
4 Borsod-Abaúj-Zemplén
5 Csongrád
6 Fejér
7 Györ-Sopron
8 Hajdú-Bihar
9 Heves
10 Komárom
11 Nógrád
12 Pest
13 Somogy
14 Szabolcs-Szatmár
15 Szolnok
16 Tolna
17 Vas
18 Veszprém
19 Zala
CZECHOSLOVAKIA
AUSTRIA
WIEN
HUNGARY
BUDAPEST
ROMANIA
BUCUREŞTI
(Bucharest)
YUGOSLAVIA
BEOGRAD
(Belgrade)
BULGARIA
SOFIJA
(Sofia)
U. S. S. R.
Ukrainian S. S. R.
Moldavian S. S. R.
KIŠIN'OV
BLACK SEA
ADRIATIC SEA
Karadeniz B.
Bratislava
Trnava
Piešťany
Nitra
Levice
Zvolen
Lučenec
Košice
Michalovce
Užgorod
Čop
Mukačevo
Chust
Rachov
Nadvornaja
Kolomyja
Černovcy
Chotin
Kamenec-Podol'skij
Soroki
Rybnica
Kotovsk
Bel'cy
Ungeny
Bendery
Bessarabka
Komrat
Kagul
Bolgrad
Izmail
Reni
Kilija
Eisenstadt
Sopron
Győr
Komárno
Tatabánya
Székesfehérvár
Szombathely
Veszprém
Siófok
Zalaegerszeg
Nagykanizsa
Kaposvár
Szekszárd
Pécs
Mohács
Dunaújváros
Kecskemét
Kiskunhalas
Baja
Szeged
Vác
Salgótarján
Eger
Miskolc
Nyíregyháza
Mátészalka
Debrecen
Cegléd
Szolnok
Szentes
Orosháza
Békéscsaba
Hódmezővásárhely
Balaton
Neusiedler See
Satu Mare
Sighetul Marmaţiei
Vişeu de Sus
Baia-Mare
Carei
Zalău
Oradea
Cluj-Napoca
Dej
Turda
Bistriţa
Vatra Dornei
Tirgu Mureş
Suceava
Botoşani
Paşcani
Iaşi
Roman
Piatra-Neamţ
Bacău
Vaslui
Miercur -Ciuc
Bîrlad
Gh. Gheorghiu-Dej
Tecuci
Focşani
Galaţi
Brăila
Tulcea
Sulina
Sf. Gheorghe
Arad
Brad
Alba-Iulia
Mediaş
Sibiu
Sf. Gheorghe
Făgăraş
Braşov
Deva
Hunedoara
Lugoj
Timişoara
Reşiţa
Petroşani
Cimpulung
Sinaia
Rimnicu-Vilcea
Tirgu-Jiu
Ploieşti
Buzău
Tirgovişte
Piteşti
Slobozia
Drobeta-Turnu-Severin
Slatina
Călăraşi
Craiova
Caracal
Roşiori-de-Vede
Giurgiu
Alexandria
Turnu-Măgurele
Calafat
Medgidia
Constanţa
Mamaia
Mangalia
Iron Gate Res.
L. Razelm
L. Sinoe
Zagreb
Varaždin
Virovitica
Sisak
Osijek
Vinkovci
Slavonski Brod
Prijedor
Banja Luka
Doboj
Tuzla
Bijeljina
Jajce
Zenica
Zvornik
Sarajevo
Split
Jablanica
Mostar
Kardeljevo
(Ploče)
Foča
Dubrovnik
O. Brač
O. Hvar
O. Korčula
O. Mljet
Kotor
Cetinje
Bar
Nikšić
Šavnik
Titograd
Ivangrad
Subotica
Sombor
Novi Sad
Kikinda
Zrenjanin
Vršac
Pančevo
Šabac
Valjevo
Kragujevac
Titovo Užice
Čačak
Kraljevo
Kruševac
Novi Pazar
Prahovo
Bor
Zaječar
Niš
Leskovac
Vranje
Priština
Peć
Prizren
Pirot
Kumanovo
Skopje
Tetovo
Gostivar
Titov Veles
Štip
Shkodër
Peshkopi
Laç
Tiranë (Tirana)
L. Scutari
Vidin
Lom
Mihajlovgrad
Vraca
Pleven
Svištov
Ruse
Tutrakan
Silistra
Razgrad
Šumen
Tolbuhin
Balčik
Varna
Loveč
Gabrovo
Veliko Tărnovo
Tărgovište
Provadija
Kazanlăk
Sliven
Karnobat
Burgas
Pomorie
Sozopol
Mičurin
Jambol
Stara Zagora
Pernik
Kjustendil
Blagoevgrad
Goce Delčev
Petrič
Pazardžik
Plovdiv
Dimitrovgrad
Haskovo
Kărdžali
Smoljan
Elhovo
Edirne
Kırklareli
Lüleburgaz
Zonguldak
Ereğli
Karasu
Dunaj
Dunav
Dunărea
Drava
Sava
Tisa
Mureşul
Olt
Prut
Dnestr
Morava
Velika Morava
Iskăr
Marica
Tundža
Kamčija

B U L G A R I A : 28 districts

1 Blagoevgrad
2 Burgas
3 Gabrovo
4 Haskovo
5 Jambol
6 Kărdžali
7 Kjustendil
8 Loveč
9 Mihajlovgrad
10 Pazardžik
11 Pernik
12 Pleven
13 Plovdiv
14 Razgrad
15 Ruse
16 Silistra
17 Sliven
18 Smoljan
19 Sofija(town)
20 Sofija
21 Stara Zagora
22 Šumen
23 Târgovište
24 Tolbuhin
25 Varna
26 Veliko Târnovo
27 Vidin
28 Vraca

G R E E C E : 9 regions

1 Aigaíoi Nísoi
2 Iónioi Nísoi
3 Ípiros
4 Kríti
5 Makedhonia
6 Pelopónnisos
7 Stereá Ellas
8 Thessalia
9 Thráki

R O M A N I A : is divided into 40 districts (judeţ)

Merchant shipping 27,800,000 GRT (1977). Chief ports: Narvik, Oslo, Bergen. Civil aviation (1976): 52.3 mn. km flown, 3,853,000 passengers carried. – **Exports:** machines and equipment, petroleum and products, vessels, non-ferrous metals, fish and fish products, iron, paper. **Imports:** fuels, ores, machines, cars and others. Chief trading partners: United Kingdom, Sweden, Fed. Rep. of Germany.

POLAND

Polska Rzeczpospolita Ludowa, area 312,683 sq.km, population 35,032,000 (1978), **socialist republic** (Chairman of the Council of State Henryk Jabłoński since 1972).

Administrative units: 49 provinces (województwo). **Capital:** Warszawa (Warsaw) 1,532,100 inhab. (1977); **other towns** (1977, in 1,000 inhab.): Łódź 818, Kraków 713, Wrocław 593, Poznań 534, Gdańsk 444, Szczecin 381, Katowice 350, Bydgoszcz 339, Lublin 292, Częstochowa 227, Białystok 207, Radom 184, Toruń 164, Kielce 162.
Population: Poles 98.5%, Ukrainians, Byelorussians, Germans. **Density** 112 persons per sq.km; average annual rate of population increase 0.82% (1970–75); urban population 55.4% (1975). 33.7% of inhabitants employed in agriculture, 25.1% in industry (1976). – **Currency:** złoty = 100 groszy.
Economy: advanced industrial and agricultural country with important mining industry. Industry is concentrated chiefly in the south-west part of the country and the Kraków region. **Principal industries:** mineral mining, engineering (Poznań, Wrocław, Katowice, Kraków, Kielce), chemical (Upper Silesian region, Łódź, Poznań), textile (cotton processing – Łódź, wool processing – the South), food processing. Metallurgy plays an important role. **Mining** (1977, in 1,000 tons, metal content): coal 186,108 (Upper and Lower Silesian Basins in the vicinity of Katowice, Bytom, Zabrze, third world producer), brown coal 37,680 (Konin, Turoszów), crude petroleum 456 (Krosno), natural gas 7.2 billion cub.m (Przemyśl), iron ore 215 (1976), copper 289 (Polkowice, Lubin), lead 55, zinc 217 (Olkusz, Chrzanów), nickel 2,500 tons (1976), silver 550 tons, cadmium, magnesite 26 (1976), salt 5,470 (1976, Wapno, Inowrocław), sulphur 4,891 (1976, second world producer, Tarnobrzeg, Grzybów). Electricity 109.4 billion kWh (1977), of which 98% are thermal power stations. **Production** (1977, in 1,000 tons): pig iron 10,080, crude steel 17,844 (Silesia, Kraków-Nowa Huta, Częstochowa, Katowice, Warszawa), zinc 228 (Silesia), lead 85, copper – smelted 260, – refined 307 (Katowice), aluminium 104 (Skawina), locomotives, railway carriages, agricultural machines, tractors 59,000 units, passenger cars 279,000 units (Warszawa), vessels 492,000 GRT (Gdańsk, Szczecin, Gdynia), radio 2.3 mn. and television receivers 920,000 units, sulphuric acid 3,288, nitric acid 2,185 (1976), soda ash 726 (1976), nitrogenous fertilizers 1,548, phosphorus fertilizers 928, plastics and resins 559 (1976), synthetic rubber 119 (Oświęcim), coke oven coke 19,900; capacity of petroleum refineries 16 mn. tons (1976), pipeline for the import of Soviet petroleum, motor spirit 2.3 mn. tons (1976), oils 10.3 mn. tons (1976), pharmaceuticals (Warszawa), cement 21,300, chemical wood pulp 606 (1976), paper 1,325 (1976), yarn – cotton 220, woollen 107 – woven fabrics – cotton 948 mn. m, woollen 125 mn. m, linen and hemp 146 mn. m, silk 1.3 mn. m (1976) – rayon and acetate – fibres 28.7 (1976), staple fibre 66.9, woven fabrics 100.6 mn. m (1976) – synthetic – fibres 64.6 (1976), staple fibre 68.4 (1976); meat 2,658, milk 17,852, butter 270, cheese 345.7, eggs 526.4, sugar 1,900, beer 12.3 mn. hl (1976); 88.8 billion cigarettes (1976).
Agriculture: private holdings predominate, vegetable production concentrates on cereals and potatoes. Land use: arable land 49.4%, meadows and pastures 13.5%, forests 28.3%. Tractors 467,000 (1967). **Crops** (1977, in 1,000 tons): rye 6,200 (second world producer), wheat 5,310, barley 3,404, oats 2,600, maize (corn), potatoes 41,300 (third world producer), flax 50 (second world producer), hemp 10, sugar beet 15,933, rapeseed 700, hops, fruit 1,628, tomatoes 350, strawberries 150, vegetables 3,775, tobacco 100; **livestock** (1977, in 1,000 head): cattle 13,019, pigs 20,051, sheep 3,934, horses 2,062, poultry 200,000; wool 6,600 tons, honey 14,500 tons; fish catch 750,000 tons (1976); roundwood 21.6 mn. cub.m (1976).
Communications (1976): railways 26,695 km, of which 5,988 km electrified, roads 143,088 km, passenger cars 1,290,100. Navigable waterways 4,527 km, important river port Koźle on the Odra. Merchant shipping 3,448,000 GRT (1977). Chief ports: Szczecin, Gdańsk, Gdynia. Civil aviation (1976): 26 mn. km flown, 1,472,000 passengers carried. Tourism 9.6 mn. visitors (1976).
Exports: coal, coke oven coke, iron and steel, copper, sulphur, cars, vessels, engineering, chemicals, textiles and food products. **Imports:** petroleum, natural gas, raw materials (iron ore, bauxite, cotton), foodstuffs, machines and industrial equipment, industrial consumer goods. Chief trading partners: socialist countries 50% of turnover (chiefly U.S.S.R., German Dem. Rep., Czechoslovakia), Fed. Rep. of Germany, United Kingdom, France.

PORTUGAL

República Portuguesa, area 91,631 sq.km, population 9,786,000 (1977), **republic** (President Gen. Antônio dos Santos Ramalho Eanes since 1976).

Administrative units: 22 districts (distrito), of which 3 on Azores and 1 on Madeira (both autonomous territories). **Capital:** Lisboa (Lisbon) 829,900 inhab. (1975); **other towns** (1970, in 1,000 inhab.): Porto 336 (1975), Amadora 66, Coimbra 56, Barreiro 54, Vila Nova de Gaia 51, Setúbal 50, Braga 49, Funchal 38, Ponta Delgada 20. **Population:** Portuguese. **Density** 107 persons per sq.km; average annual rate of population increase 0.31% (1970–75); urban population 27.8% (1975). 29% of inhabitants employed in agriculture, 18.4% in industry (1976). – **Currency:** escudo = 100 centavos.
Economy: agricultural and industrial country, having good raw material resources. **Mining** (1976, in 1,000 tons, metal content): tungsten 1,588 tons (Panasqueira), tin 252 tons (1977), uranium 88 tons, gold 247 kg, silver, iron ore, manganese, pyrites 337 (1977), sulphur, copper, salt 523, marble 253. Electricity 13.1 billion kWh (1977). **Production** (1976, in 1,000 tons): pig iron 356, crude steel 389, vessels 252,000 GRT, cement 3,713, chemical wood pulp 559, woven cotton fabrics 61.3, traditional textile industry – embroidery, lace – canned fish 50.4, wine 523 (1977, Porto), beer 3 mn. hl; 12.6 billion cigarettes; further data for 1977: meat 382, milk 834, butter 4.3, cheese 27.3, eggs 50.6,

Agriculture: arable land 39.3%, meadows and pastures 5.8%, forests 39.7%. **Crops** (1977, in 1,000 tons): cereals 899 (chiefly wheat and maize), rice 112, potatoes 1,144, sugar beet, olives 265, fruit 1,697–grapes 1,200, oranges 100, lemons 19, bananas–vegetables 1,822, tomatoes 790, sweet chestnuts 27; **livestock** (1977, in 1,000 head): cattle 1,140, sheep 3,657, goats 700, pigs 2,120, asses 179, mules; poultry 16,800; fish catch 339,200 tons (1976); roundwood 7.5 mn. cub.m, cork 133,707 tons (1975, leading world producer).
Communications: railways 3,591 km (1976), of which 431 km are electrified, roads 31,912 km (1975), passenger cars 1,034,000 (1976). Merchant shipping 1,281,000 GRT (1977). Chief port: Lisboa. Civil aviation (1976): 34.9 mn. km flown, 1,650,000 passengers carried. Tourism 958,200 visitors (1976).
Exports: woven cotton fabrics and textile products, wine, canned fish and tomatoes, cork. **Imports:** industrial products, transport equipment, metals, petroleum, chemicals and foodstuffs. Chief trading partners: United Kingdom, Fed. Rep. of Germany, U.S.A.
Azores
Area 2,335 sq.km, population 292,200 (1975), mountainous volcanic islands in the Atlantic Ocean (the largest São Miguel 747 sq.km). **Capital:** Ponta Delgada 20,190 inhab. (1970). **Economy:** cultivation of maize, wheat, bananas and vines; raising of cattle and pigs, fishing and whaling. Roads 1,627 km (1975), sea and air transport.

ROMANIA

Republica Socialistă România, area 237,500 sq.km, population 21,953,000 (1979), **socialist republic** (President and Chairman of the State Council since 1974 Nicolae Ceauşescu).

Administrative units: 40 districts including capital district Bucureşti. **Capital:** Bucureşti (Bucharest) 1,820,829 inhab. (1977); **other towns** (1977, in 1,000 inhab.): Timişoara 272, Iaşi 269, Cluj-Napoca 266, Braşov 261, Constanţa 260, Galaţi 244, Craiova 226, Ploieşti 201, Brăila 197, Oradea 174, Arad 173, Sibiu 154, Tîrgu Mureş 133, Bacău 129, Piteşti 127, Satu Mare 105, Baia-Mare 103, Buzău 99, Reşiţa 87, Hunedoara 80, Piatra-Neamţ 80. **Population:** Romanians 88%, Hungarians 7.9%, Germans 1.6%, Gipsies 1.1%. **Density** 92 persons per sq.km; average annual rate of population increase 0.90% (1970–75); urban population 44.6% (1975). 51.6% of inhabitants employed in agriculture, 24.3% in industry (1975). – **Currency:** leu = 100 bani.
Economy: industrial and agricultural country. **Principal industries:** engineering, mining, metallurgy, chemical, textiles and food processing. Industry is concentrated in the south of the country. 50% of Romanian industry is located at Bucureşti and in the petroleum extraction region near Ploieşti. **Mining** (1977, in 1,000 tons, metal content): crude petroleum 14,652 (Ploieşti, Piteşti, Ticleni, Moineşti), oil pipelines to Bucureşti and ports, natural gas 35.1 billion cub.m (Transylvania), brown coal 19,416 (Rovinari), coal 7,368 (Petroşani, Reşiţa), iron ore 727 (1976, Munţii Poiana Ruscăi, Ocna de Fier), manganese 31 (1976), lead 45 (1976), bauxite 890 (1976, Roşia), gold (Munţii Apuseni), silver 40 tons (1976), salt 4,210 (1976). Electricity 59.9 billion kWh (1977), of which 14% are hydro-electric power stations. **Production** (1977, in 1,000 tons): pig iron 7,400 (1976), crude steel 11,500 (Galaţi, Hunedoara, Reşiţa, Roman), coke oven coke 2,472 (1976), aluminium 209 (Oradea, Slatina), lead 42 (Baia-Mare), zinc 65, tractors 59,300 units (Braşov), motor vehicles – commercial 34,700 (Braşov), passenger 68,100 units (Piteşti) – locomotives and carriages (Craiova), agricultural machinery, petroleum mining and processing equipment (Ploieşti, Bucureşti), shipyards (Galaţi, Constanţa), radio 791,000 (1976) and television receivers 512,000 units (1976), sulphuric acid 1,523, nitrogenous – 1,331 (1976) and phosphorus fertilizers 493 (1976), soda ash 814 (1976), synthetic rubber 146.5 (1976), plastics and resins 543; capacity of petroleum refineries 23.5 mn. tons (1976), motor spirit 3,900 (1976), oils 14,600 (1976), cement 13,900, chemical wood pulp 607 (1976). Textile industry (Bucureşti, Arad, Timişoara, Braşov and others – 1976, in 1,000 tons): yarn – cotton 165, woollen 55.8 – woven fabrics – cotton 677 mn. sq.m, woollen 105 mn. sq.m, silk 1 mn. sq.m (1975) – rayon and acetate staple fibre 52, synthetic fibres 28, synthetic staple fibre 78.6, shoes 98.8 mn. pairs (1977); food industry (1977, in 1,000 tons): meat 1,443, milk 5,397, butter 52, cheese 138.3, eggs 305, sugar 713, canned vegetables 567, vegetable oil 367, wine 875, beer 7.6 mn. hl (1976); 27 billion cigarettes (1976).
Agriculture: vegetable production is predominant. Land use: arable land 45.7%, meadows and pastures 19.3%, forests 27.4%. **Crops** (1977, in 1,000 tons): wheat 6,540, barley 1,626, maize 10,103, rice 50, potatoes 3,738, linseed 48, hemp 26, sugar beet 6,249, sunflower seed 807, soybeans 193, castor beans 12, fruit 2,870 – plums 545, grapes 1,500–vegetables 3,932, tomatoes 1,366, walnuts 32, tobacco 47; **livestock** (1977, in 1,000 head): cattle 6,129, horses 576, buffaloes 222, sheep 14,331, goats 444, pigs 10,193; poultry 91,503; wool 19,900 tons, raising of silkworms (135 tons of cocoons), honey 10,000 tons; fish catch 127,200 tons (1976); roundwood 20.5 mn. cub.m (1976).
Communications: railways 11,080 km, of which 1,407 km are electrified (1976), roads 77,768 km (1976). Navigable waterways – primarily the Danube – 1,659 km (1975). Merchant shipping 1,218,000 GRT (1977). Chief port Constanţa. Civil aviation (1976): 17.2 mn. km flown, 913,000 passengers carried. Tourism 3.2 mn. visitors (1976).
Exports: machines, chemical products, minerals and metals, foodstuffs. **Imports**· machines and industrial equipment, raw materials, electrotechnical and chemical products, consumer goods. Chief trading partners: U.S.S.R., Fed. Rep. of Germany, German Dem. Rep., Poland, Czechoslovakia, China. Socialist countries 47% of turnover (1977).

SAN MARINO

Repubblica di San Marino, area 60.6 sq.km, population 20,520 (1977), **republic** (headed by two Captains-Regents, appointed every 6 months).

Capital: San Marino 4,608 inhab. (1976); Italian **currency,** customs union with Italy. **The official language** is Italian. **Economy:** agriculture (wheat, maize, vines, fruit); tourism (2.4 mn. visitors in 1976) and postage stamps are the chief sources of revenue.

15a Iceland 1:10 000 000

1 : 10 000 000
0 50 100 150 200 250 Km
0 50 100 150 Mi
NORTH SEA
Skagerrak
Kattegat
BALTIC SEA
Gulf of Finland
Rižskij Zaliv
Zal. Gdańska
Zat. Pomorska
NORWAY
SWEDEN
DENMARK
FINLAND
POLAND
GERMAN DEM. REP.
FED. REP. OF GERMANY
Estonian S.S.R.
Latvian S.S.R.
Lithuanian S.S.R.
Byelorussian S.S.R.
R.S.F.S.R.
U.S.S.R.
Oslo
STOCKHOLM
Helsinki
KØBENHAVN (Copenhagen)
LENINGRAD
Tallinn
RIGA
Vilnius
MINSK
Bergen
Stavanger
Kristiansand
Göteborg
Malmö
Uppsala
Västerås
Örebro
Norrköping
Linköping
Jönköping
Helsingborg
Gävle
Turku
Tampere
Vantaa
Espoo
Gotland
Öland
Bornholm
Ahvenanmaa
Hiiumaa
Saaremaa
Tartu
Pärnu
Narva
Novgorod
Pskov
Daugavpils
Liepāja
Ventspils
Kaunas
Klaipėda
Šiauliai
Kaliningrad
Gdańsk
Gdynia
Szczecin
Rostock
Kiel
Århus
Ålborg
Odense
Grodno
Białystok
Bydgoszcz
Toruń
Vitebsk
Velikije Luki
Bobrujsk
Baranoviči
BREMEN
Groningen
12°
16°
20°
24°
28°
56°
60°

SPAIN

Estado Español, area 504,783 sq.km, population 37,109,000 (1978), including Balearic and Canary Is., **kingdom** (King Juan Carlos I since 1975).

Administrative units: 14 regions (50 provinces – continental Spain, Balearic Is., Canary Is., North African settlements – towns Ceuta and Melilla, islands Islas Chafarinas, Peñón de Vélez de la Gomera, Peñón de Alhucemas). **Capital:** Madrid 3,206,067 inhab. (1976); **other towns** (1976, in 1,000 inhab.): Barcelona 1,755, Valencia 720, Sevilla 594, Zaragoza 547, Bilbao 433, Málaga 415, Las Palmas de G. Can. 356, Valladolid 293, Palma 287, Córdoba 257, Hospitalet 242 (1970), Alicante 223, Granada 217, La Coruña 209, Vigo 197 (1970), Santa Cruz de Tenerife 190, Gijón 188 (1970), Vitoria 175, San Sebastián 170, Pamplona 167, Santander 167, Badalona 163 (1970), Oviedo 163, Sabadell 159 (1970), Jerez de la Frontera 150 (1970), Cartagena 147 (1970), Cádiz 143, Tarrasa 139 (1970), Burgos 136, Salamanca 134, Elche 123 (1970), Almería 122, Léon 116, Huelva 113, San Vicente de Baracaldo 109 (1970).
Population: Spaniards 75%, Catalans, Basques, Galicians. **Density** 74 persons per sq.km; average annual rate of population increase 0.96% (1970–75); urban population 69.4% (1975). 21.3% of inhabitants employed in agriculture, 19.2% in industry (1975). – **Currency:** peseta = 100 centimos.
Economy: industrial and agricultural country with considerable raw material resources. Heavy foreign investments have encouraged the rapid development of industry in the last few years. Developed industries: mineral mining, metallurgy, engineering, production of motor vehicles, electrotechnics, the chemical and textile industries. **Mining** (1977, in 1,000 tons, metal content): coal 11,712 (Asturia, Castilla-León), brown coal 5,784 (Teruel), crude petroleum 1,224 (Ayoluengo), iron ore 3,918 (prov. Vizcaya, Santander, Oviedo), manganese 3.6 (1974), copper 18.1 (Minas de Ríotinto, Cangas de Onís), lead 61.1 (Sierra Morena), zinc 96.1 (Reocin), tungsten 370 tons (1976), mercury 1,207 tons (second world producer, Almadén, Mieres), tin 468 tons, antimony 150 tons (1976), titanium, silver 100 tons, magnesite 300 (1976), bauxite 11 (1976), uranium 170 tons (1976, Ciudad Rodrigo), salt 3,132 (1975), potassium salt 530 (1975, Suria), pyrites 2,411 (1976, prov. Huelva), sulphur 1,006 (1973, Minas de Ríotinto). Electricity 93.4 billion kWh (1977), of which 25% are hydro-electric and 8% nuclear power stations (Zorita, Santa María de Garoña, Vandellós). **Production** (1977, in 1,000 tons). **Metallurgy:** pig iron 6,924, crude steel 10,932 (San Vicente de Baracaldo, Avilés, Mieres, Santander, Sagunto), coke oven coke 4,422 (1975), aluminium 211 (Valladolid), copper – smelted 145, refined 149 (Minas de Ríotinto, Córdoba) – lead 89 (Cartagena), zinc 157 (Avilés), tin 5,340 tons (Villagarcía de Arosa). **Engineering** predominantly in large cities: locomotives (Barcelona), passenger cars 1,021,000 units (1977, Madrid, Barcelona, Zaragoza), aircraft (Madrid and vicinity, Sevilla), tractors 38,830 (1976), vessels 1,568,000 GRT (1977, El Ferrol del Caudillo, Cartagena), weapons (Reinosa, Toledo), television receivers 722,000 (1976). **Chemical industry** (Cataluña, Barcelona and vicinity, Asturia, Madrid, Valladolid, Zaragoza – 1976, in 1,000 tons): acid – sulphuric 3,276 (1977), nitric 899, hydrochloric 133 – caustic soda 388 (1977), soda ash 524, fertilizers – nitrogenous 883, phosphorus 592, potash 566 – synthetic rubber 78, plastics and resins 1,007 (1977); capacity of petroleum refineries 74 mn. tons (in the ports, mainly near Bilbao), motor spirit 5.2 mn. tons, oils 36 mn. tons; cement 28 mn. tons (1977), chemical wood pulp 769, paper 2,200. **Textile industry** (chiefly Barcelona and vicinity, 1977, in 1000 tons): yarn – cotton 81.6, woollen 34.6 – woven fabrics – cotton 65.5, woollen 26.8 – artificial fibres 14.6, synthetic fibres 56, shoes 149 mn. pairs (1976). **Food industry** (1977, in 1,000 tons): sugar 1,201, meat 2,174, milk 6,073, cheese 127, butter 19, eggs 617, olive oil 359 (second world producer), wine 2,219 (Jerez, Málaga, Sherry), beer 17.1 mn. hl (1976), canned fish 137.2 (1975); 59 billion cigarettes, 1 billion cigars (1976). Glass industry (Bilbao, Santander), ceramics.
Agriculture is extensive, not very productive. Vegetable production predominates; large output of cereals, fruit and viniculture are of importance. Irrigation in dry areas. Land use: arable land 41.4%, meadows and pastures 21.7%, forests 30.7%. **Crops** (1977, in 1,000 tons): wheat 4,045 (Castilla, Andalucía), barley 6,707, maize (corn) 1,885 (Galicia), rye 218, oats 428, rice 379, potatoes 5,553, sweet potatoes, cotton – seed 135, lint 40 – sugar beet 8,285 (provinces Valladolid, Burgos, León), sugarcane 286, sunflower seed 381, olives 1,772, soybeans, hops, fruit 8,183 – oranges 1,706 (third world producer, Valencia), tangerines 716 (second world producer), lemons 332, almonds 134, dates 16, figs 93 (1975), bananas 371 (Canary Is.), grapes 3,494 (Mediterranean coast, Castilla-La Mancha and Andalucía), peaches, apricots 65, sweet chestnuts 45–vegetables 7,943, tomatoes 2,330, watermelons 379, melons 686, onions 1,239, garlic 199 (leading world producer), tobacco 24; alfalfa 28.4. **Animal production** is extensive (1977, in 1,000 head): cattle 4,500, horses 256, asses 242, mules 269, sheep 15,590, goats 2,231, pigs 9,008; poultry 52,000; wool (merinos) 11,100 tons, raising of silkworms (152 tons of cocoons in 1974); fish catch 1,483,200 tons (1976); roundwood 12.1 mn. cub.m (1976), cork 88,965 tons (1976).
Communications: railways 15,832 km, of which 4,883 km are electrified (1976), roads 145,997 km (1977), motorways 1,291 km (1977). Navigable waterways – only the Guadalquivir 103 km. Merchant shipping 7,186,000 GRT (1977). Chief ports: Barcelona, Bilbao, Cartagena, Valencia. Civil aviation (1976): 135.8 mn. km flown, 11,801,000 passengers carried. Tourism 34.3 mn. visitors (1977).
Exports: metals, transport equipment, engineering products, food products (citrus fruit, wine, olive oil, canned fish), chemicals, petroleum products, textiles, shoes and hides. **Imports:** machines and equipment, petroleum, chemical products, iron and steel, foodstuffs (cereals, sugar and others). Chief trading partners: U.S.A., Saudi Arabia, Fed. Rep. of Germany, France and other EEC countries.

SVALBARD

Area 62,422 sq.km, population 3,500 (1976), **autonomous territory of Norway,** including the islands of Jan Mayen and Bjørnøya. **Capital:** Longyearbyen.
Economy: mining of coal 456,000 tons (1977), exported to Norway and the U.S.S.R.; petroleum prospecting is in progress. Fishing station. **Jan Mayen, area 372 sq.km,** inhabited by radio and meteorological stations staff.

SWEDEN

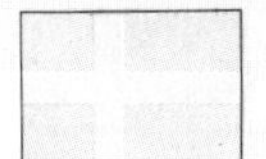

Konungariket Sverige, area 449,964 sq.km, population 8,284,000 (1978), **kingdom** (King Carl XVI Gustaf since 1973).

Administrative units: 24 provinces (län). **Capital:** Stockholm 661,258 inhab. (1977, with agglomeration 1,364,175); **other towns** (1977, in 1,000 inhab.): Göteborg 442, Malmö 240, Uppsala 140, Norrköping 120, Västerås 118, Örebro 117, Linköping 110, Jönköping 108, Borås 104, Helsingborg 101, Sundsvall 94, Eskilstuna 92, Gävle 87, Lund 77, Umeå 76, Karlstad 73, Skellefteå 73, Kristianstad 68, Luleå 67, Karlskrona 60, Solna 53. **Population:** Swedes 95%, in the North Finns, nomadic Lapps. **Density** 18 persons per sq.km; average annual rate of population increase 0.61% (1970–75); urban population 83,6% (1975). 26% of inhabitants employed in industry (1975). – **Currency:** Swedish krona = 100 öre.
Economy: highly developed industrial country with intensive, mechanized agriculture. Sweden is one of the economically most advanced countries in the world. Basic natural resources: forests, iron ore, hydro-electric energy. **Chief industries:** engineering (electrotechnics, shipbuilding), wood processing and paper industry, mineral mining, energy production. Metallurgy, the chemical industry and food processing are also highly developed. **Mining** (1976, in 1,000 tons, metal content): iron ore 15,885 (1977, Kiruna, Gällivare), copper 44.9 (Boliden, Aitik), lead 87 (1977, Laisvall), zinc 136 (1977, Åmmeberg), tungsten 194 tons, molybdenum 500 tons (1975), silver 159 tons (1977), gold 3,662 kg (Boliden), uranium, pyrites 404. Electricity 87.5 billion kWh (1977), of which 60% hydro-electric and 25% nuclear power stations (Ringhals, Oskarshamn).
Production (1977, in 1,000 tons): pig iron 2,316, crude steel 3,960 (Borlänge, Luleå, Oxelösund, Sandviken), aluminium, primary 82.7 (Kubikenborg), refined copper 54 (Rönnskär, Helsingborg), lead 45 (Landskrona), electrotechnics (Västerås), motor vehicles – passenger 235,000 units ("Volvo"), commercial 51,000 (Göteborg, Trollhättan, Södertälje) – weapons (Bofors), aircraft (Malmö, Linköping), vessels 2,127,000 GRT (Göteborg, Malmö, Landskrona – second world producer), sulphuric acid 900, nitric acid 337 (1975), plastics and resins 600, explosives; capacity of petroleum refineries 14 mn. tons (1976), motor spirit 2,500 (1976), oils 10,800 (1976), cement 2,532, sawn wood 10.8 mn. cub.m, chemical wood pulp 6,718 (1976, Husum, Örnsköldsvik, Karlsborg), paper and paper products 5,060, matches (Jönköping), leather industry (Örebro, Kumla), ornamental glass (prov. Kronoberg), woven cotton fabrics 11.6 (Borås), rayon and acetate staple fibre 30.5, meat 503, milk 3,247, butter 58, eggs 109, cheese 89; 11.3 billion cigarettes (1976).
Agriculture: predominantly animal production. Land use: arable land 7.3%, meadows and pastures 1.8%, forests 64.2%. **Crops** (1977, in 1,000 tons): wheat 1,562, barley 1,992, oats 1,399, rye, potatoes 1,346, sugar beet 2,214, rapeseed 277; **livestock** (1977, in 1,000 head): cattle 1,876, sheep, pigs 2,585; poultry 12,000; fish catch 208,600 tons (1976); roundwood 52.4 mn. cub.m (1976).
Communications: railways 12,061 km, of which 7,484 km are electrified (1976), roads 97,402 km (1977), motorways 964 km (1976), passenger cars 2,857,000 (1977). Navigable waterways 736 km (1975). Merchant shipping 7,429,000 GRT (1977). Chief ports: Göteborg, Luleå, Stockholm, Helsingborg. Civil aviation (1976): 63.5 mn. km flown, 3,909,000 passengers carried.
Exports: engineering and metal products, cars, paper, chemical wood pulp, wood and wood products, iron and steel.
Imports: fuels, machines and equipment, metals, foodstuffs, chemicals. Chief trading partners: Fed. Rep. of Germany, United Kingdom, Denmark, U.S.A., Norway, Finland.

SWITZERLAND

Schweizerische Eidgenossenschaft – Confédération Suisse – Confederazione Svizzera – Confederaziun Svizra, area 41,293 sq.km, population 6,327,000 (1977), **federal republic** (President Fritz Honegger in 1982).

Administrative units: 26 cantons. **Capital:** Bern 146,500 inhab. (1977); **other towns** (1977, in 1,000 inhab.): Zürich 383, Basel 188, Genève (Geneva) 153, Lausanne 133, Winterthur 88, St. Gallen 76, Luzern 64. **Official languages:** German (65% of inhabitants), French (18%), Italian (12%), Romansch (0.8%). **Density** 153 persons per sq.km; average annual rate of population increase 0.84% (1970–75); urban population 57.2% (1975). 22% of inhabitants employed in industry (1976). – **Currency:** franc = 100 rappen.
Economy: highly developed industrial country and a leading financial and banking centre. **Principal industries:** precision engineering (watches, electrical and optical apparatus), the chemical and textile industries, food processing. Chief industrial centres: Zürich, Basel. **Mining:** salt 312,000 tons. Electricity 45.4 billion kWh (1977), of which 3% are hydro-electric and 21% nuclear power stations (Beznau, Mühleberg). **Production** (1977, in 1,000 tons): crude steel 545 (1976), aluminium 80 (canton Valais), watches 66 mn. pieces (cantons: Neuchâtel, Bern and Solothurn; Genève, Schaffhausen), geodetic apparatus (Genève), electrical apparatus (Basel, Baden), electrochemical products (canton Valais), dyes and pharmaceuticals (Basel), cement 3,500, paper and paper products 720 (Jura, Alp region), cotton yarn 46, woven cotton fabrics 140 mn. m (eastern Switzerland), wool-processing (Solothurn, canton Thurgau), woven silk fabrics 17.6 mn. m (1976, Zürich, Basel), synthetic fibres 49.4 (1976), shoes 10 mn. pairs, meat 440, milk 3,481, cheese 114.5, butter 34.6, eggs 41.8, chocolate, wine 121, beer 4.2 mn. hl (1976); 29.4 billion cigarettes (1976).
Agriculture: predominantly livestock production, chiefly cattle raising. Land use: arable land 10%, meadows and pastures 40.1%, forests 26.5%. **Crops** (1977, in 1,000 tons): cereals 703, potatoes 870, sugar beet 545, fruit 601, grapes 167, vegetables; **livestock** (1977, in 1,000 head): cattle 2,001, pigs 2,065, sheep 368; poultry 6,000; roundwood .6 mn. cub.m (1976).
Communications (1976): railways 4,990 km (all electrified), roads 62,175 km, motorways 964 km, passenger cars ,863,615. Navigable waterways 21 km, river port Basel. Merchant shipping 253,000 GRT (1977). Civil aviation (1976): 6 mn. km flown, 5,085,000 passengers carried. Tourism 7.6 mn. visitors (1976).

NORWEGIAN SEA
Greenland (Den.)
J. Mayen (Nor.)
Svalbard (Nor.)
Longyearbyen
Nordauslandet
Vestspitsbergen
Bjørnøya (Bear I.)
BARENTS SEA
ARCTIC
Zemla Franca-Iosifa
Z. Georga
M. Arkt
O. Komsomolec
O. Oktabr'skoj Revolucii
O. Bol'ševik
Mys Želanija
Novaja Zemla
KARSKOJE MORE
O. Belyj
Dikson
Pol. Tajmyr
Krasino
O. Kolgujev
Karskije Vor.
O. Vajgač
Amderma
Jamal Pol.
Gydanskij Pol.
Obskaja Guba
Matočkin Šar
BELOJE MORE
Češskaja G.
UNITED KINGDOM
GLASGOW
Edinburgh
LIVERPOOL
Thurso
NORTH SEA
HAMBURG
BREMEN
F.R.G.
KØBENHAVN
DEN.
Frederikshavn
Bergen
Oslo
Trondheim
NORWAY
SWEDEN
STOCKHOLM
Göteborg
Malmö
Kiel
G.D.R.
BERLIN
PRAHA
POLAND
Szczecin
WARSZAWA
Gdan'sk
Kaliningrad
BALTIC SEA
Liepāja
Lithuanian S.S.R.
Latvian S.S.R.
RIGA
Estonian S.S.R.
Tallinn
Helsinki
Vaasa
FINLAND
Gulf of Bothnia
Luleå
Östersund
Bodø
Narvik
Tromsø
Pečenga
Murmansk
Kirovsk
Kandalakša
Belomorsk
Vyborg
LENINGRAD
Petrozavodsk
Onežskoje O.
Ladožskoje O.
Pskov
Novgorod
Vilnius
Byelorussian S.S.R.
Brest
MINSK
Smolensk
Kalinin
LVOV
Černovcy
KIŠIN'OV
Ukrainian S.S.R.
KIJEV
Vinnica
ODESSA
Nikolajev
Kursk
CHAR'KOV
TULA
MOSKVA (Moscow)
JAROSLAVL
Vologda
Archangelsk
Mezen'
Severnaja Dvina
Kotlas
Uchta
Syktyvkar
Pečora
Narjan-Mar
Chalmer-Ju
Vorkuta
Labytnangi
Salechard
Ber'ozovo
Novyj Port
Nadym
Dudinka
Noril'sk
Igarka
Tazovskij
Urengoj
Turuchansk
Vereščagino
Nižnaja Tunguska
Kurejka
UNION OF SOVIET SOCIALIST REPUBLICS
Or'ol
Rjazan
Saransk
GOR'KIJ
Joškar-Ola
Kirov
IŽEVSK
KAZAN'
Uljanovsk
Kama
Serov
Berezniki
PERM'
Sergino
Polunočnoje
Surgut
Chanty-Mansijsk
Nižnevartovsk
Tobol'sk
Severo-Jenisejskij
Jenisejsk
DNEPROPETROVSK
Ždanov
Sevastopol'
AZOVSKOJE MORE
ROSTOV-n. D.
VORONEŽ
Tambov
Penza
SARATOV
KUJBYŠEV
UFA
SVERDLOVSK
Magnitogorsk
ČEL'ABINSK
Tumen'
Kurgan
Petropavlovsk
OMSK
NOVOSIBIRSK
Kolpaševo
Belyj Jar
Asino
Tomsk
Ačinsk
Kansk
KRASNOJARSK
BLACK SEA
Novorossijsk
KRASNODAR
Suchumi
Batumi
Georgian S.S.R.
Stavropol'
Ordžonikidze
VOLGOGRAD
Volga
Ural
Uralsk
Orenburg
Aktubinsk
Orsk
Kazakh S.S.R.
Kustanaj
Kokčetav
Pavlodar
BARNAUL
NOVOKUZNECK
Kemerovo
Minusinsk
Abakan
Taštagol
Gorno-Altajsk
Kyzyl
Gurjev
Astrachan'
TBILISI
JEREVAN
Nachičevan
BAKU
Machačkala
CASPIAN SEA
Fort-Ševčenko
Ševčenko
Emba
Čelkar
Tobol
Arkalyk
Išim
Celinograd
Temirtau
KARAGANDA
Irtyš
Bijsk
Semipalatinsk
Ust'-Kamenogorsk
Zyr'anovsk
Araľskoje More
Aral'sk
Bajkonur
Džezkazgan
Syrdarja
Kzyl-Orda
Balchaš
O. Balchaš
O. Zajsan
TABRĪZ
Astara
Rasht
Mianeh
Qom
TEHRĀN
IRAN
ESFAHĀN
Yazd
Kermān
Zāhedān
Bam
Krasnovodsk
Turkmen S.S.R.
Bandar Torkaman
Shāhrūd
Tašauz
Nukus
Uzbek S.S.R.
Amudarja
Čardžou
Aschabad
MASHHAD
Buchara
Samarkand
TAŠKENT
FRUNZE
Kirgiz S.S.R.
ALMA-ATA
Taldy-Kurgan
Yining
Kelamayi
Chovd
Ulaangom
Uliastaj
Altaj
MONGOLIA
O. Issyk-Kul
Prževal'sk
Akesu
Kashi
Suoche
Hetian
CHINA
Tibet
Kuška
Termez
DUŠANBE
Tadzhik S.S.R.
Namangan
Fergana
Herāt
Meymaneh
AFGHANISTAN
KĀBUL
Baghlān
Qandahār
Helmand
Islāmābād
Jammu and Kashmir
Quetta
Peshāwar
PAKISTAN
LAHORE
RĀWALPINDI
Srīnagar
Gwādar
Sukkur
Indus
Chenāb
Sutlej
MULTĀN
DELHI
New Delhi
Ganga
ĀGRA
INDIA
NEPAL
Kātmāndu
1 Armenian S. S. R.
2 Azerbaijan S. S. R.
3 Moldavian S. S. R.
1 : 45 000 000
0 250 500 750 1 000 Km
0 200 400 600 Mi
Kalančak
Arm'ansk
Novoalek
Skadovsk
Karkinitskij Zaliv
Krasnop
Razdol'noje
Vojkovo
Černomorskoje
Okťabr'skoje
O. Donuzlav
O. Sasyk
Jevpatorija
Saki
Simferopol'
Kača
Bachč
Sevastopol'
BLACK
M. Saryč
U.S.S.R.
60° 70° 80° 90° 50° 40° 30° 46° 34°

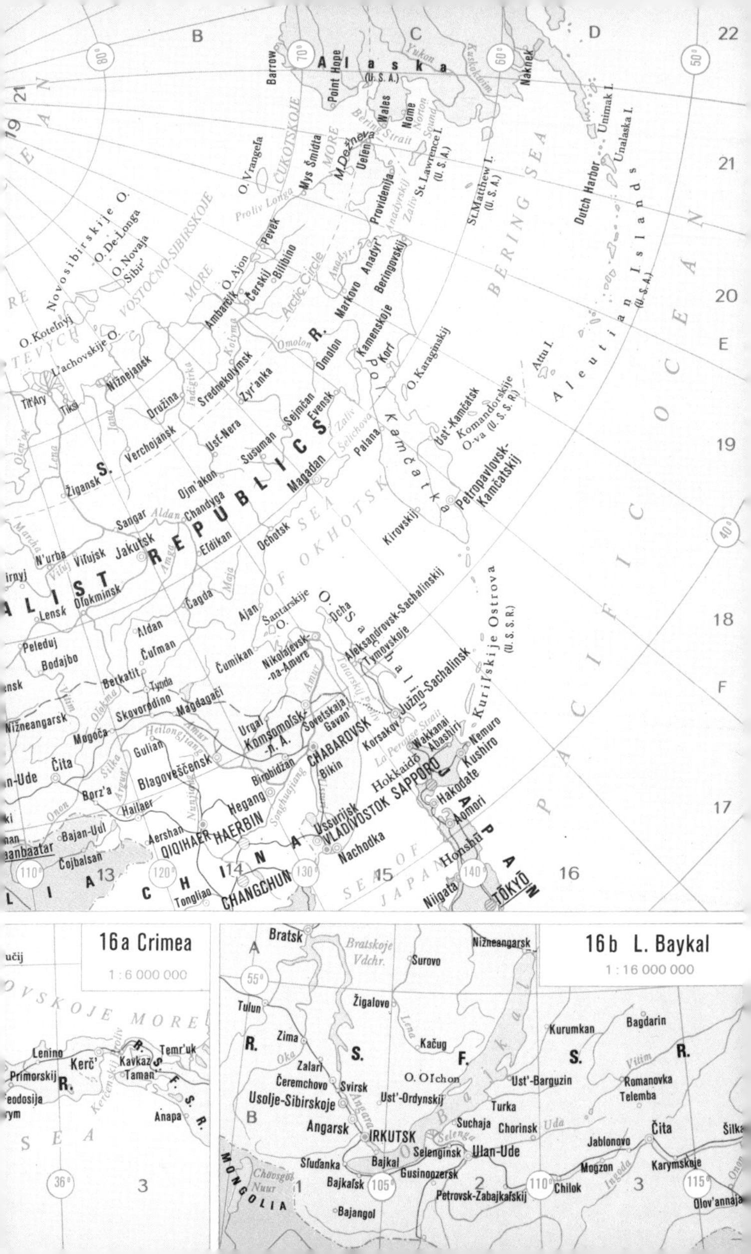

16a Crimea
1 : 6 000 000
16b L. Baykal
1 : 16 000 000

Exports: machines and apparatus, watches, chemical products (dyes, pharmaceuticals), textiles, foodstuffs (dried and condensed milk, cheese, chocolate). **Imports:** fuels (petroleum, coal), raw materials for the textile and food industries. Chief trading partners: Fed. Rep. of Germany, France, Italy, U.S.A.

VATICAN CITY, STATE OF

Stato della Città del Vaticano, area 0.44 sq.km, population 723 (1977), **papal state** (Pope John Paul II since 1978).

The smallest state in the world as to area; religious and political centre of the Roman Catholic Church, seat of the Pope, Cardinals and the highest ecclesiastical officials. **Official languages** are Latin and Italian. – **Currency:** Italian lira. Revenue from tourism and postage stamps.

WEST BERLIN

Berlin (West), area 480 sq.km, population 1,927,000 (1978), administrated by the Senate, headed by the Ruling Burgomaster (Dr Richard von Weizsäcker since 1981).

Density 4,015 persons per sq.km. After the Second World War Berlin was divided into 4 sectors on the basis of the Potsdam Agreement (1945). West Berlin consisted of 3 sectors under the control of the U.S.A., the United Kingdom and France. West Berlin contains the **administrative districts** of Kreuzberg, Neukölln, Schöneberg, Steglitz, Tempelhof, Zehlendorf, Charlottenburg, Spandau, Tiergarten, Wilmersdorf, Wedding, Reinickendorf.

Economy: highly developed industries, dependent on imported raw materials. Leading industries: electrotechnical, engineering and textiles. The extensive exchange of goods between West Berlin and Fed. Rep. of Germany is dependent upon specific autobahns and air corridors for transport across the territory of the German Dem. Rep.

YUGOSLAVIA

Socialistička Federativna Republika Jugoslavija, area 255,804 sq.km, population 22,014,000 (1978), **socialist federal republic** (President Sergej Krajgher since 1981).

Administrative units: 6 socialist republics – Bosna i Hercegovina, Crna Gora, Hrvatska, Makedonija, Slovenija, Srbija (2 autonomous regions of Vojvodina and Kosovo). **Capital:** Beograd (Belgrade) 1,345,000 inhab. (1976, with agglomeration); **other towns** (1971, in 1,000 inhab.): Zagreb 566, Skopje 313, Sarajevo 244, Ljubljana 174, Split 152, Novi Sad 141, Niš 133, Rijeka 132, Maribor 97, Osijek 94, Banja Luka 90, Subotica 89, Kragujevac 71, Priština 70, Bitola 66, Zrenjanin 60, Titograd 55. **Population:** Serbs, Croats, Slovenians, Montenegrians. **Density** 86 persons per sq.km; average annual rate of population increase 0.91% (1970–75); urban population 38.5% (1975). 43.5% of inhabitants employed in agriculture, 18% in industry (1975). – **Currency:** dinar = 100 para.

Economy: developed industrial and agricultural country. Rich resources of raw materials, forests and hydro-electric energy. **Principal industries:** mining of minerals, engineering, metallurgy, textiles, wood-working and food processing. Manufacturing of motor vehicles, shipbuilding, electrotechnics, radioelectronics and the chemical industry are under development. **Mining** (1977, in 1,000 tons, metal content): coal 504, lignite 38,592 (Slovenija - Trbovlje, Zagorje; Srbija), crude petroleum 3,948 (Lendava, Gojilo, Stružec), natural gas 1.9 billion cub.m, iron ore 1,558 (Ljubija, Vareš), manganese 6.6 (1976), chromium 700 tons (1976), copper 116.3 (Bor, Majdanpek), lead 127.2, zinc 107.2 (Mežica), mercury 140 tons, antimony 2,550 tons, gold 4,886 kg (1976), silver 146 tons, bauxite 2,040 (Rovinj, Mostar, Titograd), magnesite 391 (1976, Štip), salt 267 (1976), asbestos 13 (1976). Electricity 48.6 billion kWh (1977), of which 50% is from hydro-electric power stations.

Production (1977, in 1,000 tons): pig iron 2,112, crude steel 2,580 (Zenica), coke oven coke 1,786 (1976), copper – smelted 93, refined 121 (Bor) – lead 129.6 (Mežica), zinc 98.9 (Celje), aluminium 200.4 (Lozovac, Kidričevo), motor vehicles – commercial 15,900 units, passenger 233,500 (Maribor, Kragujevac) – tractors 41,800 (Kruševac), agricultural machinery (Subotica, Novi Sad), railway carriages (Kraljevo), locomotives (Slavonski Brod), vessels 436,000 GRT (Rijeka, Split, Trogir), electrotechnics (Beograd), sulphuric acid 937 (Bor), nitric acid 662, nitrogenous fertilizers 389 (1976), phosphorus fertilizers 213 (1976, Kosovska Mitrovica); capacity of petroleum refineries 12 mn. tons (1976), motor spirit 1,900 (1976), oils 8,000 (1976), cement 8,004, yarn – cotton 121, woollen 46 – woven fabrics – cotton 383.9 mn. sq.m, woollen 78.7 mn. sq.m, silk 42.1 mn. sq.m – rayon and acetate staple fibre 50 (1976), woven rayon and acetate fabrics 37.1 mn. sq.m (1976), leather shoes 53.4 mn. pairs, chemical wood pulp 438 (1976), paper and paper products 623 (1976), sugar 765, meat 1,256, milk 3,909, butter 12.3, cheese 135.6, eggs 191, wine 638, beer 8.7 mn. hl (1976), alcoholic spirits; 42.2 billion cigarettes (1976).

Agriculture: arable land 31.3%, meadows and pastures 24.7%, forests 35.5%. **Crops** (1977, in 1,000 tons): maize (corn) 9,856, wheat 5,622, barley 650, oats, rice, potatoes 2,854, hemp, sugar beet 5,286, sunflower seed 481, sesame, olives, soybeans 66, hops, fruit 2,853, apples 367, plums 757 (leading world producer), grapes 1,291 (Dalmacija, Danubeland), vegetables 3,692, tomatoes 440, watermelons 600, tobacco 62 (Bosna i Hercegovina, Makedonija), walnuts 32.9; **livestock** (1977, in 1,000 head): cattle 5,641, buffaloes 65, horses 812, asses, sheep 7,484, pigs 7,326; poultry 53,779; wool 5,700 tons, honey 6,100 tons, raising of silkworms (12 tons of cocoons in 1977); fish catch 58,800 tons (1976); roundwood 14 mn. cub.m (1976).

Communications (1976): railways 9,967 km, roads 113,513 km, of which 41,136 km asphalted, passenger cars 1,732,100. Navigable waterways 2,001 km. Merchant shipping 2,285,000 GRT (1977). Chief ports: Rijeka, Split, Dubrovnik. Civil aviation (1976): 30.6 mn. km flown, 2,647,000 passengers carried (1976). Important tourism 5.6 mn. visitors (1977).

Exports: ores and non-ferrous metals, machines and equipment, transport equipment (vessels, cars, railway carriages), foodstuffs (meat, canned fish, fruit, wine), electrotechnical products, furniture, textiles and leather goods. **Imports:** machinery and industrial equipment, fuels, raw materials and semi-finished products, foodstuffs. Chief trading partners: U.S.S.R., Fed. Rep. of Germany, Italy, U.S.A.

UNION OF SOVIET SOCIALIST REPUBLICS

Soyuz Sovyetskikh Sotsialisticheskikh Respublik, area 22,274,900 sq.km (22,402,200 sq.km incl. the White Sea /Beloje More 90,000 sq.km and the Sea of Azov /Azovskoje More 37,300 sq.km), **population 260,100,000** (1978); of this, the European part has approx. 5,443,900 sq.km with population 191.2 mn. and the Asiatic part about 16,831,000 sq.km with population 67.7 mn. (1977); **Union of Soviet Socialist Republics** (President of the Presidium of the Supreme Soviet of the U.S.S.R. Leonid Ilyich Brezhnev since 1977).

Administrative units: 15 federal Soviet Socialist Republics (S.S.R.) divided into 3,150 districts and 593 urban districts (1978). In addition to this, some of the republics include Autonomous Soviet Socialist Republics (A.S.S.R., 20), Autonomous Regions (8), Autonomous Areas (only in the Russian Soviet Federal Socialist Republic, 10), Territories and Regions. **Capital:** Moskva (Moscow) 7,911,000 inhab. (1978); **other towns** (1978, in 1,000 inhab.): Leningrad 4,480, Kijev 2,133, Taškent 1,733, Baku 1465, Char'kov 1,428, Gor'kij 1,322, Novosibirsk 1,324, Minsk 1,273, Kujbyšev 1,221, Sverdlovsk 1,204, Dnepropetrovsk 1,061, Tbilisi 1,052, Odessa 1,051, Omsk 1,042, Čel'abinsk 1,019, Doneck 997, Perm' 985, Jerevan 982, Kazan' 980, Ufa 962, Volgograd 943, Rostov-na-Donu 935, Alma-Ata 890, Saratov 866, Riga 827, Voronež 790, Zaporožje 784, Krasnojarsk 781, L'vov 655, Krivoj Rog 648, Jaroslavl' 592, Karaganda 580, Krasnodar 560, Iževsk 549, Vladivostok 548, Irkutsk 543, Novokuzneck 543, Chabarovsk 536, Barnaul 531, Frunze 522, Tula 515, Kišin'ov 492, Toljatti 479 (1977), Dušanbe 476, Ždanov 474 (1977), Vilnius 470, Astrachan' 466 (further 1977), Ivanovo 461, Kemerovo 454, Nikolajev 447, Uljanovsk 447, Orenburg 446, Vorošilovgrad 445, Penza 443, R'azan' 442, Tallinn 442, Makejevka 437, Tomsk 423, Kalinin 401, Nižnij Tagil 399, Magnitogorsk 398.
Population: the most numerous nationalities at the 1970 census were: 129 mn. Russians, 40.8 mn. Ukrainians, 9.2 mn. Uzbeks, 9.1 mn. Byelorussians, 5.9 mn. Tatars, 5.3 mn. Kazakhs, 4.4 mn. Azerbaijanians, 3.6 mn. Armenians, 3.2 mn. Georgians, 2.7 mn. Lithuanians, 2.7 mn. Moldavians, 2.2 mn. Jews, 2.1 mn. Tadzhiks, 1.8 mn. Germans, 1.7 mn. Chuvashes, 1.5 mn. Kirgiz, 1.5 mn. Turkmenians, 1.4 mn. Latvians, 1.3 mn. Mordovians, 1.2 mn. Bashkirs, 1.2 mn. Poles. **Density** 11.6 persons per sq.km; average annual rate of population increase 0.99% (1970–75); urban population 62% (1978). 18.9% of the economically active inhabitants employed in agriculture. – **Currency:** rouble = 100 kopeks.
Economy: well-developed industrial and agricultural country. It produces about 20% of the total world industrial output and holds the leading position in many branches of mining and processing: mining of crude petroleum, peat, iron ore, manganese, lead, mercury, asbestos, potash, silver; in the production of pig iron and crude steel, coke oven coke, cement, bricks, cotton and woollen yarn, woven woollen and linen fabrics, milk, butter, flour, sugar, roundwood etc. The U.S.S.R. is the second greatest producer in the world of coal, lignite and brown coal, natural gas, copper, zinc, chromium, tungsten, vanadium, diamonds, magnesite and phosphates. The size of the cultivated area is the largest in the world, and the U.S.S.R. is the world's leading producer of wheat, barley, rye, oats, potatoes, sunflower seed, flax fibre, sugar beet and it has the largest sheep population. Arable land constitutes 10%, meadows and pastures 17% and forests 28% of the total land area. The U.S.S.R. is the second greatest producer in the world of hemp, cotton lint, tomatoes, sheepskins, wool-grease and honey.
Agriculture (1977, in 1,000 tons) – **crops:** wheat 92,042 (Ukraine, Plain of Kuban', a district called the Black Earth Central Zone – Tambov, Voronež, Kursk; North Kazakhstan), barley 52,653 (southern European Russia), maize (corn) 10,993 (Ukraine, North Caucasia, Moldavia), rye 8,471 (central European Russia), oats 18,379 (western Siberia, central European Russia), rice 2,200 (Central Asian republics), millet 2,000, sorghum 150, potatoes 83,400 (European Russia, Byelorussia), sunflower seed 5,870 (southern European Russia), linseed 340, flax fibre 517 (Byelorussia, Baltic shore, central European Russia), hemp 54, cotton – seed 5,694, lint 2,716 (Uzbekistan), sugar beet 93,300 (Ukraine), tomatoes 4,500, grapes 3,170 (Moldavia, Transcaucasia, Central Asia), oranges 200 (Georgia), tea 55 (Georgia), tobacco 318. **Livestock** (1977, in 1,000 head): horses 5,996, asses 444, cattle 110,346, buffaloes 393, camels 239, pigs 63,055 (second world population), sheep 139,834, goats 5,539, poultry 880,900, eggs 3,360,000 tons, cow-hides 734,996 tons, sheepskins 120,000 tons, wool-grease 458,000 tons, honey 188,000 tons. Fish catch 10,134,000 tons (second world catch). Roundwood 389 mn. cub.m (leading world producer).
Mining (1977, in 1,000 tons, metal content): coal 499,768 – Donbas Basin (Doneck and Vorošilovgrad region), Kuzbas Basin (Kemerovo region), Central Kazakhstan (Karaganda), lignite and brown coal 160,032, peat, combustible shale, crude petroleum 551,500 (the Ural-Volga area – Tatar A.S.S.R., Bashkir A.S.S.R., Kujbyšev region; T'umen' region in western Siberia, Baku etc.), international oil pipelines supply Czechoslovakia, Hungary, German Dem. Republic and Poland; natural gas 347 billion cub.m (Doneck-Dnepr district, North Caucasia, Volga Basin, T'umen' region), an extensive gas pipeline system supplies industrial regions in the U.S.S.R. and in a number of European countries (Czechoslovakia, German Dem. Republic, Fed. Rep. of Germany, Austria, Italy, France, Yugoslavia), uranium, iron ore 131,418 (Fe content – Krivoj Rog, Kursk, the Ural Mts.), manganese 2,904 (Nikopol', Čiatura), copper 1,110 (Kazakhstan, the Ural Mts.), bauxite 6,700 (1976 – the Ural Mts., Kazakhstan, eastern Siberia), zinc 1,040 (Kazakhstan, Kuzbas Basin, North Caucasia), lead 625 (Kazakhstan, North Caucasia, the Ural Mts.), nickel 168 (northern Siberia, the Ural Mts., Kol'skij Pol. = Kola Peninsula), chromium 910 (the Ural Mts., Kazakhstan), tin, antimony 7,900 tons (Kazakhstan, Central Asia), molybdenum 9,700 tons (Transcaucasia, Central Asia, eastern Siberia), tungsten 10,350 tons (Kazakhstan, Central Asia, eastern Siberia), vanadium 3,200 tons (1976, the Ural Mts., Kazakhstan), mercury 2,200 tons (Central Asia), asbestos 2,290 (1976, the Ural Mts.), mica (eastern Siberia), gold (eastern Siberia, the Ural Mts.), silver 1,550 tons (Kuzbas Basin, Central Asia), platinum (the Ural Mts., northern Siberia), diamonds 9.9 mn. carats (1976, Jakut A.S.S.R., the Ural Mts.), magnesite 1,850 (the Ural Mts.), cobalt, phosphates 24,200 (Kola Pen., Kazakhstan), sulphur 2,500 (the Ural Mts., Ukraine), potash 8,500 (the Ural Mts., Byelorussia, Turkmenistan), salt 14,317.
Electricity 1,150 billion kWh, of which 147 billion kWh hydro-electric power stations and 34.8 billion kWh nuclear power stations. Largest power stations: Krivoj Rog 3 mn. kW (thermal), Krasnojarsk 6 mn. kW, Bratsk 4.6 mn. kW, Sajano-Šušenskaja 6.3 mn. kW – partially in operation (hydro-electric), Novovoronežskaja 2.5 mn. kW (nuclear).
Metallurgy: iron – Doneck-Dnepr district (Krivoj Rog), the Ural Mts.; non-ferrous metals – the Ural Mts., Kazakhstan, North Caucasia, Kola Peninsula. Production (1977, in 1,000 tons): pig iron 107,368, crude steel 146,678, aluminium 1,640, copper 1,100, lead 510, zinc 1,020; coke oven coke 86,800.

17 a Ural Region 1:10 000 000

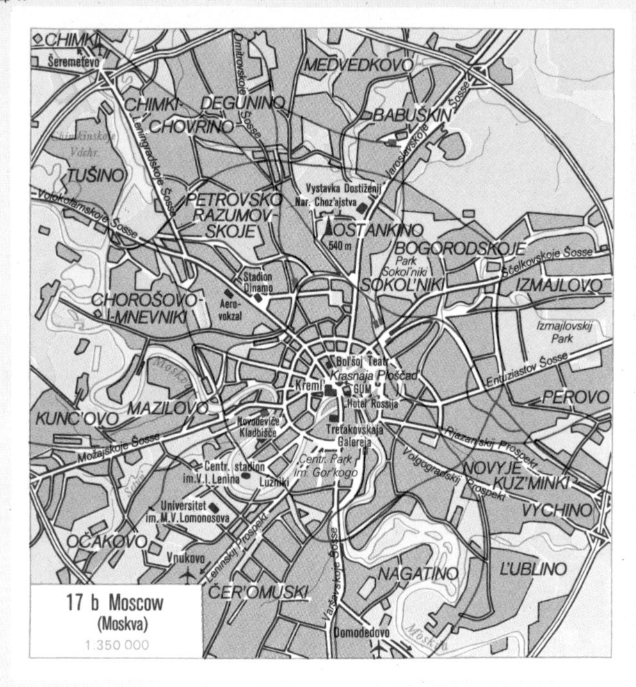

Engineering (1977, in 1,000 units): metal-working machines 237.5 (Moskva, Leningrad), motor vehicles – passenger 1,280 (Toljatti, Iževsk, Moskva, Gor'kij), commercial 733.7 (Moskva), buses 74.6 (L'vov) – locomotives 4.2 (Vorošilovgrad, Char'kov, Kolomna), railway carriages 73.3 (Nižnij Tagil, Ždanov), tractors 546.3 (Volgograd, Char'kov, Minsk), grain combines 105.5 (Rostov-na-Donu), shipbuilding (Leningrad, Nikolajev, Cherson), radio receivers 8,652, television receivers 7,073, refrigerators 5,798.

Chemical industry – Moskva and surroundings, the Ural Mts., Doneck-Dnepr district, Leningrad, Gor'kij, regions near raw material deposits. Production (1977, in 1,000 tons): sulphuric acid 21,104, synthetic resins and plastics 2,765, caustic soda 2,658, nitrogenous fertilizers 9,114, phosphate fertilizers 6,024, chemical wood pulp 5,472.

Building industry – Moskva and surroundings, Leningrad, Char'kov, Krivoj Rog, Volgograd, Kujbyšev, the Ural Mts. Production (1977): cement 127 mn. tons, bricks 46 billion pieces. **Wood and paper industry:** production of furniture – Moskva, Ivanovo, Leningrad, Riga, Užgorod, Kijev; paper 5.4 mn. tons, of which 1.4 mn. tons newsprint.

Textile industry: cotton yarn 1.6 mn. tons (1976), woollen yarn 436,700 tons; woven fabrics – cotton 7,461 mn. sq.m, woollen 989 mn. sq.m, silk 43.6 mn. sq.m (1976), linen 817 mn. sq.m. Leather footwear 736 mn. pairs. **Food industry** – in cities and agricultural production regions (1977, in 1,000 tons): meat 14,800, milk 94,300, butter 1,408, cheese 1,363, flour 42,863, sugar 8,885, margarine 1,168, wine 30,726,000 hl, beer 61,861,000 hl; 378 billion cigarettes.

Communications: railways 139,762 km (1977), roads 1,405,000 km, of which hard surfaced 713,000 km (1977), motor vehicles – passenger 3.8 mn., commercial 4.5 mn. (1975). Navigable waterways 143,109 km (1977). Merchant shipping 21.4 mn. GRT (1976). Chief ports: Novorossijsk, Odessa (with Iljičovsk), Leningrad, Nachodka, Astrachan', Archangel'sk, Murmansk, Baku. Civil aviation 92.9 mn. passengers carried (1977). Length of oil and oil products pipelines 61,897 km (1977), gas pipelines 103,500 km (1976). Tourism 4,399,799 visitors (1977).

Foreign trade: with 118 countries (1977). Total turnover 63.4 billion roubles, of which exports 33.3 billion, imports 30.1 billion roubles.–**Exports:** fuels, raw materials, ores 48.1%; machines, equipment and transport equipment 18.8%; chemical products, building materials and other products 18.4%; agricultural raw materials and their products, consumer goods. **Imports:** machines, industrial and transport equipment 38%, raw materials, rolled iron, food materials, foodstuffs and consumer goods. Chief trading partners: German Dem. Rep., Poland, Czechoslovakia, Bulgaria, Hungary, Fed. Rep. of Germany, U.S.A., Japan, Finland, Yugoslavia.

RUSSIAN SOVIET FEDERAL SOCIALIST REPUBLIC (R.S.F.S.R.), Rossiskaya Sovietskaya Federativnaya Sotsialisticheskaya Respublika, **area 17,075,400 sq.km, population 136,500,000** (1978); 16 Autonomous Soviet Socialist Republics (A.S.S.R.): Bashkir A.S.S.R., Buryat A.S.S.R., Checheno-Ingush A.S.S.R., Chuvash A.S.S.R., Daghestan A.S.S.R., Kabardino-Balkar A.S.S.R., Kalmyk A.S.S.R., Karelian A.S.S.R., Komi A.S.S.R., Mari A.S.S.R., Mordovian A.S.S.R., North Ossetian A.S.S.R., Tatar A.S.S.R., Tuva A.S.S.R., Udmurt A.S.S.R., Yakut A.S.S.R.; **capital:** Moskva 7,911,000 inhab. (1978). **Population:** more than 100 nationalities live there; the most numerous are (1970): Russians 82.8%, Tatars 3.7%. – **Economy:** the R.S.F.S.R. produces about two thirds of the total industrial and one half of the agricultural output of the Soviet Union. 90% of the total coal and 80% of petroleum reserves are found here as well as 60% of iron ore, 80% of peat, 90% of wood, 70% of hydro-electric resources, the majority of precious ores and gems. About 60% of the total cultivated area of the U.S.S.R. lies here.

UKRAINE, Ukrainska Radyanska Sotsialistichna Respublika (Ukrainian S.S.R.), **area 603,700 sq.km, population 49,500,000** (1978); **capital:** Kijev 2,133,000 inhab. (1978). **Population** (1970): Ukrainians 74.9%, Russians 19.4%, – **Economy:** raw materials: high quality coal, iron and manganese ores, petroleum, chemical raw materials, important metallurgical, engineering and chemical production, intensive agriculture (sugar beet, wheat, maize /corn/).

BYELORUSSIA, Belaruskaya Sovietskaya Sotsialistychnaya Respublika, **area 207,600 sq.km, population 9,500,000** (1978); **capital:** Minsk 1,272,700 inhab. (1978). **Population** (1970): Byelorussians 81%, Russians 10.4%, Poles 4.3%. – **Economy:** raw materials – forest products, peat, phosphides; engineering and food industry, cultivation of potatoes and flax; cattle breeding.

AZERBAIJAN, Azerbaijchan Soviet Sotsialistik Republikasy, **area 86,600 sq.km, population 5,866,000** (1978), 1 autonomous republic – Nakhichevan A.S.S.R.; **capital:** Baku 1,465,000 inhab. (1978). **Population** (1970): Azerbaijanis 73.8%, Russians 10%, Armenians 9.4%.–**Economy:** important mining of petroleum, chemical, engineering, textile and food industries, cultivation of cotton, subtropical products and sheep breeding.

GEORGIA, Sakartvelos Sabchota Sotsialisturi Respublica, **area 69,700 sq.km, population 5,041,000** (1978), 2 autonomous republics – Abkhaz A.S.S.R. and Adzhar A.S.S.R.; **capital:** Tbilisi 1,052,000 inhab. (1978). **Population** (1970): Georgians 66.8%, Armenians 9.7%, Russians 8.5%.–**Economy:** mineral resources – manganese ore and coal, metallurgical, engineering, textile and food industries. Main crops: tea, tobacco, cotton, citrus fruit, grapes.

ARMENIA, Haikakan Sovetakan Sotsialistakan Respublika, **area 29,800 sq.km, population 2,955,600 inhab.** (1978); **capital:** Jerevan 982,000 inhab. (1978). **Population** (1970): Armenians 88.6%, Azerbaijanis 5.9%, Russians 2.7%. **Economy:** mining and processing of copper ore, engineering, chemical and food industries; cultivation of cotton, vines, fruit.

MOLDAVIA, Respublika Sovietike Sochialiste Moldovenyaske, **area 33,700 sq.km, population 3,315,000** (1978); **capital:** Kišin'ov 492,000 inhab. (1978). **Population** (1970): Moldavians 64.6%, Ukrainians 14.2%, Russians 11.6%, Gagauzians 3.5%.–**Economy:** cultivation of vines, fruit, sugar beet, tobacco, cereals. Food industry.

ESTONIA, Eesti Nõukogude Sotsialistik Vabariik, **area 45,100 sq.km, population 1,459,000** (1978); **capital:** Tallinn 442,000 inhab. (1978). **Population** (1970): Estonians 68.2%, Russians 24.7%.–**Economy:** mining of combustible shales and peat; engineering, textile industry, production of cement. Cultivation of potatoes, flax, barley, fodder crops. Cattle breeding.

LATVIA, Latvijas Padomju Socialistiska Republika, **area 63,700 sq.km, population 2,530,000** (1978); **capital:** Riga 827,000 inhab. (1978). **Population** (1970): Latvians 56.8%, Russians 29.8%, Byelorussians 4%.–**Economy:** mining of peat; engineering (electrotechnical), food industry. Cultivation of flax, sugar beet, potatoes; livestock breeding.

LITHUANIA, Lietuvas Taryu Socialistine Respublika, **area 65,200 sq.km, population 3,364,000** (1978); **capital:** Vilnius 469,800 inhab. (1978). **Population** (1970): Lithuanians 80.1%, Russians 8.6%, Poles 7.7%, Byelorussians 1.5%.–**Economy:** mining of peat, engineering and food industry. Cultivation of flax and potatoes; livestock breeding.

KAZAKHSTAN, Kazak Soviettik Sotzialistik Respublikasy, **area 2,717,300 sq.km, population 14,700,000** (1978); **capital:** Alma-Ata 890,000 inhab. (1978). **Population** (1970): Kazakhs 32.6%, Russians 42.4%.–**Economy:** mining and metallurgy of non-ferrous metals, extraction of coal and petroleum. Cultivation of wheat, fruit, cattle breeding.

TURKMENISTAN, Tiurkmenostan Soviet Sotsialistik Respublikasy, **area 488,100 sq.km, population 2,722,000** (1978); **capital:** Aščhabad 309,000 inhab. (1978). **Population** (1970): Turkmenians 65.6%, Russians 14.5%, Uzbeks 8.3%. **Economy:** extraction and processing of petroleum, mining of sulphur, textile industry. Deserts and dry steppes cover about 90% of the land area; cultivation of cotton and rice on irrigated land. Karakul sheep breeding.

UZBEKISTAN, Ozbekistan Soviet Sotsialistik Respublikasy, **area 447,400 sq.km, population 14,800,000** (1978); 1 autonomous republic – Karakalpak A.S.S.R.; **capital:** Taškent 1,733,000 inhab. (1978). **Population** (1970): Uzbeks 65.5%, Russians 12%, Tatars 4.9%, Kazakhs 4%.–**Economy:** mining of petroleum, coal, copper, sulphur; heavy engineering, chemical, textile and food industries. Cultivation of cotton and breeding of Karakul sheep.

TADZHIKISTAN, Respublikai Sovieth Sotsialistii Tojokiston, **area 143,100 sq.km, population 3,689,000** (1978); **capital:** Dušanbe 476,000 inhab. (1978). **Population** (1970): Tadzhiks 56.2%, Uzbeks 23%, Russians 11.9%.–**Economy:** mining of coal, petroleum, polymetallic ores; processing of agricultural products. Cultivation of cotton, rice, fruit and vines; sheep breeding.

KIRGIZIA, Kyrgyz Sovietik Sotsialistik Respublikasy, **area 198,500 sq.km, population 3,511,000** (1978); **capital:** Frunze 522,000 inhab. (1978). **Population** (1970): Kirghizians 43.8%, Russians 29.2%.–**Economy:** mining of petroleum, coal, mercury, non-ferrous metals; engineering. Cultivation of cotton, sugar beet, fruit; cattle and sheep breeding.

1 Abkhaz A. S. S. R.
2 Adzhar A. S. S. R.
3 Checheno-Ingush A. S. S. R.
4 Kabardino-Balkar A. S. S. R.
5 Nakhichevan A. S. S. R.
6 North Ossetian A. S. S. R.
7 Adygei Aut. Reg.
8 Karachayevo-Cherkessk Aut. Reg.
9 Nagorno-Karabakh Aut. Reg.
10 South Ossetian Aut. Reg.
11 Trans-Carpathian Reg.
18 a Caucasus and Transcaucasia
1 : 12 000 000
1 : 10 000 000
0 50 100 150 200 250 Km
0 50 100 150 Mi
Lithuanian S.S.R.
Byelorussian S.S.R.
Ukrainian
Moldavian S.S.R.
ROMANIA
R. S. F. S. R.
Georgian S. S. R.
Armenian S.S.R.
Azerbaijan S. S. R.
Daghestan A.S.S.R.
TURKEY
IRAN
Klaipėda
Šiauliai
Daugavpils
Velikije Luki
Nelidovo
Ržev
Panevežys
Tauragė
Utena
Druja
Nevel'
Žarkovskij
Sovetsk
Ukmergė
Novopolock
Polock
Veliž
Vladimirskij
Syčovka
Kaliningrad
Kaunas
Glubokoje
Tupik
Gagarin
Demidov
Vitebsk
Jarcevo
V'az'ma
Čern'achovsk
Kapsukas
Vilnius
Lepel'
Smolensk
Smorgon'
Suwałki
Alytus
Vilejka
Molodečno
Orša
Juchnov
Kętrzyn
Olsztyn
Druskininkai
Borisov
Gorki
Ełk
Lida
Grodno
MINSK
Mogil'ov
Roslavl'
Kirov
Szczytno
Kričov
Ludinovo
Ostrołęka
Białystok
Volkovysk
Osipoviči
Ostrów Mazowiecka
Kost'ukoviči
Žukovka
Slonim
Baranoviči
Hajnówka
Sluck
Bobrujsk
Br'ansk
Karačev
WARSZAWA
Soligorsk
Žlobin
Klincy
Uneča
Navl'a
Siedlce
Brest
Ber'oza
Okt'abr'skij
Gomel'
Kobrin
Pinsk
Luninec
Svetlogorsk
Novozybkov
Dobruš
Rečica
David-Gorodok
Mozyr'
Radom
Puławy
Kamen'-Kaširskij
Novgorod-Severskij
Družba
Lublin
Ostrowiec Świętokrz.
Kovel'
Černigov
Ščors
Šostka
Rylsk
Chełm
Sarny
Ovruč
Gluchov
Tarnobrzeg
Zamość
Vladimir-Volynskij
Černobyl'
Bachmač
Hrubieszów
Gorodnica
Korosten'
Oster
Konotop
Belopolje
Luck
Kijevskoje Vdchr.
Nežin
Červonograd
Dubno
Rovno
Novograd-Volynskij
Radomyšl'
Sumy
Jarosław
Rava-Russkaja
KIJEV
Priluki
Romny
Lebedin
Przemyśl
Brody
Šepetovka
Žitomir
Borispol'
Sanok
LVOV
Fastov
Pir'atin
Gad'ač
Achtyrka
Starokonstantinov
Berdičev
Lubny
Mirgorod
Sambor
Drogobyč
Stryj
Berežany
Ternopol'
Belaja Cerkov'
Kazatin
Mironovka
Turka
Zolotonoša
Poltava
Kaluš
Bučač
Chmel'nickij
Čerkassy
Užgorod
Ivano-Frankovsk
Čortkov
Vinnica
Smela
Kremenčugskoje Vdchr.
Mukačevo
Nadvornaja
Kolomyja
Žmerinka
Tal'noje
Špola
Kremenčug
Beregovo
Chust
Kamenec-Podol'skij
Gajsin
Tul'čin
Uman'
Novomoskovsk
Chotin
Mogil'ov-Podol'skij
Aleksandrija
Satu-Mare
Baia-Mare
Rachov
Černovcy
Kirovograd
P'atichatki
Carei
Dneprodzeržinsk
Novoukrajinka
Botoşani
Soroki
Balta
Dolinskaja
KRIVOJ ROG
Pervomajsk
Nikopol'
Bel'cy
Kotovsk
Rybnica
Voznesensk
Iaşi
Ungeny
Novyj Bug
Marganec
Paşcani
KIŠIN'OV
Tiraspol'
Ber'ozovka
Michajlovka
Roman
Snigir'ovka
Vaslui
Bendery
Nikolajev
Kachovskoje Vdchr.
Bacău
Bessarabka
Cherson
Novaja Kachovka
Bîrlad
Belgorod-Dnestrovskij
ODESSA
Kagul
Skadovsk
Tecuci
Tatarbunary
Karkinitskij Zaliv
Krasnoperekopsk
Galaţi
Bolgrad
Izmail
Kilija
Džankoj
Černomorskoje
Nižnegorskij
Krymskij Poluo.
Sulina
Jevpatorija
Gvardejskoje
Simferopol'
Sevastopol'
Jalta
M. Saryč
B L A C
Tkvarčeli
Beslan
Ordžonikidze
Alagir
Očamčira
Levaši
Kutaisi
Sačchere
Ciatura
Cchinvali
Poti
Zestafoni
Gori
Telavi
Chudat
Kusary
Macharadze
TBILISI
Zakataly
Batumi
Achalciche
Rustavi
Čiteli-Ckaro
Samur
Chačmas
Ardahan
Šeki
Kuba
Artvin
Alaverdi
Mingečaurskoje Vdch.
Leninakan
Tauz
Mingečaur
Sumgait
Kirovakan
Oltu
Artik
Kirovabad
Geokčaj
Kars
K'urdamir
Sarıkamış
Oktember'an
JEREVAN
Oz. Sevan
Kuščinskij
Kazi-Magomed
BAKU
Aras
Stepanakert
Agdam
Ali-Bajramly
Eleşkirt
Karaköse
Iljičevsk
Fizuli
Saljany
Bank
Nachičevan'
Kafan
Džalilabad
Džul'fa
Port-Il'ič
Lenkoran'
Khvoy
Marand
Ahar
Ardabil
Astara
TABRĪZ
CASPIAN SEA
55°
25°
30°
35°
50°
40°
45°

Vladimir
Pavlovo
Sergač
Šumerla
Kanaš
Aľmetjevsk
Murom
Oka
5
45°
6
50°
55°
40°
Orechovo-
-Zujevo
Gus'-Chrustaľnyj
Vyksa
Arzamas
Alatyr'
Buinsk
Kujbyševskoje
Vdchr.
Bugul'ma
Okťabr'skij
Jegorjevsk
Melenki
Lukojanov
Dimitrovgrad
Nurlat
Kasimov
Temnikov
Uljanovsk
Suchodol
Bugurusłan
Zarajsk
Mordovian
A.S.S.R.
Saransk
Jazykovo
R'azan'
Šilovo
Sasovo
Sura
Inza
TOLJATTI
Novomoskovsk
Kovylkino
Ruzajevka
Baryš
KUJBYŠEV
Kineľ
Otradnyj
Novokujbyševsk
Skopin
R'ažsk
Zemetčino
Nižnij Lomov
Lunino
Syzran'
Čapajevsk
Neftegorsk
Buzuľuk
Don
Moršansk
Penza
Kuzneck
Soročinsk
Volovo
Pervomajskij
Sosnovka
Kamenka
Saratovskoje
Vdchr.
Boľšaja-
Glušica
Jefremov
Lev Tolstoj
Mičurinsk
Chvalynsk
Chop'or
Serdobsk
Jelec
Lipeck
Tambov
Kirsanov
Petrovsk
Pugačov
Bol. Irgiz
Kotovsk
Voľsk
Pereľub
Sosna
Gr'azi
Inžavino
Rtiščevo
Balakovo
Atkarsk
Volga
Marks
Uraľsk
Usman'
Arkadak
SARATOV
Engeľs
Jeršov
Ozinki
Uvarovo
Puškino
Ramon'
VORONEŽ
Ertiľ
Borisoglebsk
Balašov
Kalininsk
R.
S.
Krasnyj Kut
Boľšoj Uzeń
Ural
F.
Anna
Novochop'orsk
Povorino
Krasnoarmejsk
Rovnoje
Novouzensk
Čapajev
Staryj Oskol
Bobrov
Ur'upinsk
Jelan'
Aleksandrov Gaj
Buturlinovka
Georgiu Dež
Aleksejevka
Novoanninskij
Kotovo
Pallasovka
Malyj Uzeń
Furmanovo
Pavlovsk
Michajlovka
Kamyšin
Nikolajevsk
Valujki
Rossoš'
Kalač
Volgogradskoje
Bykovo
Vodochranilišče
Džanybek
Frolovo
Medvedica
Kantemirovka
Bogučar
Novaja Kazanka
Kazakh S.S.R.
Oz. Eľton
Kup'ansk
Serafimovič
Dubovka
Svatovo
Kletskij
Urda
Starobeľsk
Bokovskaja
VOLGOGRAD
Volžskij
Kapustin Jar
Šungaj
Rubežnoje
Kalač-na-Donu
Severodoneck
Millerovo
Oz. Baskunčak
Achtubinsk
Verchnij Baskunčak
Stachanov
Vorošilovgrad
Kamensk-
-Šachtinskij
Morozovsk
C
Kommunarsk
Ciml'anskoje
Vodochranilišče
Jenakijevo
Sverdlovsk
Cagan-Aman
Charabali
Krasnyj Luč
Šachty
Konstantinovsk
Koteľnikovo
Novošachtinsk
Ciml'ansk
Sovetskoje
Jenotajevka
ROSTOV-
-na-Donu
Novočerkassk
Volgodonsk
Sarpa
Volnovacha
Taganrog
Manyč
Veselovskoje
Vdchr.
Zimovniki
Kalmyk A.S.S.R.
Batajsk
Astrachan'
Azov
Kamyzak
Kirovskij
Proletarsk
Jejsk
Kuščovskaja
Saľsk
Oz. Manyč
Gudilo
Elista
Jaškuľ
Mumra
Jegorlyk
Starominskaja
Gorodovikovsk
CASPIAN SEA
Divnoje
Krasnogvardejskoje
Kaspijskij
45°
-Achtarsk
Pavlovskaja
Tichoreck
Novoaleksandrovsk
Arzgir
Komsomoľskij
Timaševsk
Korenovsk
Kropotkin
Svetlograd
Kuma
Slav'ansk-na-Kubani
KRASNODAR
Stavropoľ
Blagodarnyj
7
Armavir
Neftekumsk
Daghestan
Kubań
Nevinnomyssk
Budennovsk
Kočubej
Krymsk
Belorečensk
Labinsk
Aleksandrovskoje
Novorossijsk
Mineraľnyje
Vody
Laba
Gelendžik
Majkop
Čerkessk
Georgijevsk
Kizľar
Apšeronsk
Kamennomostskij
Šedok
P'atigorsk
Prochladnyj
Terek
Mozdok
Tuapse
Zelenčukskaja
Kislovodsk
Majskij
Groznyj
Gudermes
Chasavjurt
Machačkala
8
4
Nalčik
Beslan
3
Kaspijsk
D
Dombaj
Tyrnyauz
Bujnaksk
Soči
Gagra
6
Alagir
Ordžonikidze
Izberbaš
1
Georgian S.S.R.
Levaši
Inguri
Suchumi
Tkvarčeli
Rioni
Derbent
Očamčira
Sačchere
10
A.S.S.R.
Cchinvali
Chudat
Kutaisi
Telavi
4
Micha-Cchakaja
5
Čiatura
Gori
45°
6
Kusary
Poti
Zestafoni
TBILISI
Zakataly
Kuba
Macharadze
Šeki
40°
E
A
B

ASIA

The Asian continent lies in the northern hemisphere, although in the south-east some Indonesian islands belong to the southern hemisphere. On its western side Asia is linked to Europe, which is in fact a gigantic peninsula of Asia (Eurasia). In the south-west the boundary with Africa runs along the Isthmus of Suez (120 km long) on the Sinai Peninsula. The name "Asia" derives from the Accadian word "Asu", meaning "Land of the East, the Dawn".

Asia covers an **area of 44,413,000 sq.km** (including the islands) and takes up 29.72% of the world's land surface, making it the largest continent. It has **2,476 mn. inhabitants** (1978, including the Asian part of the U.S.S.R.), i.e. 58.9% of the world's population with a density of 56 persons per sq.km.

Geographical position – northernmost point of the continent: cape Mys Čel'uskin (U.S.S.R.) 77°43′ N.Lat. (of the entire continent: cape Mys Arktičeskij /island O. Komsomolec/ in Severnaja Zeml'a 81°16′ N.Lat.); southernmost: cape Tg. Buru on the Malay Peninsula 1°25′ N.Lat. (island P. Roti in the Indonesian Lesser Sunda Is. 11° S.Lat.); westernmost: cape Baba B. in Asia Minor 26°03′ E.Long. and easternmost: cape Mys Dežneva on the peninsula Čukotskij Poluostrov (U.S.S.R.) 169°40′ W.Long. (island O. Ratmanova in the Diomede Is. in the Bering Strait 169°02′ W.Long.).

The coast line of Asia is very varied and is 69,000 km long (excluding the offshore islands). Largest peninsulas: Arabian (area: 2,780,000 sq.km), India (1,850,000 sq.km), Indo-China (1,450,000 sq.km), Asia Minor (580,000 sq.km), Tajmyr (420,000 sq.km), and Kamčatka (275,000 sq.km). There are numerous islands; the largest include: The Greater Sunda Is. (area: 1,548,600 sq.km), the Japanese Is. (377,458 sq.km), the Philippines (297,413 sq.km), the Lesser Sunda Is. (91,860 sq.km) and the Moluccas (Maluku, 74,500 sq.km).

The complex **geological and tectonic structure** of Asia is the main reason for its varied relief. The basic geological structure is formed by Primary continental tables: the Siberian table in the north and the Chinese table in the east, the Indian in the south and the Arabian in the south-west. **The surface.** The vertical features of the Asian continent differ enormously. The average height above sea level is 960 m, i.e a figure higher than that for any other continent, with the exception of Antarctica. About 26% of Asia is lowland, not higher than 200 m, but 14% of the land is above 2,000 m. Huge mountain ranges cross the continent in two zones which link up in the Pamir in Central Asia (7,719 m). The first zone, a system of folded ranges of the Alpine-Himalayan type, stretches from Asia Minor across Central Asia, and through Indo-China as far as Sumatra and Java. This includes: The Armenian Plateau (5,165 m), the Caucasus (Bol'. Kavkaz, 5,642 m), the Iranian Mountains (5,670 m) and the Hindu Kush (7,708 m) which adjoins the Pamir. The Karakoram Range (Godwin Austen, 8,611 m) runs in a south-easterly direction from the Pamir, as well as the mighty range of the Himalayas (11 summits above 8,000 m), with the highest mountain in the world, Mt. Everest (8,848 m). The Kunlunshan (7,723 m) lies to the east of the Pamir. The second zone of mountains stretches in a north-easterly direction from the Pamir as far as the peninsula of Čukotskij Pol.: Tien Shan (7,439 m), Altai (4,506 m), and the lower Sayan Mts., Stanovoj Chr. and Chr. Čerskogo (3,147 m). The Pacific zone of Tertiary mountains lies in eastern Asia, with high volcanic activity (130 volcanoes) and frequent earthquakes. – Greatest plains: Western Siberian (Zapadno-Sibirskaja Nizm.), Indo-Gangetic, Great Plain of China. The Dead Sea (−394 m) is the lowest point and lake O. Bajkal the deepest crypto-depression (−1,286 m below sea level).

The river system. Over 40% of central and south-west Asia has no outlet to the sea. The Siberian rivers carry most water and flow into the Arctic ocean (the longest is the Ob'-Irtyš, 5,410 km, which has a drainage basin of 2,975,000 sq.km and a mean annual discharge of 12,600 cub.m per sec). Monsoon rivers are an important source of water for agriculture. They include Asia's longest river, the Yangtze (Changjiang, 5,520 km, drainage basin 1,942,000 sq.km, and a mean annual discharge of 31,000 cub.m per sec). The rivers in the tropical belt, especially on the islands, have ample water throughout the year. The largest among the many **lakes** is the Caspian Sea covering 371,000 sq.km (depth 1,025 m).

Large parts of Asia have a continental **climate;** the very low winter temperatures and hot inland summers cause oscillations in the atmospheric circulation. The north of Asia is influenced by the cold Arctic air and has heavy frosts. The lowest absolute temperature, −78°C, was measured at Ojm'akon in Siberia. Almost two thirds of Asia lie in the temperate belt and the Asian Plateau is a region of extreme drought. The south of Asia has a tropical climate: the south-west is dry with high temperatures; the highest absolute temperature, 53°C, was measured at Jacobābād in Pakistan; in the south-east, in particular on the islands, it is hot and damp with only minor variations in temperature. The monsoon rains are a life-giving force bringing an average precipitation of 2,500 mm per year. Maximum rainfall was recorded at Cherrapunji (India) 11,013 mm a year and the minimum at Al-'Aqabah (Jordan) 24 mm annually.

Mean January and July temperatures in °C (and annual rainfall in mm) are as follows: Verchojansk −48.9 and 15.3 (142), Sapporo −6.2 and 19.4 (1,063), Beijing −4.7 and 25.6 (610), Ulaanbaatar −26.7 and 17.2 (101), Alma-Ata −7.4 and 23.3 (575), İstanbul 5.6 and 23.1 (679), Baghdād 9.4 and 34.7 (156), Tehrān 2.1 and 29.7 (250), Multān 13.5 and 36.1 (161), Bombay 23.9 and 27.4 (1,878), Calcutta 19.6 and 28.8 (1,588), Cherrapunji 11.7 and 20.4 (11,013), Hong Kong 15.6 and 27.9 (2,177), Tōkyō 3.2 and 25.0 (1,575), Manila 24.6 and 26.9 (2,123), Krung Thep 26.1 and 28.6 (1,420), Ar-Riyād 14.2 and 33.6 (97), Aden 25.4 and 32.5 (41), Colombo 26.2 and 27.2 (2,236), Singapore 26.1 and 27.4 (2,413), Djakarta 25.9 and 26.5 (1,779), Kupang 26.8 and 25.3 (1,413).

The vegetation of Asia falls into two zones. First, the extensive Holarctic realm in the north, west and south-west which includes, from north to south, tundra, tundra with woodland, typical taiga with coniferous forest, mixed forest, broad-leaved deciduous forest, the Sino-Japanese region of evergreen plants, Central Asian wooded steppe, grassland steppe, semi-desert with scrub, Mediterranean evergreen maquis, and the North-African-Indian desert region with xerophilous scrub. Second, the smaller Paleo-tropical realm in the south and south-east of Asia, which includes semi-desert with sparse thorn forest, dry deciduous tropical forest and grassland savanna in India and monsoon tropophilous woodland, subtropical forest, evergreen tropical rain forest and mangrove swamp, etc., in India and Southeastern Asia. – **The fauna** belongs to two zoogeographical realms: the larger is Paleoarctic and poorer in species (e.g. polar bear, ermine, reindeer, wolf, tiger, stag, sheep, forest and water fowl, pheasant, vulture, bustard, numerous rodents, fresh water fish, insects, etc.). The smaller Indo-Malaysian is richer in species and older in evolution (e.g. monkeys, rare orangutans, Indian elephant, rhinoceros, buffalo, leopard, tiger, tapir, pheasant, peacock, python, crocodile, flying gurnard, and large numbers of insects and other fishes).

LONGEST RIVERS

Name	Length in km	River basin in sq.km
Changjiang /Yangtze/	5,520	1,942,000
Ob' (-Irtyš)	5,410	2,975,000
Huanghe /Yellow/	4,845	772,000
Mekong /Lancangjiang/	4,500	810,000
Irtyš	4,422	1,595,680
Amur (-Šilka, -Onon)	4,416	1,855,000
Lena	4,400	2,490,000
Jenisej	4,092	2,580,000
Indus /Sindh/	3,190	960,000
Nižn'aja Tunguska (-Nepa)	2,989	473,000
Brahmaputra /Yaluzangbujiang/	2,960	935,000
Syrdarja (-Naryn)	2,860	136,000
Salween /Nujiang/	2,820	325,000
Euphrates /Al-Furăt, Fırat/	2,760	765,000
Talimuhe	2,750	1,210,000
Ganga-Padma /Ganges/	2,700	1,125,000
Vil'uj	2,650	454,000
Amudarja (-P'andž)	2,620	227,000
Kolyma	2,513	647,000

LARGEST LAKES

Name	Area in sq.km	Greatest Depth in m	Altitude in m
Caspian Sea+	371,000	1,025	−28
Aral'skoje More+	64,115	67	53
O. Bajkal	31,500	1,620	455
O. Balchaš	18,200	26	340
D. -y. Reẕā'īyeh+	7,500	15	1,275
O. Issyk-Kul'+	6,280	702	1,609
Tônlé Sab	5,700	10	15
O. Tajmyr	4,560	26	6
Qinghai /Kukunuoer/+	4,460	38	3,250
O. Chanka	4,200	10	68
Hongzehu	3,780	.	3
Dongtinghu	3,750	.	25
Van Gölü+	3,738	25	1,662
Uvs-Nuur+	3,350	.	759
Poyanghu	3,150	.	15
Chövsgöl Nuur	2,620	.	1,645
Luobubo /Lop Nor/+	2,600	.	780
Dead Sea+	980	399	−394

+Salt Lake

LARGEST ISLANDS

Name	Area in sq.km	Name	Area in sq.km	Name	Area in sq.km
Borneo (Kalimantan)	746,546	Mindanao	98,692	Hainandao	33,670
Sumatera	433,800	Hokkaidō	78,073	Timor	33,615
Honshū	227,414	Sachalin	76,400	Seram	18,625
Sulawesi (Celebes)	179,416	Sri Lanka (Ceylon)	65,607	Shikoku	18,256
Djawa (Java)	126,700	Kyūshū	36,554	Halmahera	17,800
Luzon	106,983	Taiwan	35,961	Flores	15,600

HIGHEST MOUNTAINS

Name (Country)	Height in m	Name (Country)	Height in m
Mt. Everest /Sagarmatha, Zhumulangmafeng/ (Nepal, China)	8,848	Himalchuli (Nepal)	7,893
Godwin Austen /K2/ (China-Pakistan)	8,611	Nuptse (Nepal)	7,879
Kānchenjunga (Nepal-India)	8,598	Masherbrum (Pak.)	7,821
Lhotse (Nepal-China)	8,511	Nanda Devi (India)	7,816
Makālu (Nepal-China)	8,481	Rakaposhi (Pak.)	7,789
Lhotse-Shar (Nepal–China)	8,383	Disteghil Sar (Pak.)	7,785
Dhaulāgiri (Nepal)	8,172	Batura (Pak.)	7,785
Manāslu (Nepal)	8,156	Kamet (India-China)	7,756
Cho Oyu (China-Nepal)	8,153	Namuchabawashan (China)	7,755
Nanga Parbat (Pak.)	8,126	Ulugh Muztagh (China)	7,723
Annapurna (Nepal)	8,078	Gonggeershan (China)	7,719
Gasherbrum (Pak.-China)	8,068	Tirich Mīr (Pak.)	7,708
Phalchan Kangrī /Broad Pk./ (Pak.-China)	8,047	Pik Kommunizma (U.S.S.R.)	7,495
Gosainthan (China)	8,013	Nowshāk (Afghan.)	7,492
		Pik Pobedy (U.S.S.R.)	7,439

ACTIVE VOLCANOES

Name (Country)	Altitude in m	Latest eruption
V.Kl'učevskaja Sopka U.S.S.R.)	4,750	1974
G.Kerintji (Indon.)	3,805	1968
G.Rindjani (Indon.)	3,726	1966
G.Semeru (Indon.)	3,676	1976
Kronockaja Sopka (U.S.S.R.)	3,528	1923
Korjakskaja Sopka (U.S.S.R.)	3,456	1957
G.Slamet (Indon.)	3,428	1967

FAMOUS NATIONAL PARKS

Name (Country)	Area in sq.km
Kronockij Zap. (U.S.S.R.)	9,770
Altajskij Zap. (U.S.S.R.)	8,638
Issik-Kulskij Zap. (U.S.S.R.)	7,816
Taman Negara (Malaysia)	4,360
Gunung Leuser (Indon.)	4,165
Sagarmatha (Mt.Everest, Nepal)	1,243
Fuji-Hakone (Jap.)	948

ARCTIC OCEAN
NORWEGIAN SEA
NORTH SEA
BARENTS SEA
KARSKOJE MORE
MORE LAPTEVYCH
BALTIC SEA
G. of Bothnia
BLACK SEA
CASPIAN SEA
Aral'skoje M.
O. Balchaš
O. Bajkal
Persian Gulf
G. of Oman
RED SEA
Gulf of Aden
ARABIAN SEA
Bay of Bengal
ANDAMAN SEA
INDIAN OCEAN
Equator
Tropic of Cancer
British Is.
London
Scandinavia
Oslo
Stockholm
Helsinki
Svalbard
Spitsbergen
Nordkinn
Murmansk
Kol'skij Pol.
Novaja Zeml'a
Zeml'a Fr. Iosifa
Severnaja Zeml'a
M. Čel'uskin
Dikson
Pol. Tajmyr
O. Tajmyr
Alps
Wien
Praha
Berlin
Warszawa
Riga
Leningrad
Moskva
Kijev
Carpathian Mts
Bucurešti
Istanbul
Srednerusskaja Vozvyš.
Volgograd
Astrachan'
Ural'skije Gory
Narodnaja
Salechard
Pol. Jamal
Dudinka
Kamen'
Srednesibirskoje Ploskogor'je
Zapadno-Sibirskaja Nizmennost'
Omsk
Novosibirsk
Irkutsk
Sayan Mts.
G. Munku Sardyk
G. Belucha
Altai
Changajn Nu.
Zhuangaerp.
Ulaanbaatar
Asia Minor
Ankara
Baba B.
Toros Dağ.
Cyprus
Bol'.Kavkaz
G. El'brus
Büyük Ağri D.
Baku
Plato Ust Urt
Turanskaja Nizmen.
Kyzylkum
Karakumy
Aschabad
Taškent
Alma-Ata
O. Issyk-Kul'
Tien Shan
Tulufanwadi
Luobubo
Pik Pobedy
Pik Kommunizma
Talimupendi
Suez Canal
Sinai Pen.
Dead S.
Bayrūt
Euphrates
Tigris
Mesopotamia
Baghdād
Tehrān
Q.-ye Damāvand
Elburz Mts.
Dasht-e Kavir
Zagros
Plateau of Iran
Pamir
Hindu Kush
Godwin Austen
Kābul
Nānga Parbat
Karakoram Ra.
Kunlunshan
Aerjinshan.
Qiliansh.
Chaidamup.
Qinghai
TIBET
Himalayas
Mt. Everest
Dhaulāgiri
Lasa
Gonggash
An-Nafūd
Arabian Pen.
Ar-Riyāḍ
Dubayy
J. ash-Shām
Ar-Rab' al-Khālī
Hadūr Shu'ayb
Aden
Ra's al-Hadd
Karāchi
Jacobābād
Indus
Thar Desert
Delhi
Indo-Gangetic Plain
Ganga
Brahmaputra
Cherrapunji
INDIA
Calcutta
Narmada
Bombay
Godāvari
Krishna
Deccan
West Ghāts
East Ghāts
Madras
Sri Lanka
C. Comorin
Lakshadweep
Maldives
Chagos Arch.
Andaman Is.
Nicobar Is.
Rangoon
Irrawaddy
Salween
Str. of Malacca
Malay
Sumatra
Kep. Mentawai
Suqutrā
Somali Pen.
R. Hafun
Madagascar
Helmand
Sutlej
Amudarja
Syrdarja
Emba
Ural
Volga
Don
Dnepr
Dunaj
Kama
Pečora
Sev. Dvina
Ladožskoje O.
Onežskoje O.
Ob'
Irtyš
Tobol
Išim
Jenisej
Niž. Tunguska
Podkam. Tunguska
Angara
Lena
Chatanga
Selenga
Ili
Huanghe
Yangtze
4810
2655
2472
1894
1701
1640
455
1620
3491
4506
5642
5165
-28
-132
53
-154
7439
7495
7708
8611
8126
7723
5934
5670
4420
-394
3018
3760
8172
8848
7590
3053
2695
2524
3835
5824
5803
1 : 65 000 000
0 300 600 900 1 200 1 500 Km
0 200 400 600 800 1 000 Mi

19a Hong Kong and Macau
1:2 000 000
Zhangmutou
Tangtouxia
Danshui
Longgang
Pinghu
Aotou
Nantou
Baoan
Mirs Bay
Zhongshan
Hau Hoi Wan
Tai Po
Yuen Long
1112
Sha Tin
Tuen Mun
Tsuen Wan
Sai Kung
541
Tangjia
Sanxiang
NEW KOWLOON
Xijiang
(U.K.)
KOWLOON
Zhuhai
VICTORIA
Lan Tao
Taio
Hong Kong
970
Macau (Port.)
Zhujiangkou
Soko Is.
Pok Liu Chao
Po Toi Group
Modaomen
Sanzao
Danganquandao
Wanshanqundao
Sanzaodao
Dangangqundao
Dongao
SOUTH CHINA SEA
19b Malay Peninsula
1:15 000 000
Kangar
THAILAND
Kota Bharu
Alor Setar
1 Johore
2 Kedah
3 Kelantan
4 Melaka
5 Negri Sembilan
6 Pahang
7 Perak
8 Perlis
9 Pinang
10 Selangor
11 Trengganu
Butterworth
Kuala Trengganu
Pinang
Taiping
2190
Langsa
2182
G. Tahan
Ipoh
Kuala Lipis
Telok Anson
Kuantan
Bindjai
Tebingtinggi
KUALA LUMPUR
Pekan
MEDAN
Tandjungbalai
Kelang
Kep. Anambas (Indon.)
2475
Seremban
Pematangsiantar
D. Toba
Melaka
2078
Bagansiapi-api
Tarutung
Kluang
Rantauprapat
Muar
Batu Pahat
Sibolga
Johore Bahru
SINGAPORE
Padangsidimpuan
Bengkalis
Strait of Malacca
SINGAPORE
P. Nias
Tandjungpinang
Natal
Pakanbaru
Kep. Riau
G. Tamalau
2912
Lubuksikaping
Kep. Lingga
P. Pini
Equator
MALAYSIA
INDONESIA
Sumatra
19c Cyprus
1:4 500 000
Akr. Apostólou
Rizokárpason
Aiyialoúsa
Akr. Kormakíti
Lápithos
Kirínia
Leonárison
Attila Line since 27.7.1976
Kól. Mórfou
1025
Levkónoikon
Tríkomon
Mórfou
Levkosia (Nicosia)
Kól. Ammókhostou
Akr. Akámas
Lévka
Ássia
Pedias
Ammókhostos
Pólis
Messaoria
Dherínia
1951
Xilotimbou
Ólimbos
Akrotíri
Páno Lévkara
Lárnax
Árso̧s
Akr. Pidálion
Néa Páfos
Kól. Lárnakos
Episkopí
Lemesós
Palaikhóri
Akr. Gátas
CYPRUS
MEDITERRANEAN SEA
Dežneva
Čukotskij Pol.
St. Lawrence I.
Aleutian Is.
Pevek
BERING SEA
Kolymskij Chr.
Komandorskije O.
Kl'učevskaja Sopka 4750
Magadan
Petropavlovsk-Kamč.
SEA OF OKHOTSK
Nikolajevsk
O. Sachalin
Kuril'skije O.
Tatarskij Proliv
Sichote-Alin
2077
Hokkaido
Vladivostok
Paektusan 2744
SEA OF JAPAN
Honshū
Tōkyō
3776
Fuji-san
Pyŏngyang
Korea
Korea Str.
Shikoku
YELLOW SEA
Kyūshū
Shanghai
EAST CHINA SEA
Nansei-shotō
7507
2120
Wui Shan
Formosa Str.
3997
Yü Shan
Taiwan
PHILIPPINE SEA
Philippines
Luzon
Manila
Mindoro
10830
5559
Panay
Palawan
Negros
SULU SEA
Mindanao
Mt. 2965 Apo
4101
Mt. Kinabalu
Sulu Arch.
CELEBES SEA
Borneo
Sulawesi
Kep. Sula
Buru
3455
B. Rantekombola
BANDA SEA
JAVA SEA
Flores
3676
3726
PACIFIC OCEAN

ASIA

Country	Area in sq.km	Population	Year	Density per sq.km
Afghanistan	647,497	17,855,000	1978	28
Bahrain	662	282,000	1978	426
Bangladesh	143,998	84,655,000	1978	588
Bhutan	47,000	1,232,000	1977	26
Brunei (U.K.)	5,765	190,000	1977	33
Burma	678,033	32,334,000	1978	48
China	9,560,980	879,250,000	1978	92
Cyprus	9,251	640,000	1977	69
Egypt – Sinai Peninsula	58,824	50,000	1976	1
Gaza Strip (Egypt)	378	410,000	1977	1,085
Hong Kong (U.K.)	1,046	4,514,000	1977	4,320
India	3,287,590	634,200,000	1978	193
Indonesia (inc. E.Timor)	1,919,270	145,350,000	1978	76
Iran	1,648,100	35,213,000	1978	21
Iraq	438,446	12,171,480	1977	28
Israel	20,770	3,709,000	1978	179
Japan	372,487	115,100,000	1979	309
Jordan	97,740	2,874,000	1977	29
Kampuchea	181,035	7,895,000	1977	44
Korea, Democratic People's Republic of	120,538	17,028,000	1978	141
Korea, Republic of	98,484	37,019,000	1978	376
Kuwait	17,818	1,199,000	1978	67
Laos	236,800	3,427,000	1978	14
Lebanon	10,400	3,056,000	1978	294
Macau (Port.)	16	279,000	1977	17,438
Malaysia	329,749	12,826,000	1977	39
Maldives	298	143,046	1978	480
Mongolia	1,565,000	1,531,000	1977	1
Nepal	140,797	13,421,000	1978	95
Oman	212,457	817,000	1977	4
Pakistan	803,942	76,770,000	1978	95
Philippines	297,413	46,351,000	1978	156
Qatar	11,000	157,000	1977	14
Saudi Arabia	2,149,690	9,522,000	1977	4
Singapore	597	2,334,000	1978	3,910
Sri Lanka	65,610	14,082,000	1978	215
Syria	185,180	8,103,000	1978	44
Taiwan	35,981	16,793,000	1977	467
Thailand	514,121	45,100,000	1978	88
Turkey	779,452	43,210,000	1978	55
Turkey-Asiatic part	755,688	38,296,000	1977	51
Union of Soviet Socialist Republics – Asiatic part	16,831,000	67,691,000	1977	4
United Arab Emirates	83,600	862,000	1978	10
Vietnam	332,560	49,120,000	1978	148
Yemen (Y.A.R.)	195,000	5,642,000	1977	29
Yemen, South (D.Y.)	287,683	1,853,000	1978	6

Asia has a **population of 2,476 mn.** (1978), which is roughly 58.9% of the total population of the world, with a density of 56 persons per sq.km. Its geographical distribution is very uneven. Large desert regions, tundra and forest zones are almost uninhabited. The fertile agricultural lands of China, India, Bangladesh and elsewhere, and the industrial areas of Japan, are among the most densely populated regions in the world. Extreme density of population is found in the smallest countries (1977): Macau 17,438, Hong Kong 4,320, Singapore 3,910 (1978); among other countries (1978): Bangladesh 588, Bahrain 426, Rep. of Korea 376 and Japan 309 (1979).

The average **population increase** is highest (1970–75) in Kuwait 7.13%, Iraq 3.36%, the Philippines 3.34% and Jordan 3.29%. The highest birth rate is in the two Yemens (49.6 per 1,000), Bangladesh and Saudi Arabia (49.5 per 1,000), the highest death rate in Bangladesh (28.1 per 1,000). The average life expectancy in southern Asia is 48.5 years and 62.5 years in eastern Asia (1970–75). There are several hundred nations and nationalities in Asia. The country with the highest population is China. Asia has an ancient tradition of urban settlement. The highest percentages of urban dwelling are in Japan (75.1%) and Israel (92.6%) – 1975. There are more than 60 **towns** with over 1 mn. people. The largest are (in 1,000 inhab., 1970–76): Shanghai 10,820 (agglom. 12,000), Tōkyō 8,392 (agglom. 11,684), Beijing (Peking) 8,000, Sŏul 6,889, Bombay 5,970, Djakarta 4,576.

Economy: Asia is an important source of raw materials, primarily petroleum, ores, textile raw materials, skins and hides, oil-producing crops and fruit. In the majority of Asian countries agriculture remains the most important industry. 60% of the working population worked in agriculture (1977), tilling 455 mn. hectares of arable land. Only one third of the world total of arable land is in Asia, which is far less than its share of the world's population. Industry is located unevenly on the continent, and in the majority of countries industrial production is in its infancy. Exceptions to this include Japan, a leading industrial world power, Israel and the Republic of Korea.

AFGHANISTAN

Jumhouri Demokratike Chalghie Afghānistān, area 647,497 sq.km, population 17,855,000 (1978), **republic** (Chairman of the Revolutionary Council and Prime Minister Babrak Karmal since 1979).
Administrative units: 28 provinces. **Capital:** Kābul 377,715 inhab. (1976, with agglom. 587,643); **other towns** (1976, in 1,000 inhab.): Qandahār 149, Baghlān 118, Herát 116. – **Population:** Afghans 60%, Tadzhiks 25–30%, Uzbeks etc. Average annual rate of population increase 2.54% (1970–75); urban population 12.2% (1975). 79% of the economically active inhabitants engaged in agriculture (1977). – **Currency:** afgháni = 100 puls.
Economy:predominantly agricultural country. Arable land 13.11% of the land surface. **Agriculture** (1977, in 1,000 tons): wheat 2,640, rice 461, barley 407, maize (corn) 817, millet, sesame, cotton – seed 87, lint 43 – sugar beet 100, vegetables, fruit, grapes 43, raisins 60; livestock (1977, in 1,000 head): cattle 3,800, sheep 22,000 (of which more than one third are karakul sheep), camels 290, goats 3,000; hides and skins, wool; fish catch 1,500 tons (1976). **Mining** (1976): coal, natural gas 2.5 billion cub.m, salt 73,000 tons. Textile and food industries. – **Communications:** roads 17,973 km (1974). Civil aviation: 4.2 mn. km flown, 102,000 passengers carried (1976). – **Exports:** fruit, raisins, natural gas, karakul skins, wool, cotton, carpets. Chief trading partners: U.S.S.R., India, Japan, the United Kingdom.

BAHRAIN

Doulat al-Bahrain, area 662 sq.km, population 282,000 (1978), **emirate** (Amir Shaikh Isa bin Sulman Al-Khalifa since 1961).

Capital: Al-Manāmah 114,000 inhabitants (1978). – **Currency:** Bahrain dinar = 1,000 fils.
Economy: dates 16,000 tons (1977), pearl fishery, fish catch 1,500 tons (1976). Mining of petroleum 2,808,000 tons (1977), natural gas 2.8 billion cub.m; petroleum refineries, production of aluminium 122,100 tons (1976); electricity 682 mn. kWh (1976). – **Communications:** roads 450 km, passenger cars 23,800. – **Exports:** petroleum and petroleum products, aluminium, dates, pearls. Chief trading partners: United Kingdom, U.S.A., Japan, Saudi Arabia.

BANGLADESH

People's Republic of Bangladesh, area 143,998 sq.km, population 84,655,000 (1978), **republic, member of the Commonwealth** (President Justice Abdus Sattar since 1981).
Administrative units: 19 districts. **Capital:** Dacca 1,730,000 inhab. (1974); **other towns** (1974): Chittagong 889,760, Khulna 437,000, Nārāyanganj 270,680, Rājshāni 132,909. – **Population:** Bengals. **Density** 588 persons per sq.km; average annual rate of population increase 1.7% (1970–75); urban population 6.4% (1975). 84.6% of the economically active inhabitants engaged in agriculture (1977). – **Currency:** taka = 100 poisha.
Economy: predominantly an agricultural country. Arable land 66% of the land area. **Agriculture – crops** (1977, in 1,000 tons): rice 19,300, wheat 259, sugarcane 6,504, jute 1,026 (third world producer), bananas 600, tea 34, tobacco 64, vegetables, pineapples 141; **livestock** (1977, in 1,000 head): cattle 26,500, buffaloes 445, sheep 1,200, goats 8,000, fish catch 640,000 tons (1976). **Mining:** coal, crude petroleum, natural gas 0.9 billion cub.m (1976). Electricity: 1,710 mn. kWh (1976). **Production** (1976, in 1,000 tons): jute products 498, sugar 110. Textile and food industries. – **Communications:** railways 2,874 km (1974), roads 10,875 km (1974). Merchant shipping 244,000 GRT (1977). – **Exports:** jute and jute products, hides and skins, tea. Chief trading partners: India, U.S.A., EEC countries, Japan.

BHUTAN

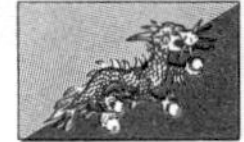

Druk-yul, area 47,000 sq.km, population 1,232,000 (1977), **monarchy** (King Jigme Singhye Wangchuk since 1972).
Capital: Thimbu 15,000 inhab., winter seat Paro. – **Currency:** ngultrum = 100 chetrums.
Economy: arable land and permanent crops only 5.4%, forests 63.8%. Agriculture in the valleys (1977, in 1,000 tons); rice 275, maize (corn) 57, millet, barley; cattle raising in the mountains (1977, in 1,000 head): cattle 202, sheep 39, horses. Mining of coal, handicrafts. – **Exports:** wood, rice, coal, animal products.

BRUNEI

The Sultanate of Brunei, area 5,765 sq.km, population 190,000 (1977), **sultanate under the protection of United Kingdom** (Sultan Hassanal Bolkiah Muizzaddin Waddaulah since 1967).
Capital: Bandar Seri Begawan 36,987 inhab. (1971). – **Population:** Malays, Chinese, Dayaks and others. Average annual rate of population increase 1.9% (1970–75). – **Currency:** Brunei dollar = 100 cents.
Economy (1977, in 1,000 tons): rice 10, bananas 2, coconuts, natural rubber 1, rare timber; raising of buffaloes 17,000 head, pigs 15,000 head (1977). Fish catch 1,565 tons (1976). **Mining** of petroleum 10.1 mn. tons (1976), natural gas 7.8 billion cub.m (1976). Petrochemical industry. – **Communications:** roads 1,257 km (1972), passenger cars 24,600 (1976). – **Exports:** petroleum and petroleum products, natural gas, natural rubber, rare timber.

BURMA

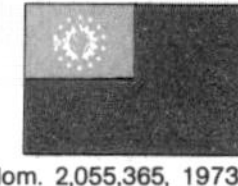

Pyidaungsu Socialist Thammada Myanma Naingngandaw, area 678,033 sq.km, population 32,334,000 (1978), **socialist republic** (President U San Yu since 1981).
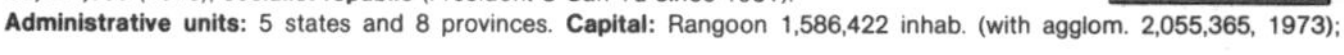
Administrative units: 5 states and 8 provinces. **Capital:** Rangoon 1,586,422 inhab. (with agglom. 2,055,365, 1973);

ARCTIC OCEAN
NORWEGIAN SEA
NORTH SEA
BARENTS SEA
KARSKOJE MORE
MORE LAPTEVYCH
BALTIC SEA
BLACK SEA
CASPIAN SEA
Aral'skoje M.
O. Balchaš
O. Bajkal
Persian Gulf
G. of Oman
ARABIAN SEA
Gulf of Aden
INDIAN OCEAN
Bay of Bengal
ANDAMAN SEA
G. of Thailand
Tropic of Cancer
Equator
Svalbard (Nor.)
Vestspitzbergen
Zemľa F. Iosifa
Novaja Zemľa
Severnaja Zemľa
Mys Čeľuskin
Novosibirskije
NORWAY
SWEDEN
FINLAND
UNION OF SOVIET SOCIALIST
TURKEY
CYPRUS
EGYPT
IRAQ
IRAN
SAUDI ARABIA
YEMEN
SOUTH YEMEN
OMAN
SOMALIA
AFGHANISTAN
PAKISTAN
INDIA
NEPAL
SRI LANKA
MALDIVES
BURMA
THAILAND
CHINA
MONGOL
Aberdeen
Belfast
GLASGOW
LONDON
U.K.
AMSTERDAM
BRUXELLES
KØBENHAVN
PARIS
BERLIN
Bonn
MÜNCHEN
Bern
WIEN
PRAHA
WARSZAWA
BEOGRAD
BUDAPEST
BUCUREŞTI
SOFIJA
Bergen
Oslo
Malmö
STOCKHOLM
Helsinki
Narvik
RIGA
Tallinn
Vilnius
LENINGRAD
Petrozavodsk
Murmansk
Archangeľsk
Amderma
Narjan-Mar
MOSKVA
MINSK
KIJEV
LVOV
VORONEŽ
CHAR'KOV
ODESSA
Sevastopoľ
ROSTOV-n.-D.
GOR'KIJ
KAZAN'
Kirov
PERM'
KUJBYŠEV
UFA
SARATOV
VOLGOGRAD
SVERDLOVSK
ČEĽABINSK
Magnitogorsk
Novorossijsk
KRASNODAR
Groznyj
TBILISI
BAKU
JEREVAN
Astrachan'
Gurjev
Dikson
Chalmer-Ju
Salechard
Dudinka
Noriľsk
Chatanga
Tiksi
Turuchansk
Tura
Niž. Tunguska
Surgut
Chanty-Mansijsk
OMSK
Tomsk
NOVOSIBIRSK
KRASNOJARSK
Ust'-Kut
Bratsk
Nižneudinsk
BARNAUL
Bijsk
Abakan
Kyzyl
IRKUTSK
Ulan-Ude
Semipalatinsk
Ust'-Kamenogorsk
Bajkonur
KARAGANDA
Ulaanbaatar
Uliastaj
Chovd
Dalandzadgad
ALMA-ATA
FRUNZE
TAŠKENT
Samarkand
DUŠANBE
Tašauz
Krasnovodsk
Aschabad
MASHHAD
Yining
WULUMUQI (Urumchi)
Kashi
Hetian
LANZHOU
Lasa
CHENGDU
CHONGQING
KUNMING
Gejiu
İZMİR
İSTANBUL
ANKARA
HALAB
BAYRŪT (Beirut)
DIMASHQ
AMMĀN
AL-MAWSIL
TABRĪZ
Hamadān
TEHRĀN
BAGHDĀD
AL-BASRAH
ESFAHĀN
Ābādān
Shīrāz
Al-Kuwayt
Ad-Dammām
Al-Manāmah
AR-RIYĀD
JUDDAH
Makkah
Ad-Dawhah
Abū Zaby
Dubayy
Masqat
Bandar 'Abbās
Bam
Zāhedān
San'ā
Aden
Socotra (South Yemen)
Herāt
KĀBUL
Qandahār
Quetta
Islāmābād
Srinagar
RĀWALPINDI
LAHORE
KARĀCHI
New Delhi
DELHI
ĀGRA
KĀNPUR
ALLĀHĀBĀD
AHMADĀBĀD
INDORE
SURAT
NĀGPUR
BOMBAY
PUNE
HYDERĀBĀD
Cuttack
CALCUTTA
Mangalore
Calicut
BANGALORE
MADRAS
MADURAI
Trivandrum
COLOMBO
Lakshadweep (India)
Male
Kātmāndu
Thimbu
Dangori
DACCA
Myitkyinā
MANDALAY
RANGOON
Chiang Mai
KRUNG THEP (Bangkok)
Phnum Pénh
Andaman Is. (India)
Nicobar Is. (India)
Banda Atjeh
Pinang
MEDAN
Sumatera
Kep. Mentawai
Volga
Kama
Don
Dnepr
Danube
Ural
Pečora
Ob'
Irtyš
Jenisej
Angara
Lena
Amudarja
Syrdarja
Indus
Sutlej
Ganga
Brahmaputra
Godāvari
Krishna
Huanghe
1 BAHRAIN
2 BANGLADESH
3 BHUTAN
4 DEM. PEOPLE'S REP. OF KOREA
5 ISRAEL
6 JORDAN
7 KUWAIT
8 LEBANON
9 QATAR
10 REP. OF KOREA
11 SINGAPORE
12 SYRIA
13 UNITED ARAB EMIRATES
1 : 65 000 000
0 300 600 900 1 200 1 500 Km
0 200 400 600 800 1 000 Mi

20 a Southern India and Sri Lanka 1 : 15 000 000

20 b Levantine Countries 1 : 12 500 000

other towns: Mandalay 418,008, Kanbe 253,600, Moulmein 171,977, Insein 143,625, Bassein 126,045, Pegu 124,643. –
Population: Burmese, minorities: Karens, Shans, Kachins and others. **Density** 48 persons per sq.km; average annual rate of population increase 2.37% (1970–75); urban population 22.2% (1975). 54% of the economically active inhabitants engaged in agriculture (1977). – **Currency:** kyat = 100 pyas.
Economy: predominantly an agricultural country with extraction industry. Arable land 14.8%, forests 66.9% of the land area. **Agriculture** (1977, in 1,000 tons): rice 9,460, millet 60, sugarcane 1,660, groundnuts 530, cotton, jute 4[illegible], vegetables, fruit, tobacco 75, teak, natural rubber 16; livestock (1977, in 1,000 head): cattle 7,696, buffaloes 1,78[illegible], pigs 1,800, sheep 206, goats 587, fish catch 501,600 tons. **Mining** (1977, in 1,000 tons): crude petroleum 1,29[illegible], natural gas, lead, zinc, copper, silver, tungsten 370 tons (1976), antimony 559 tons, tin 760 tons, salt, precious stones. Food and textile industries. Production of woven cotton fabrics 73 mn. m, electricity 968 mn. kWh (1976).
Communications: railways 4,324 km (1974), roads 22,339 km (1974), passenger cars 37,700 (1976). Civil aviation 5 mn. km flown, 396,000 passengers carried (1976). – **Exports:** rice, teak, ores, vegetable oil, natural rubber, cotton. Chief trading partners: Japan, India, United Kingdom, U.S.A., Fed. Rep. of Germany, China.

CHINA

Chung-Hua Jen-Min Kung-Ho Kuo – People's Republic of China, area (the UN data) **9,560,980 sq.km, population 879,250,000** (1978), **people's republic** (Chairman of the State Council Chao Ts'-Yang since 1980).

Administrative units: 22 provinces (incl. Taiwan), 5 autonomous regions, 3 self-administrated municipalities. **Capital** Beijing (Peking) 8 mn. inhab. (1976); **other towns** (1970, in 1,000 inhab.): Shanghai 10,820 (with agglom. 12,000), Tianjin (Tientsin) 4,280, Shenyang (Mukden) 4,000, Guangzhou (Canton) 3,000, Wuhan 2,226, Chongqing 2,16[illegible], Nanjing 1,750, Haerbin 1,670, Lüda 1,650, Xi'an 1,600, Lanzhou 1,450, Taiyuan 1,350, Qingdao 1,300, Chengdu 1,25[illegible], Changchun 1,200. – **Population:** about 94% Chinese (Han), minorities Chuang, Uighur, Hui, Yi, Tibetans, Mongolians and others. **Density** 92 persons per sq.km; average annual rate of population increase 1.66% (1970–75); urban population 23.3% (1975). 62% of the economically active inhabitants work in agriculture. – **Currency:** renminpi juan = 10 chiao = 100 fen.
Economy: agricultural and industrial country. The main industrial region is eastern China (40% of industrial production – provinces Jiangsu, Anhui, Zhejiang, Fujian and self-administered city of Shanghai). Large raw material resources and heavy industry in north-east China (provinces Liaoning, Jilin, Heilongjiang). **Agriculture:** arable land 13.4%, meadows and pastures 22.2%, forests 15% of the land area. There are two or three harvests annually in the east and south-east of China. **Crops** (1977, in 1,000 tons): rice 131,472 (leading world producer), wheat 40,003 (third world producer), millet 20,007 (leading world producer), barley 15,401 (second world producer), maize (corn) 33,615 (second world producer), oats 1,900, sweet potatoes 117,164 (leading world producer), potatoes 41,646 (second world producer), beans 2,331, peas 4,900, sugarcane 46,240, sugar beet 8,240, cotton lint 2,342, jute 1,450 (leading world producer), flax, hemp, sisal, ramie, soybeans 12,955 (second world producer), cotton seed 4,684, sesame 361 (second world producer), coconuts 56, palm kernels 42, palm oil 162, tung oil 63 (two thirds of world production), rape seed, poppy, groundnuts 2,781 (second world producer), tea 336 (second world producer), tobacco 1,022 (leading world producer), oranges 81[illegible], tangerines 252, grapefruits 127, lemons 62, pineapples 874 (leading world producer), grapes 172, bananas, apples 49[illegible], pears 1,041 (second world producer), tomatoes 3,466, spices. **Livestock** (1977, in 1,000 head): horses 6,800, cattle 65,129, buffaloes 30,321 (second world population), camels 1,050, pigs 243,300 (highest world population), sheep 76,000, goats 62,196 (second world population), asses 11,500, mules 1,530, chickens 1,332,840 (highest world population), silkworms; eggs 3,862,000 tons (leading world producer), cow hides 358,620 tons, wool 62,000 tons, raw silk 15,814 tons (second world producer), honey 247,500 tons (leading world producer); fish catch 6,880,000 tons (third largest world catch); roundwood 195.2 mn. cub.m, natural rubber 29,000 tons.
Mining (1978, in 1,000 tons): coal 618,000 (provinces Shānxī, Hebei, Liaoning), crude petroleum 104,050 (Heilongjiang, Shandong, northern China, north-west China), uranium, natural gas 20 billion cub.m (provinces Hebei, Hubei – 1976), combustible shale (Liaoning), iron ore 32,500 (1976), manganese 300, antimony 12, cobalt, bauxite 1,1[illegible], copper 155 (1977), lead 150 (1977), magnesite 1,000, mercury 700 (1977), molybdenum 1.5, nickel, gold, silver, tin 20 (1977), tungsten 11.3 (leading world producer – provinces Jiangxi, Guangdong, Hunan), zinc 145 (197[illegible]), phosphates 3,750, potash 450, sulphur 130, asbestos 150, graphite, vanadium, salt 30,000. Electricity 256.55 billion kWh (1978). **Production** (1978, in 1,000 tons): crude steel 31,780, cement 65,240, fertilizers 8,693, metal-working machines 183,000 units, motor vehicles 149,100 units, woven cotton fabrics 11,029 billion sq.m, synthetic staple fibre 37 (1976), newsprint 1,100 (1976), sugar 4,823 (1977), meat 16,344, cow milk 3,846, buffalo milk 1,1[illegible] (1977), butter 93.3 (1977), cheese 190.6 (1977).
Communications: railways 36,000 km (1976), roads 550,000 km (1964), motor vehicles 705,000 (of which 35,0[illegible] passenger cars, 1975). Merchant shipping 4,245,000 GRT (1976). – **Exports:** agricultural products – rice, soybeans, fruit, tea, meat and meat products, eggs, sugar, edible oils, tung oil, cotton, silk; textile products, raw materials for the power industry, ores – especially tungsten and ore concentrates. Chief trading partners: Japan, Hong Kong, Canada, U.S.A., Fed. Rep. of Germany, U.S.S.R.

TAIWAN

Ta Chunghwa Min-Kuo, area 35,981 sq.km, population 16,793,000 (1977), **republic** under the control of the Kuomintang Nationalist Party (President Chiang Ching-kuo since 1978).

Capital: T'aipei 2,127,625 inhab. (1978); **other towns** (in 1,000 inhab.): Kaohsiung 1,041, T'aichung 571. – **Currency:** Taiwan dollar = 100 cents.
Economy: arable land 8.9%, forests 64.5% of the land area. **Agriculture** (1976, in 1,000 tons): rice 2,713, sugar cane 8,728, manioc, soybeans, sweet potatoes 1,851, pineapples 278.8, bananas 213.4, tea 24.8; livestock (197[illegible]

in 1,000 head): cattle 253.3, pigs 3,700, silkworms; fish catch 810,600 tons (1976). **Mining** (1976, in 1,000 tons): coal 3,236, crude petroleum 247, gold 838 kg, silver 3,109 kg, salt 497. Electricity 26,877 mn. kWh (1976). **Industry:** textile and chemical industries, engineering. **Production:** woven cotton fabrics 631 mn. m, sugar 879,800 tons (1976). **Communications:** railways 4,300 km (1974), roads 17,172 km (1976), passenger cars 122,517 (1975). Merchant shipping 1,449,957 GRT (1975). – **Exports:** textile products, plastic products, television and radio receivers, synthetic fibres.

CYPRUS

Kypriaki Dimokratia – Kıbrıs Cumhuriyeti, area 9,251 sq.km, population 640,000 (1977), **republic, member of the Commonwealth** (President Dr. Spyros Kyprianou since 1977).

Administrative units: Since 27 July 1976 Cyprus has been divided into two parts: southern with Greek and northern with Turkish speaking population by Attila Line. **Capital:** Levkosia (Nicosia) 117,100 inhab. (1974); **other towns** (1974, in 1,000 inhab.): Lemesós 80.6, Ammókhostos 39.4. – **Population:** Cyprian Greeks 79%, Cyprian Turks 18%, Armenians, Englishmen and others. **Density** 69 persons per sq.km; average annual rate of population increase 1.5% (1970–75); urban population 41.9% (1975). 35.4% of the economically active inhabitants engaged in agriculture. – **Currency:** Cyprus pound = 1,000 mils.

Economy: Mediterranean agriculture and extraction of raw materials. Arable land and permanent crops 46.7%. **Agriculture:** crops (1977, in 1,000 tons): grapes 180, oranges 160, lemons 32, grapefruit 65, olives 20, wheat 66, vegetables, early potatoes; livestock (1977, in 1,000 head): cattle 35, pigs 155, sheep 490, goats 399; fish catch 1,100 tons (1976). **Mining** (1976, in 1,000 tons): pyrites, copper 7.6, chromium 4.5, asbestos 34, sulphur. **Production:** wine 530,000 hl (1976), olive oil, raisins, cigarettes.

Communications: roads 9,719 km (1975), passenger cars 69,100 (1976). Civil aviation: 4.9 mn. km flown, 225,000 passengers carried (1976). Merchant shipping 2,788,000 GRT (1977). Tourism 47,084 visitors (1975). – **Exports:** citrus fruit, potatoes, tobacco and cigarettes, wine, copper and concetrates, pyrites. Chief trading partners: United Kingdom, Greece, U.S.S.R., Italy, Fed. Rep. of Germany, France.

INDIA

Bharat, area 3,287,590 sq.km, population 634,200,000 (1978), **federal republic, member of the Commonwealth** (President Neelam Sanjiva Reddy since 1977).

Administrative units: 22 federal states and 9 territories. **Capital:** New Delhi 301,801 inhab. (1971); **other towns** (1971, in 1,000 inhab., +with agglom.): Bombay 5,970, Delhi 3,288 (+3,647), Calcutta 3,149 (+7,031), Madras 2,469 (+3,170), Hyderābād 1,607 (+1,796), Ahmadābād 1,585 (+1,741), Bangalore 1,541 (+1,654), Kānpur 1,154 (+1,275), Nāgpur 866 (+931), Pune (Poona) 856 (+1,135), Lucknow 749 (+814), Āgra 592 (+635), Vārānasi (Benares) 584 (+607), Madurai 549, Indore 543, Allāhābād 514. – **Population:** Hindustani, Bihari, Marathi, Bengali, Santhali, Telugu and others. **Density** 193 persons per sq.km; average annual rate of population increase 2.43% (1970–75); urban population 21.4% (1975). 65.3% of the economically active inhabitants engaged in agriculture (1977). – **Currency:** Indian rupee = 100 paise.

Economy: agricultural and industrial country; arable land 49.5%, meadows and pastures 3.8%, forests 20.6%. **Agriculture. Crops** (1977, in 1,000 tons): rice 74,000 (second world producer), wheat 29,082, millet 10,000 (second world producer), sorghum 10,400 (second world producer), maize (corn) 6,800, sugarcane 154,023 (leading world producer), potatoes 7,287, important legumens, peas 245, lentils 431, soybeans 120, groundnuts 5,500 (leading world producer), chick-peas 5,366 (leading world producer), sesame 450 (leading world producer), coconuts 4,347, copra 320.3, cotton – seed 2,342, lint 1,171, linseed 431, jute 1,262 (second world producer), hemp 58, tea 561 (leading world producer), coffee 103, oranges 1,004, lemons 451, bananas 3,553 (second world producer), pineapples 110, apples 719, cashew nuts 150, sweet chestnuts 165, tobacco 414, vegetables, tomatoes 715. **Animal production** (1977, in 1,000 head): horses 900, cattle 181,092 (highest world population), buffaloes 60,398 (highest world population), camels 1,050, pigs 8,732, sheep 40,352, goats 70,060 (highest world population), silkworms; cowhides 795,600 tons (second world producer), goat skins 70,200 tons (leading world producer), grease wool 32,700 tons (1977), raw silk 2,600 tons (1977). Fish catch 2,540,000 tons (1977). Roundwood 134.4 mn. cub.m.

Mining (1977, in 1,000 tons, metal content): coal 100,110 (West Bengal – Rānīganj and Bihār), brown coal 3,628, crude petroleum 10,185 (Punjab), natural gas 1.1 billion cub.m (1976), uranium, iron ore 26,520 (Bihār, Orissa, Madhya Pradesh), manganese 665.3, lead 14.4, zinc 20.5, chromium 169.7, copper 33.2, gold 1,994 kg, mica 18.6 (1973, leading world producer – Bihār - Hazāribāgh, Andhra Pradesh, Rājasthān), asbestos 20, bauxite 1,512, magnesite 403, salt 5,329, phosphates 734, diamonds. Electricity 990.96 mn. kWh (1977), of which 37,176 mn. kWh hydroenergy. **Industry.** Important textile and food industries, metal-working. **Production** (1977, in 1,000 tons): pig iron 9,966, crude steel 9,310, coke oven coke 9,650, motor vehicles – commercial 41,200, passenger 48,000 units – merchant vessels 60,000 GRT, radio receivers 1,813,000 units, cement 19,173, naphtha 2,023, motor spirit 1,370, kerosene 2,488, jet fuel 1,020, sulphuric acid 2,016, caustic soda 517, nitrogenous fertilizers 1,999.7, phosphate fertilizers 670, woven fabrics – cotton 6,902 mn. m, – silk 4,160,000 sq.m (1971) – rayon and acetate 1,457.8 mn. m – woven woollen fabrics, sugar 5,019, flour 1,958, meat 841, cow milk 8,424, buffalo milk 16,500, butter 570 (second world producer), cheese 1,608; 63,996 mn. cigarettes.

Communications: railways 60,301 km (1975), roads 1,337,000 km (1973), motor vehicles – passenger 805,400, commercial 690,400 (1977). Merchant shipping 5,482,000 GRT (1977). Civil aviation: 80 mn. km flown, 5,168,000 passengers carried (1977). Tourism 640,422 visitors (1977). – **Exports:** woven cotton fabrics and products 5%, jute and jute products 4%, tea 5.7%, hides, skins and furs 5.1%, spices, fruit, vegetables, tobacco, sugar, iron ore and concentrates .6%, handicraft products 7.8%. Chief trading partners: U.S.A., Japan, United Kingdom, Fed. Rep. of Germany.

BLACK SEA
İSTANBUL
Karadeniz Boğ.
SEA OF MARMARA
Zonguldak
Ereğli
Kastamonu
Samsun
Ordžonikidze
Kutaisi
Poti
Batumi
TBILISI
Leninakan
Sakarya
Kocaeli
Bandırma
Çanakkale
Çanakkale Boğ.
Bursa
Edremit
Balıkesir
Kütahya
Eskişehir
Karabük
Çankırı
Kırıkkale
ANKARA
Merzifon
Çorum
Amasya
Ordu
Giresun
Trabzon
Rize
Çoruh
Kelkit
Tokat
Zile
Sivas
Yozgat
Erzincan
Erzurum
Kars
Aras
Karaköse
Fırat (Euphrates)
Murat
Van G.
Khvoy
Van
Tatvan
Bitlis
Batman
Kurtalan
Reẕā'īyeh
Lésvos
Khios
Manisa
Akhisar
Gediz
Uşak
İZMİR
Nazilli
Afyon
Kırşehir
Kızıl
Tuz G.
Kayseri
Malatya
Elâzığ
Sámos
Aydın
Denizli
Isparta
Burdur
Konya
Aksaray
Niğde
Ereğli
Diyarbakır
Muğla
Beyşehir G.
Karaman
Tarsus
Adana
Seyhan
Maraş
Dicle (Tigris)
Cizre
Dhodhekánisos
Antalya
Antalya Kör.
Göksu
İçel
Gaziantep
Fırat
Urfa
Mardin
Birecik
Al-Qāmishlī
Dijlah
Ródhos
İskenderun
İskenderun Kör.
Iráklion
Kríti
HALAB
Ar-Raqqah
Irbīl
AL-MAWSIL
Al-Lādhiqīyah
B. al-Asad
Al-Furāt
Levkosía
Ammókhostos
CYPRUS
Lemesós
Bāniyās
Hamāh
Dayr az-Zawr
Kirkūk
Hims
Tudmur
Tikrīt
MEDITERRANEAN SEA
Tarābulus
LEBANON
An-Nabk
Abū Kamāl
BAYRŪT
Zahlah
Sāmarrā'
Saydā
Al-Hadīthah
Al-Kāzimīyah
DIMASHQ (Damascus)
Ar-Ramādī
Hefa
Dar'ā
Ar-Rutbah
Karbalā'
Tel Aviv-Yafo
'AMMĀN
JORDAN
IRAQ
AL-ISKANDARĪYAH (Alexandria)
Matrūh
Dumyāt
Būr Sa'īd
Ghazzah
Yerushalayim (Jerusalem)
Al-Hillah
An-Najaf
Al-'Alamayn
Al-'Arīsh
Dead Sea
Tantā
Suez Canal
Al-Ismā'īlīyah
Oron
Badanah
AL-JĪZAH
As-Suways (Suez)
Ma'ān
ISRAEL
As-Samāwah
Sīwah
Al-Fayyūm
AL-QĀHIRAH (Cairo)
Elat
Ra's an-Naqb
Al-Jawf
Sakākah
Banī Suwayf
Abū Zanīmah
Al-'Aqabah
EGYPT
Al-Bawītī
G. of Suez
Al-Minyā
At-Tūr
Tabūk
SAUDI ARABIA
Nile
Mallawī
Ras Gharib
Nabq
Ash-Sharmah
Qasr al-Farāfirah
Asyūt
Al-Ghurdaqah
RED SEA
Taymā'
Hā'il
Zibā'
Būr Safājah
Sawhāj
BLACK SEA
Elhovo
BULGARIA
Mičurin
Kırklareli
Edirne
Lüleburgaz
GREECE
Çorlu
İSTANBUL
Tekirdağ
Keşan
Alexandroúpolis
SEA OF MARMARA
Gökçeada
Gelibolu
Mudanya
Kocaeli
Gemlik
Bandırma
Zonguldak
Bartın
Ayancık
Bafra
Samsun
Çarşamba
Ordu
Tirebolu
Ereğli
Kastamonu
Karabük
Tosya
İskilip
Merzifon
Sakarya
Düzce
Bolu
Çankırı
Amasya
Giresun
Gümüşhane
Göynük
Beypazarı
Çorum
Yeşil
Tokat
Şebinkarahisar
Bursa
Bilecik
Sakarya
ANKARA
Kalecik
Sungurlu
Zile
Sivas
Zara
Erzincan
Ezine
Edremit
Balıkesir
Eskişehir
Kırıkkale
Yozgat
Tavşanlı
Polatlı
Sivrihisar
Divriği
İliç
Soma
Kütahya
Kulu
Kırşehir
Şarkışla
Arapkir
Lésvos
Bergama
Akhisar
Emirdağ
Şereflikoçhisar
Bünyan
Keban
Khios
Manisa
Gediz
Uşak
Afyon
Tuz G.
Kayseri
Pınarbaşı
Cihanbeyli
Nevşehir
Köyyeri
Malatya
İZMİR
Salihli
Ödemiş
Çivril
Tire
Nazilli
Dinar
Akşehir
Ilgın
Aksaray
Niğde
Saimbeyli
Elbistan
Adıyaman
Sámos
Söke
Aydın
Büyükmenderes
Eğridir
Beyşehir G.
Konya
Beyşehir
Akköy
Denizli
Burdur
Isparta
Ulukışla
Maraş
Besni
Dhodhekánisos
Milâs
Bucak
Ereğli
Seyhan
Kozan
Fırat (Euphrates)
Muğla
Karaman
Urfa
Bodrum
Antalya
Manavgat
Tarsus
Ceyhan
Osmaniye
Gaziantep
Birecik
Göksu
İçel
Adana
İskenderun
Kilis
Fethiye
Kemer
Antalya Kör.
Alanya
Erdemli
Silifke
Ródhos
Kaş
Finike
Anamur
İskenderun Kör.
Antakya
HALAB
Al-Bāb
SYRIA
Lindhos
Idlib
Al-Lādhiqīyah
Kárpathos
MEDITERRANEAN SEA

CASPIAN SEA
Zaliv Kara-Bogaz-Gol
S.
R.
BAKU
Sumgait
Krasnovodsk
Nebit-Dag
Kizyl-Arvat
Bacharden
Aščhabad
Tašauz
Urgenč
Amudarja
Dargan-Ata
Leninabad
Navoi
Kattakurgan
Zeravšan
Buchara
Kagan
Samarkand
Šachrisabz
Karši
Kamaši
DUŠANBE
Denau
Kurgan-T'ube
Čardžou
Kerki
Termez
Bajram-Ali
Mary
Iolotan'
Tedžen
Sandykači
Tachta-Bazar
Kuška
Astara
Enzeli
Rasht
Safid
Shahsavār
Bandar Torkaman
Gorgān
Gasan-Kuli
Atrak
Bojnūrd
Shīrvān
Qūchān
MASHHAD
Sarakhs
Neyshābūr
Sabzevār
Shāhrūd
Bābol
Sārī
Qazvīn
TEHRĀN
Abhar
Karaj
Rey
Semnān
Garmsar
Kāshmar
Torbat-e Heydarīyeh
Qom
Daryācheh-ye Namak
Jandaq
Ferdows
Tabas
Arāk
Kāshān
Golpāyegān
Ardestān
Bīrjand
ESFAHĀN
Nā'īn
Ardakān
Rīz
Dezfūl
Masjed Soleymān
Shahrezā
Īzad Khvāst
Ābādeh
Yazd
Bāfq
Zarand
Naft-e Safīd
Rāmhormoz
Deh Bīd
Anār
Kermān
Bandar Chomejnī
Behbehān
Gachsārān
Ābādān
Rafsanjān
Shīrāz
Daryācheh-ye Bakhtegān
Sa'īdābād
Kāzerūn
J.-ye Khārk
Neyrīz
Bam
Fasā
Būshehr
Jahrom
Kahnūj
Kangān
Lār
Nāy Band
Bandar 'Abbās
Qeshm
Bandar-e Lengeh
Strait of Hormuz
Jāsk
Al-Kuwayt
Persian Gulf
Al-Qaṭīf
BAHRAIN
Ad-Dammām
Al-Manāmah
Abqaiq
Al-Mubarraz
QATAR
Al-Hufūf
Ad-Dawhah
Al-Khasab
Ra's al-Khaymah
Ash-Shāriqah
Ujmān
OMAN
Dubayy
Al-Fujayrah
Suḥār
Abu Zaby
Al-'Ayn
Al-Buraymī
UNITED ARAB EMIRATES
Habshān
Tarīf
Masqaṭ
Maṭraḥ
Nazwā
'Ibrī
Fahūd
Ṣūr
Ra's al-Hadd
Gulf of Oman
Tropic of Cancer
ARABIAN SEA
Al-Maṣīrah
Ra's ad Daqm
OMAN
Khūryān Mūryān
Ṣalālah
Mirbāṭ
IRAN
Shūsf
Hāmūn-e Sāberī
Zābol
Daryācheh-ye Sīstān
Zāhedān
Mīrjāveh
Khāsh
Īrānshahr
Sūrān
Bampūr
Hāmūn-e Jaz Mūrīān
Chāh Bahār
AFGHANISTAN
Sheberghān
Mazār-e Sharīf
Sar-e Pol
Meymaneh
Morghāb
Bāmiān
Qal'eh-ye Now
Harīrūd
Chaghcharān
Herāt
Ghūrīān
Tarīn-Kowt
Farāh
Mūs'a Qal'eh
Qalāt
Gereshk
Lashkar Gāh
Qandahār
Chaman
Khāsh
Helmand
Zaranj
Nushki
Chāgai
Nok Kundi
Dālbandin
Khārān
Hāmūn-i-Māshkel
PAKISTAN
Bazdār
Turbat
Pasni
Ormāra
Gwādar
1 : 18 000 000
0 100 200 300 400 Km
0 50 100 150 200 250 Mi
50°
55°
60°
65°
40°
35°
30°
25°
20°
6
7
8
9
A
B
C
D
E
F
Turkey 1 : 12 500 000
Oltu
Kars
Sarıkamış
Aras
Erzurum
Eleşkirt
Karaköse
Doğubayazıt
Mākū
Malazgirt
Erciş
Muradiye
Khvoy
Tatvan
Van
Bitlis
Van Gölü
Siirt
Çölemerik
Cizre
Zākhū
Dahūk
Qāmishlī
IRAQ
Rawāndūz
IRAN

INDONESIA

Republik Indonesia, area 1,919,270 sq.km (incl. East Timor), **population 145,350,000** (1978), **republic** (President Gen. I. Suharto since 1968).

Administrative units: 26 provinces, in 1976 annexed East Timor (area 14,925 sq.km, 704,000 inhab.). **Capital:** Djakarta 4,576,009 inhab. (1971); **other towns** (1971, in 1,000 inhab.): Surabaja 1,556, Bandung 1,202, Semarang 647, Medan 636, Palembang 583, Udjung Pandang (Makasar) 435, Malang 422, Bandjarmasin 421, Surakarta 414. – **Population:** Malay Indonesians (over 140 ethnic groups; most numerous are the Javanese and the Chinese among the minorities). **Density** 76 persons per sq.km; average annual rate of population increase 2.6% (1970–75); urban population 19.2% (1975). 61.2% of the economically active inhabitants engaged in agriculture (1977). – **Currency:** Indonesian rupiah = 100 sen.

Economy: agricultural country, and mining of minerals. Arable land and permanent crops 9.8%, meadows and pastures 5.2%, forests 63.7% of the land area. **Agriculture: Crops** (1977, in 1,000 tons): rice 23,235 (third world producer), manioc 12,169, sweet potatoes 2,453, soybeans 527, groundnuts 673, sugarcane 15,076, sisal, cotton, coconuts 6,882 (second world producer), copra 950 (second world producer), palm kernels 95, palm oil 450, sesame, fruit, tea 74, coffee 180, tobacco 84; **livestock** (1977, in 1,000 head): cattle 6,114, buffaloes 2,458, pigs 2,516, sheep 3,286, goats 6,112, horses 649; fish catch 1,488,000 tons (1976). Natural rubber 850,000 tons (second world producer), roundwood 141.4 mn. cub.m.

Mining (1977, in 1,000 tons, metal content): coal 231, crude petroleum 82,998, natural gas 8.8 mn. cub.m, tin 25.1 (third greatest producer in the world), nickel 14, manganese, bauxite 1,301, gold 256 kg, diamonds, phosphates, salt 786. Electricity 4,380 mn. kWh, of which 2,200 mn. kWh from hydro-electric power plants. **Industry:** Important processing of tin, petroleum refineries, food and textile industries. **Production** (1977, in 1,000 tons): tin 24 (second world producer), motor spirit 2,453, kerosene 4,690, jet fuel 113, nitrogenous fertilizers 457.6, sugar 1,100, 67,218 mn. cigarettes (1976); radio 1 mn. and television receivers 482,000 units.

Communications: railways 8,956 km (1972), roads 95,544 km (1974), motor vehicles – passenger 479,300, commercial 327,100. Merchant shipping 1,272,000 GRT (1977). Civil aviation: 72.3 mn. km flown, 3,781,000 passengers carried. Tourism 456,718 visitors (1977). – **Exports:** petroleum and petroleum products, wood, natural rubber, palm oil, coffee, tin concentrates, tea, tobacco, spices. Chief trading partners: Japan, U.S.A., Fed. Rep. of Germany, Netherlands.

IRAN

al Jumhouriya al Islamia Írán, area 1,648,100 sq.km, population 35,213,000 (1978), **republic** (President Hojatoislam Mohammad Ali Chamenei since 1981).

Administrative units: 14 provinces (ustáns), 9 governorates (farmandarikol). **Capital:** Tehrān 4,498,159 inhab. (1976), **other towns** (1976, in 1,000 inhab.): Esfahān 672, Mashhad 670, Tabrīz 599, Shīrāz 416, Ahvāz 329, Ābādān 296, Kermānshāh 291. – **Population:** two thirds are Iranian Persians; Azerbaijanis, Kurds, Arabs. **Density** 21 persons per sq.km; average annual rate of population increase 2.98% (1970–75); urban population 44.1% (1975). 40.7% of the economically active inhabitants engaged in agriculture (1977). – **Currency:** rial = 100 dinars.

Economy: agricultural and industrial country with important mining of petroleum. Arable land and land under permanent crops 10%, forests 10.9% of the land area. **Agriculture. Crops:** (1977, in 1,000 tons): wheat 6,200, barley 1,600, rice 1,650, millet 25, sorghum 10, maize (corn) 90, potatoes 580, beans 111, peas 30, lentils 30, groundnuts, soybeans 103, sunflower, sesame, olives, castor beans, cotton – seed 315, lint 180 – sugar beet 4,800, sugarcane 895, vegetables, oranges 68, lemons 32, apricots 72, grapes 917, raisins 63, almonds, dates 300 (third world producer), hazelnuts, pistachios 46.6 (leading world producer), tomatoes 256, onion 333; **livestock** (1977, in 1,000 head): horses 350, cattle 6,650, sheep 35,441, goats 14,375, asses 1,800; cowhides 22,943 tons, sheepskins 29,310 tons, grease wool 18,300 tons, eggs 132,400 tons (all 1977).

Mining (1977, in 1,000 tons, metal content): crude petroleum 282,608 (chiefly in south-west Iran), natural gas 22.5 billion cub.m, iron ore 670, manganese 15, copper 6, lead 40, zinc 61.5, bauxite, magnesite 10, salt 700. Electricity 18,000 mn. kWh, of which 4,000 mn. kWh hydro-energy. The most developed **industries** are petrochemical and chemical, textile, leather and food processing. **Production** (1977, in 1,000 tons): naphtha 700, motor spirit 4,706 (of which aviation spirit 400), kerosene 4,742, jet fuel 1,200, coke oven coke 400, sulphuric acid 351 (1976), phosphate fertilizers 110.4, nitrogenous fertilizers 177.9, cement 6,000 (1976), woven cotton fabrics 550 mn. m, woven woollen fabrics 23 mn. m, synthetic continuous filaments 8.5, meat 515, sugar 732, flour 4,705, milk 1,256.

Communications: railways 4,700 km (1975), roads 50,000 km (1974), motor vehicles – passenger 932,700, commercial 204,000 (1977). Merchant shipping 1,195,000 GRT (1977). Civil aviation: 44 mn. km flown, 3,005,000 passengers carried (1977). Tourism 695,500 visitors (1977). – **Exports:** petroleum and petroleum products, cotton and cotton goods, carpets, dates, raisins, skins, pistachios. Chief trading partners: Japan, Fed. Rep. of Germany, United Kingdom, U.S.A., Italy, France, Netherlands, India.

IRAQ

al Jumhouriya al'Iraqia, area 438,446 sq.km, population 12,171,480 (1977), **republic** (President Saddám Hussain since 1979).

Administrative units: 18 provinces (muhafadhas). **Capital:** Baghdād 2,183,800 inhab. (1970); **other towns** (1970, in 1,000 inhab.): Al-Basrah 371, Al-Mawsil (Mosul) 293, Kirkūk 208, An-Najaf 179. – **Population:** about 75% Arabic speaking Iraqis, 10–15% Kurds, 2% Turks. Average annual rate of population increase 3.35% (1970–75); urban population 64.8% (1976). 42% of the economically active inhabitants engaged in agriculture. – **Currency:** Iraqi dinar = = 1,000 fils.

Economy: agricultural country with developing industry. Arable land 11.7% and forests 3.5% of the land area. **Agriculture** (1977, in 1,000 tons): rice 199, wheat 696, barley 458, cotton – lint 13, seed 24 – dates 375 (second world producer), sesame, olives, oranges 46, lemons 30; livestock (1977, in 1,000 head): cattle 2,550, buffaloes 218, camels 228, sheep 11,400, goats 3,600; grease wool 18,200 tons (1977); fish catch 21,800 tons (1976). **Mining** (1977, in 1,000 tons): crude petroleum 122,390 (Kirkūk, Al-Mawsil), natural gas 1.9 billion cub.m, sulphur 610 (1976), salt 60. Electricity 5,000 mn. kWh (1977). Petrochemical, textile and food **industries. Production** (1977, in 1,000 tons): naphtha 500, motor spirit 670, kerosene 610, nitrogenous fertilizers 125.5, sulphuric acid, woven cotton fabrics 68 mn. m; 7,200 mn. cigarettes; sugar 25, canned fruit and vegetables, vegetable oils.

Communications: railways 2,394 km (1973), roads 10,824 km (1973), motor vehicles – passenger 150,400, commercial 74,500 (1977). Merchant shipping 1,135,000 GRT (1977). Civil aviation: 13.9 mn. km flown, 672,000 passengers carried. Tourism 721,577 visitors (1977). – **Exports:** petroleum and petroleum products, dates, skins, wool. Chief trading partners: United Kingdom, U.S.S.R., Italy, France, Brazil.

ISRAEL

Medinat Israel, area 20,770 sq.km, population 3,709,000 (1978), **republic** (President Yitzhak Navon since 1978).

Administrative units: 6 districts (mechuza). **Capital:** Yerushalayim (Jerusalem) 376,000 inhab. (1977); **other towns** (1977, in 1,000 inhab.): Tel Aviv-Yafo 343 (with agglom. 976), Hefa (Haifa) 228. **Population:** Jews 82%, Arabs 15%. **Density** 179 persons per sq.km; average annual rate of population increase 1.26% (1970–75); urban population 75.1% (1975). 7.6% of the economically active inhabitants engaged in agriculture. – **Currency:** Israeli pound = 100 agorot.

Economy: industrial and agricultural country. Arable land 16.5%, meadows and pastures 39.5%, forests 5.6% of the land area. **Agriculture** (1977, in 1,000 tons): wheat 230, sugar beet 343, oranges 897, grapefruits 499 (second world producer), bananas 54, peaches, cotton, vegetables, grapes 73; livestock (1977, in 1,000 head): cattle 335, sheep 218, goats 142; eggs 97,972 tons (1977). Fish catch 24,400 tons (1977). Electricity 11,108 mn. kWh. Engineering and electronics, chemical, textile and food **industries. Production** (1977, in 1,000 tons): motor spirit 950, kerosene 850, nitrogenous fertilizers 50.6, phosphate fertilizers 46.6, potash fertilizers 683.2, woven cotton fabrics 11.6 (1975), woven woollen fabrics, meat, cheese 49.5, butter 5.5, margarine 30.1; 4,751 mn. cigarettes; sugar 35, wine 366,000 hl.

Communications: railways 786 km (1975), roads 10,657 km (1973), motor vehicles – passenger 312,700, commercial 107,600. Merchant shipping 405,000 GRT (1977). Civil aviation: 31.2 mn. km flown, 1,039,000 passengers carried. Tourism 893,883 visitors (1977). – **Exports:** polished gems, citrus fruit, products of engineering, chemical and textile industries.

JAPAN

Nippon (or **Nihon), area 372,487 sq.km, population 115,100,000** (1979), **empire** (Emperor Hirohito since 1926).

Administrative units: 47 prefectures. **Capital:** Tōkyō 8,392,425 inhab. (with agglom. 11,683,613 – 1976); **other towns** (1976, in 1,000 inhab.): Ōsaka 2,750, Yokohama 2,659, Nagoya 2,080, Kyōto 1,462, Kōbe 1,364, Sapporo 1,277, Kitakyūshū 1,064, Kawasaki 1,026, Fukuoka 1,022, Hiroshima 842, Sakai 758, Chiba 675, Sendai 597, Amagasaki 536, Okayama 523, Higashiōsaka 501, Kumamoto 484, Hamamatsu 476, Kagoshima 472, Shizuoka 452, Nagasaki 447, Himeji 439, Niigata 426, Funabashi 423, Gifu 408, Kurashiki 398, Yokosuka 396, Wakayama 394, Nishinomiya 390, Sagamihara 386, Toyonaka 385, Matsuyama 379, Matsudo 355, Utsunomiya 350. – **Population:** Japanese, about 650,000 Koreans, 46,600 Chinese, 15,000 Ainu (1973). **Density** 309 persons per sq.km; average annual rate of population increase 1.26% (1970–75); urban population 76% (1975). 13.2% of the economically active inhabitants engaged in agriculture (1977). – **Currency:** yen = 100 sen.

Economy: economically highly developed industrial and agricultural country. Arable land 13.3%, forests 67% of the land area. **Agriculture: Crops:** (1977, in 1,000 tons): rice 17,000, wheat 236, barley 206, oats, maize (corn), soybeans 111, groundnuts 63, beans 144, sugarcane 1,800, sugar beet 2,300, potatoes 3,742, sweet potatoes 1,279, hemp, flax, vegetables – tomatoes 965, onion 1,132, fruit – apples 923, pears 533, oranges 289, tangerines 3,704 (leading world producer), grapefruits 48, strawberries 175, sweet chestnuts 66, tobacco 174; **livestock** (1977, in 1,000 head): cattle 3,875, pigs 7,901, extensive raising of silkworms; eggs 1,878,000 tons, raw silk 16,000 tons (1977); fish catch 10,733,300 tons (largest world catch). Roundwood 34.5 mn. cub.m.

Mining (1977, in 1,000 tons, metal content): coal 18,246, brown coal 10,000, crude petroleum 592, natural gas 3 billion cub.m, uranium, iron ore 487, pyrites, copper 81.4, lead 54.8, zinc 275.7, chromium 5.9, tungsten 973 tons, gold 37,920 kg, silver 299 tons, molybdenum 109 tons, mercury, salt 1,053. Electricity 511,776 mn. kWh, of which hydro-energy 88,373 mn. kWh, nuclear energy 34,079 mn. kWh (1976).

Industry: chief producer in the world in metallurgy (Kitakyūshū, Muroran), many branches of the engineering (Ōsaka, Yokohama, Kōbe, Nagasaki, Hiroshima), chemical (Kōbe, Tōkyō, Ōsaka) and textile industries. **Production** (1977, in 1,000 tons): coke oven coke 42,945, pig iron 87,699 (second world producer), crude steel 102,405 (third world producer), aluminium 1,192.9, copper 1,127 (second world producer), lead 287, zinc 839 (second world producer), television receivers 15,210,000 (leading world producer), merchant vessels 4,921,000 GRT (leading world producer), of which tankers 771,000 GRT (second world producer), motor vehicles – passenger 5,431,000 (second world producer), commercial 3,096,600 (leading world producer), cement 73,138 (second world producer), naphtha 20,511, motor spirit 23,072, kerosene 20,518, jet fuel 3,101, sulphur 1,078, sulphuric acid 6,392, hydrochloric acid 487.4, nitric acid 646, caustic soda 2,785, nitrogenous fertilizers 1,446, phosphate fertilizers 696, woven cotton fabrics 2,266 mn. sq.m, woven silk fabrics 155.6 mn. sq.m (leading world producer), woven woollen fabrics 347 mn. sq.m, synthetic fabrics

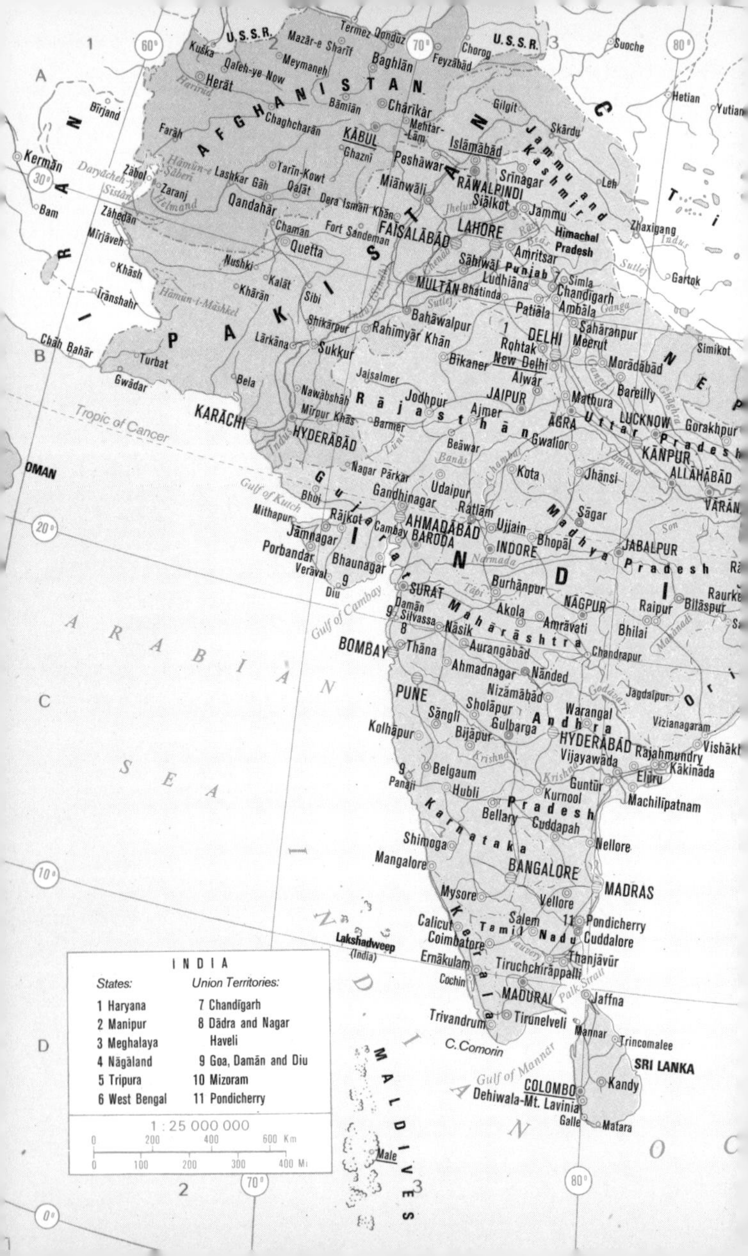
INDIA
States:
1 Haryana
2 Manipur
3 Meghalaya
4 Nāgāland
5 Tripura
6 West Bengal
Union Territories:
7 Chandīgarh
8 Dādra and Nagar Haveli
9 Goa, Damān and Diu
10 Mizoram
11 Pondicherry
1 : 25 000 000
0 200 400 600 Km
0 100 200 300 400 Mi
AFGHANISTAN
PAKISTAN
INDIA
ARABIAN SEA
MALDIVES
SRI LANKA
OMAN
U.S.S.R.
KĀBUL
Islāmābād
New Delhi
DELHI
BOMBAY
MADRAS
KARĀCHI
COLOMBO
Tropic of Cancer
Gulf of Kutch
Gulf of Cambay
Gulf of Mannar
Palk Strait
Lakshadweep (India)
Male
C. Comorin
Jammu and Kashmir
Himachal Pradesh
Punjab
Rājasthān
Gujarāt
Madhya Pradesh
Uttar Pradesh
Mahārāshtra
Andhra Pradesh
Karnataka
Kerala
Tamil Nadu

22 a Plain of Bengal

1 : 15 000 000

Rohtak Meerut Morādābād
DELHI New Delhi Rāmpur
Birendranagar
NEPAL
CHINA
Haryana
Bareilly
Nepālganj Pokhara
Aligarh Shāhjahānpur
Kātmāndu Namche Bazar
Rājasthān
Alwar Mathura
Bhaktapur
Lalitpur
Sikkim
Thimbu
Punakha
AGRA Farrukhābād
Darjeeling Gangtok
BHUTAN
Fīrozābād LUCKNOW
Amlekhganj Dhankuta
Siliguri
Uttar Pradesh
Faizābād Gorakhpur
Birātnagar
Cooch Behār
Assam
Etāwah
Chambal
KĀNPUR
Darbhanga
B Gwalior
Saidpur Dhubri
Brahmaputra
Rangpur
Meghalaya
Muzaffarpur
Shivpuri
ALLĀHĀBĀD Jaunpur
PATNA
Katihār
Bārān
Jhānsi Bānda
Ganges
VĀRĀNASI Arrah
Bihar Monghyr
Bhāgalpur
Bengal
BANGLADESH
Mymensingh
Guna Lalitpur
Mirzāpur Sāsarām
Gaya
Rājshāhi Sirājganj
Bīna-Etāwa
Satna Rewa
Bihar
Deoghar
Berhampore
Madhya Pradesh
INDIA
Sāgar
Murwāra
Daltonganj
Pābna
DACCA
Agartala
Dhānbād
Durgāpur
Comilla
Bhopāl
Asansol
Burdwān Jessore
JABALPUR
Rānchī
Bhātpāra
Chāndpur
Jamshedpur
West
HOWRAH KHULNA
Barisāl
Raurkela
CALCUTTA
Kharagpur
Orissa
Bay of Bengal

Yaluzangbujiang
Rikaze Jiangzi
Thimbu
Sikkim
BHUTAN
Gangtok
Arunachal Pradesh
Murkong Selek
Itanagar
Dangori
Putao
Brahmaputra
Dibrugarh
Assam
Gauhāti
Shillong
Kohīma
Mymensingh
Imphāl
Myitkyinā
Tengchong
Lijiang
Xichang
Huize
GUIYANG
CHINA
Dukou
Qujing Anshun
Duyun
Dushan
Guilin
Dali
Chuxiong
KUNMING
Yishan
Liuzhou
Wuzhou
Lancangjiang
Baise
Xijiang
Agartala
Aijal
DACCA
BANGLADESH
Mengzhi
Gejiu
NANNING
Yulin
KHULNA
CALCUTTA
CHITTAGONG
Ye-u
Shwebo
Lashio
Lao-Cai
Pingxiang
Beihai
Zhanjiang
Lang-Son
BURMA
MANDALAY
Jinghong
Hon-Gai
Gulf
Sg. Da
Sg. Hong
HA-NOI
HAI-PHONG
Pakokku
Chauk
Myingyan
Phong Saly
Hoa-Binh
Nam-Dinh
Qiongzhouhaixia
HAIKOU
Meiktila
Sittwe
Magwe
Taunggyi
Mekong
Louangphrabang
Thanh-Hoa
of
Hainandao
Irrawaddy
Pyinmana
Chiang Rai
Xiangkhoang
Vinh
Tonkin
Prome
Toungoo
Chiang Mai
Ha-Tinh
Yaxian
Salween
Viangchan (Vientiane)
Myanaung
Lampang
Nong Khai
M. Khammouan
Henzada
Pegu
Udon Thani
Paracel Is. (Viet.)
Hue
Phitsanulok
LAOS
Bassein
Martaban
Khon Kaen
Savannakhet
Khemmarat
DA-NANG
RANGOON
Moulmein
THAILAND
Ubon-Ratchathani
Pakxé
Moktama Kwe
Ye
Nakhon Sawan
Nakhon Ratchasima
VIETNAM
Nam Tok
Qui-Nhon
THON BURI
Koko Kyunzu (Burma)
Tavoy
Siĕm Réap
Tônlé Sab
Stoeng Tréng
KRUNG THEP (Bangkok)
Bătdâmbâng
Krâchéh
Nha-Trang
Andaman Is.
Myeik Kyunzu
Myeik
Chanthaburi
KAMPUCHEA
Da-Lat
Cam-Ranh
Phan-Rang
Loc-Ninh
ANDAMAN
(India)
Port Blair
Tenasserim
Phnum Pénh
Mekong
TH.-PHO HO CHI MINH (Saigon)
Gulf
Kâmpôt
My-Tho
Vung-Tau
SEA
Chumphon
Krŏng Preă Seihanŭ
of
Rach-Gia
Vinh-Long
Kra Buri
Can-Tho
Thailand
Quan-Long
Côn Dao
Surat Thani
Mui Bai-Bung
Nicobar Is. (India)
Phuket
Nakhon Si Thammarat
SOUTH CHINA SEA
Ko Phuket
Trang
Songkhla
Hat Yai
Pattani
Kota Bharu
Alor Setar
Pinang
MALAYSIA
Banda Atjeh
Lhokseumawe
Kuala Trengganu
Taiping
INDONESIA
Langsa
Sumatera
MEDAN
Ipoh
Kuantan
Kep. Bunguran (Indon.)

882.4 mn. sq.m, cotton yarn 441, woollen yarn 130.2, synthetic staple fibre 715.4 (second world producer), newsprint 2,370, milk 5,713, butter 48, meat 2,397, beer 42,972,000 hl, sugar 566, flour 3,960; 303,461 mn. cigarettes.
Communications: railways 27,161 km, of which 12,836 km are electrified (1975), main roads 260,000 km, motor vehicles – passenger 19.8 mn., commercial 11.6 mn. (1977). Merchant shipping 40,036,000 GRT (1977, world's second largest fleet). Civil aviation: 296.6 mn. km flown, 36,177,000 passengers carried. Tourism 890,688 visitors.
Exports: 62% machines and equipment – cars, vessels, products of heavy engineering, radio and television receivers, 17.5% metals and metal products, textile products – cotton and silk fabrics and others, chemical products. Chief trading partners: U.S.A., Saudi Arabia, Australia, Indonesia, Iran, Fed. Rep. of Germany, Canada, U.S.S.R., China.

JORDAN

Al Mamlaka al Urduniya al Hashemiyah, area 97,740 sq.km (incl. the territory occupied by Israel), **population 2,874,000** (1977), **kingdom** (King Hussein Ibn Talal since 1952).
Administrative units: 8 districts (liwa) and the Desert Area. **Capital:** Ammān 671,560 inhab. (1976); **other towns** (1976, in 1,000 inhab.): Az-Zarqā' 251.4, Irbid 130.9. – **Currency:** Jordan dinar = 1,000 fils.
Economy: developing agricultural country with nomadic cattle raising. Arable land and permanent crops 13.9% of the land area. **Agriculture** (1977, in 1,000 tons): wheat 53, barley, maize (corn), olives 23, grapes, citrus fruit, dates, vegetables; livestock (1977, in 1,000 head): cattle 36, camels 19, sheep 820, goats 490. **Mining** (1976, in 1,000 tons): phosphates 1,768, potassium salt, salt, sulphur, manganese, copper. Electricity 501 mn. kWh (1976). Processing of fruit and vegetables, chemical and textile industries. – **Communications:** railways 478 km (1977), roads 5,972 km (1974), passenger cars 41,500 (1976). Civil aviation: 10.6 mn. km flown, 417,000 passengers carried. Tourism 1,063,294 visitors (1976). **Exports:** phosphates, citrus and other fruit, vegetables, olive oil, chemical products. Chief trading partners: U.S.A., Lebanon, Fed. Rep. of Germany and other EEC countries.

KAMPUCHEA

Cambodia, area 181,035 sq.km, population 7,895,000 (1977), **republic** (Chairman of the People's Revolutionary Council Heng Samrin since 1979).
Administrative units: 4 self-administered cities and 17 provinces. **Capital:** Phnum Pénh 200,000 inhab. (1977). – **Currency:** riel = 100 sen.
Economy: agricultural country. Arable land 16.8%, forests 74% of the land area. **Agriculture** (1977, in 1,000 tons): rice 1,800, maize (corn) 80, sweet potatoes, soybeans, groundnuts 15, sugarcane 195, cotton, sesame, bananas 90, coconuts 47 (1976), pineapples 11 (1976), citrus fruit, spices, tobacco; livestock (1977, in 1,000 head): cattle 1,150, pigs 800; fish catch 84,700 tons (1976). Natural rubber 18,000 tons (1977), teak and other timber. Electricity 150 mn. kWh (1976). Developing chemical, textile and food industries. – **Communications:** railways 272 km (1971), roads 15,029 km (1973), passenger cars 27,200 (1972). Civil aviation: 800,000 km flown, 113,000 passengers carried (1976). – **Exports:** rice, natural rubber, rare timber, pepper, fruit, vegetables. Chief trading partners: Vietnam, U.S.S.R.

KOREA, DEMOCRATIC PEOPLE'S REPUBLIC OF

Chosun Minchu-chui Inmin Konghwa-guk, area 120,538 sq.km, population 17,028,000 (1978), **people's democratic republic** (Chairman of the Presidium of the Supreme People's Assembly Kim IL Sung since 1972).
Administrative units: 2 statutory cities and 9 provinces. **Capital:** P'yŏngyang 1,500,000 inhab. (1976); **other towns** (1976, in 1,000 inhab.): Ch'ŏngjin 320, Hŭngnam 260, Kaesŏng 240. **Population:** Koreans. **Density** 141 persons per sq.km; 48.5% of the economically active inhabitants engaged in agriculture (1977). – **Currency:** won = 100 yun.
Economy: agricultural and industrial country. Arable land 17.8% of the land area. **Agriculture** (1977, in 1,000 tons): rice 4,610, wheat 310, barley 340, maize (corn) 1,820, millet 418, sorghum 120, sweet potatoes 360, soybeans 310, tobacco 41, fruit, vegetables; livestock (1977, in 1,000 head): cattle 820, pigs 1,580, sheep 275, goats 210, silkworms; fish catch 800,000 tons (1976). Raw silk 2,250 tons (1977). **Mining** (1976, in 1,000 tons, metal content): coal 40,000, iron ore 3,800, tungsten 2,700 tons, graphite 75, magnesite 1,700 (third world producer), copper, lead 120, silver, tin 576 tons, zinc 150 (1977), salt, phosphates. Electricity 26,000 mn. kWh (1975). – **Production** (1976): pig iron 3 mn. tons, crude steel 3 mn. tons, cement 6 mn. tons, nitrogenous fertilizers 270,000 tons, tractors, lorries, textile and food industries. – **Communications:** railways 11,000 km, roads 20,000 km. Merchant shipping 89,000 GRT (1977). – **Exports:** ores, canned fish and fruit, agriculture products. Chief trading partners: U.S.S.R., China, Japan.

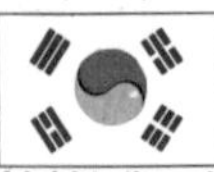

KOREA, REPUBLIC OF

Han Kook, area 98,484 sq.km, population 37,019,000 (1978), **republic** (President Chon Doo Hwan since 1980).
Administrative units: 2 statutory cities and 9 provinces. **Capital:** Sŏul 6,889,470 inhab. (1975); **other towns** (1975, in 1,000 inhab.): Pusan 2,454, Taegu 1,311, Inch'ŏn 797, Kwangju 607, Taejŏn 506. **Population:** Koreans. **Density** 376 persons per sq.km; average annual rate of population increase 2% (1970–75); urban population 48.5% (1975). 42.3% of the economically active inhabitants engaged in agriculture. – **Currency:** won = 100 chon.
Economy: industrial and agricultural country with advanced consumer industries. Arable land 24.7%, forests 67.2% of the land area. **Agriculture** (1977, in 1,000 tons): rice 8,340 (highest world yield 6,780 kg per ha), wheat 90, barley 814, sweet potatoes 1,844, soybeans 340, maize (corn), millet, legumens, tobacco 138; livestock (1977, in 1,000 head): cattle 1,553, pigs 1,953, goats 232, silkworms; fish catch 2,406,700 tons (1976). Raw silk 4,480 tons.

Mining (1977, in 1,000 tons, metal content): coal 17,316, iron ore 367.6, tungsten 3,564 tons (1976), graphite 37,400 tons (1971), gold 515 kg (1976), lead, manganese, molybdenum 162 tons (1976), silver, tin, zinc 69, salt. Electricity 24,426 mn. kWh (1976). **Production** (1976, in 1,000 tons): cement 11,873, shipbuilding 689,000 GRT, radio 6,578,000 and television receivers 2,291,000 units, sulphuric acid 626, hydrochloric acid 43.8, nitrogenous 510.3 and phosphate fertilizers 214.8, woven fabrics – cotton 340 mn. sq.m – silk 20.3 mn. sq.m – rayon and acetate 46.5 mn. sq.m – newsprint 199.
Communications: railways 3,144 km (1975), roads 45,514 km (1976), passenger cars 96,100 (1976). Merchant shipping 2,495,000 GRT (1977). Civil aviation: 38.6 mn. km flown, 2,075,000 passengers carried (1976). – **Exports:** textile, machinery and chemical products, fish and fish products, wood, skins, ores. Chief trading partners: Japan, U.S.A., Fed. Rep. of Germany, Hong Kong.

KUWAIT

Dowlat al Kuwayt, area 17,818 sq.km, population 1,199,000 (1978), **emirate** (Emir Shaikh Jabir al-Ahmad al-Jabir al-Sabah since 1978).

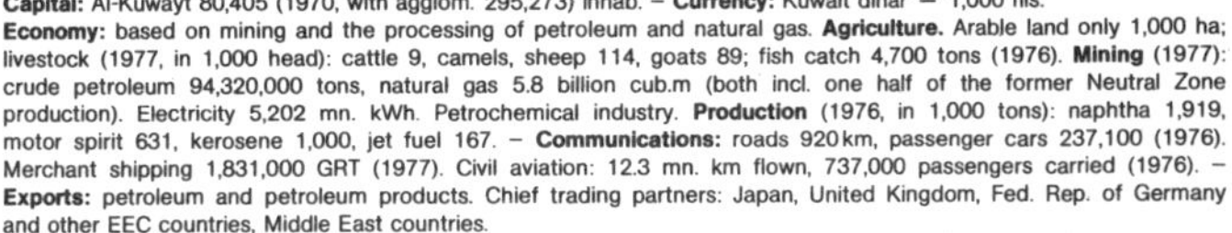

Capital: Al-Kuwayt 80,405 (1970, with agglom. 295,273) inhab. – **Currency:** Kuwait dinar = 1,000 fils.
Economy: based on mining and the processing of petroleum and natural gas. **Agriculture.** Arable land only 1,000 ha; livestock (1977, in 1,000 head): cattle 9, camels, sheep 114, goats 89; fish catch 4,700 tons (1976). **Mining** (1977): crude petroleum 94,320,000 tons, natural gas 5.8 billion cub.m (both incl. one half of the former Neutral Zone production). Electricity 5,202 mn. kWh. Petrochemical industry. **Production** (1976, in 1,000 tons): naphtha 1,919, motor spirit 631, kerosene 1,000, jet fuel 167. – **Communications:** roads 920 km, passenger cars 237,100 (1976). Merchant shipping 1,831,000 GRT (1977). Civil aviation: 12.3 mn. km flown, 737,000 passengers carried (1976). – **Exports:** petroleum and petroleum products. Chief trading partners: Japan, United Kingdom, Fed. Rep. of Germany and other EEC countries, Middle East countries.

LAOS

Sa Thalanalath Pasathipatay Pasason Lao; area 236,800 sq.km, population 3,427,000 (1978), **people's democratic republic** (President Prince Souphanouvong since 1975).
Administrative units: 16 provinces (khueng). **Capital:** Viangchan (Vientiane) 176,637 inhab. (1973); **other towns** (1973, in 1,000 inhab.): Savannakhet 51, Pakxé 45, Louangphrabang 44. – **Currency:** kip = 10 bi = 100 at.
Economy: developing agricultural country. Arable land 4.1%, forests 63.3%. **Agriculture** (1977, in 1,000 tons): rice 700, maize (corn) 35, sweet potatoes 20, manioc 14, groundnuts, coffee, cotton, tobacco, rare timber, natural rubber; livestock (1977, in 1,000 head): cattle 518, buffaloes 1,234, pigs 1,512, goats; fish catch 20,000 tons (1976). **Mining:** tin 576 tons (1977). Electricity 250 mn. kWh (1976). Handicraft production. – **Communications:** roads 7,395 km (1974), passenger cars 14,100 (1974). Civil aviation: 800,000 km flown, 45,000 passengers carried (1976). – **Exports:** timber, tin, coffee.

LEBANON

al-Jumhouriya al-Lubnaniya, area 10,400 sq.km, population 3,056,000 (1978), **republic** (President Dr. Elias Sarkis since 1976).
Administrative units: 5 provinces (mohafazat). **Capital:** Bayrūt 474,870 inhab. (1970, with agglom. 1,010,000), **other important town:** Tarābulus 157,300 inhab. – **Currency:** Lebanese pound = 100 piastres.
Economy: arable land 33.5%. **Agriculture** of Mediterranean type (1977, in 1,000 tons): grapes 100, oranges 188, tangerines 18, lemons 81, olives 46, vegetables, cereals 38; livestock (1977, in 1,000 head): cattle 84, sheep 237, goats 330; fish catch 2,500 tons (1976). Electricity 1,250 mn. kWh (1976). Food **industry,** production of carpets and jewellery. Processing of imported petroleum. – **Communications:** railways 415 km (1969), roads 7,350 km (1970), passenger cars 220,200 (1974). Merchant shipping 227,000 GRT (1977). Civil aviation: 48.5 mn. km flown, 1,050,000 passengers carried. – **Exports:** fruit, vegetables, tobacco, skins, jewellery, textile products.

MALAYSIA

Persekutuan Tanah Malaysia – Federation of Malaysia, area 329,749 sq.km, population 12,826,000 (1977), **federation, member of the Commonwealth** (Sultan Haji Ahmad Shah since 1979).
Administrative units: West Malaysia: 11 states, East Malaysia: Sabah, Sarawak. **Capital:** Kuala Lumpur 451,977 inhab. (1970, with agglom. 750,000); **other towns** (1970, in 1,000 inhab.): Pinang (George Town) 270, Ipoh 248, Johore Bahru 136, Kelang 113; Sabah – Kota Kinabalu 41; Sarawak – Kuching 64. **Population:** mainly Malays and Chinese. **Density** 39 persons per sq.km; average annual rate of population increase: West Malaysia 2.5%, Sabah 3.1%, Sarawak 2.5% (1976); urban population 28% (1975). 50.1% of the economically active inhabitants engaged in agriculture. – **Currency:** Malaysian ringgit = 100 sen.
Economy: plantational agriculture and mining industry. Arable land: West Malaysia 4.7%, Sabah 1.1%, Sarawak 19.5%; forests: West Malaysia 52.1%, Sabah 82.1%, Sarawak 75.8%. **Agriculture** (1977, in 1,000 tons): rice 1,984, maize (corn), sweet potatoes, coconuts 994, copra 151, palm kernels 341.5 (second world producer), palm oil 1,643 (leading world producer), manioc 345, pineapples 284, pepper; livestock (1977, in 1,000 head): cattle 423, buffaloes 298, pigs 1,479, goats 373; fish catch 618,700 tons (1977). Natural rubber 1,595,000 tons (leading world producer), roundwood 36.9 mn. cub.m, rare timber.

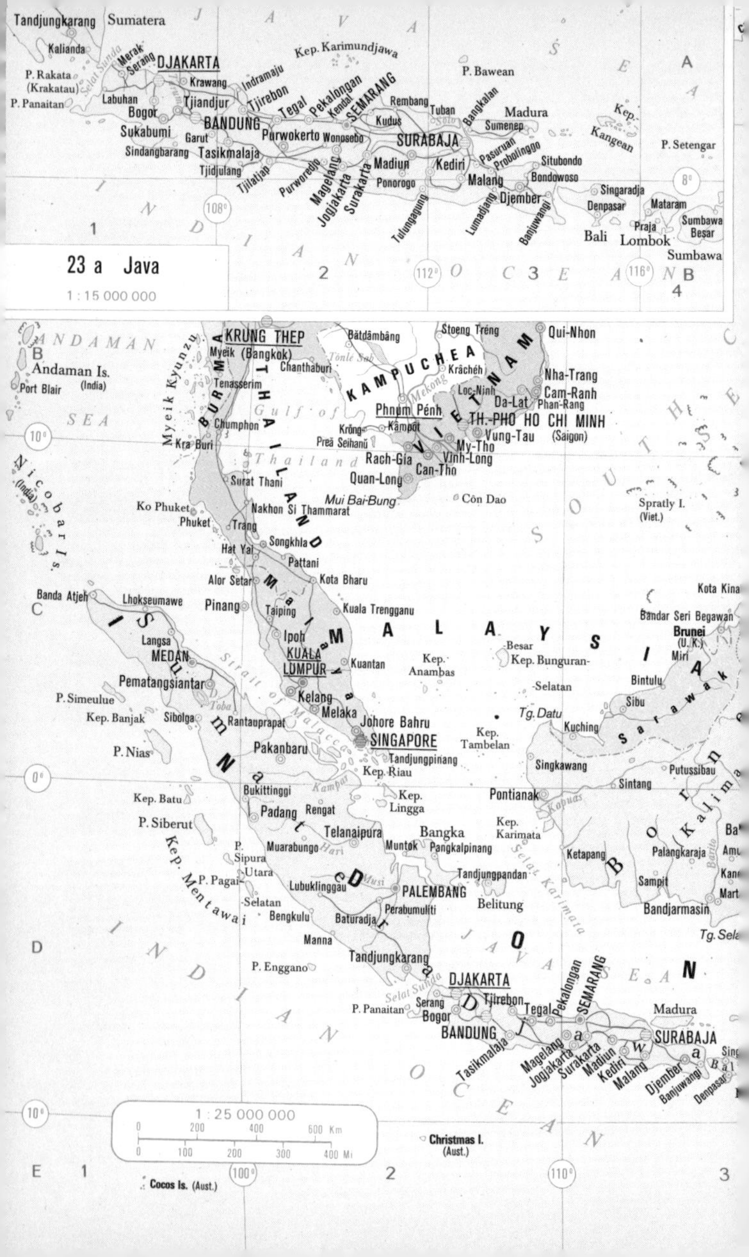
23 a Java
1 : 15 000 000
Tandjungkarang
Sumatera
Kalianda
P. Rakata (Krakatau)
P. Panaitan
Merak
Serang
DJAKARTA
Krawang
Labuhan
Bogor
Tjiandjur
Sukabumi
BANDUNG
Garut
Sindangbarang
Tasikmalaja
Tjidjulang
Tjilatjap
Indramaju
Tjirebon
Tegal
Pekalongan
Kendal
SEMARANG
Purwokerto
Wonosobo
Purworedjo
Magelang
Jogjakarta
Surakarta
Kep. Karimundjawa
Rembang
Kudus
Tuban
SURABAJA
Madiun
Ponorogo
Tulungagung
Kediri
Malang
Bangkalan
Madura
Sumenep
Pasuruan
Probolinggo
Lumadjang
Djember
Banjuwangi
Situbondo
Bondowoso
P. Bawean
Kep. Kangean
P. Setengar
Singaradja
Denpasar
Bali
Mataram
Praja
Lombok
Sumbawa Besar
Sumbawa
JAVA SEA
INDIAN OCEAN
108°
112°
116°
8°
ANDAMAN SEA
Andaman Is. (India)
Port Blair
Nicobar Is. (India)
Myeik Kyunzu
BURMA
Myeik
Tenasserim
Chumphon
Kra Buri
KRUNG THEP (Bangkok)
THAILAND
Chanthaburi
Bătdâmbâng
Tônlé Sab
KAMPUCHEA
Stoeng Trêng
Krâchéh
Mekong
Phnum Pénh
Kâmpôt
Krŏng
Preă Seihanŭ
Gulf of Thailand
VIETNAM
Qui-Nhon
Nha-Trang
Cam-Ranh
Phan-Rang
Da-Lat
Loc-Ninh
TH.-PHO HO CHI MINH (Saigon)
Vung-Tau
My-Tho
Vinh-Long
Can-Tho
Rach-Gia
Quan-Long
Mui Bai-Bung
Côn Dao
Spratly I. (Viet.)
SOUTH CHINA SEA
Surat Thani
Ko Phuket
Phuket
Nakhon Si Thammarat
Trang
Songkhla
Hat Yai
Pattani
Alor Setar
Kota Bharu
Pinang
Taiping
Kuala Trengganu
Ipoh
KUALA LUMPUR
Kuantan
Kelang
Melaka
Malaya
MALAYSIA
Johore Bahru
SINGAPORE
Strait of Malacca
Kep. Anambas
Kep. Bunguran-Besar
-Selatan
Tg. Datu
Kep. Tambelan
Kota Kina
Bandar Seri Begawan
Brunei (U.K.)
Miri
Bintulu
Sibu
Kuching
Sarawak
Banda Atjeh
Lhokseumawe
Langsa
MEDAN
Pematangsiantar
Toba
P. Simeulue
Kep. Banjak
Sibolga
Rantauprapat
P. Nias
Sumatera
Pakanbaru
Tandjungpinang
Kep. Riau
Kampar
Bukittinggi
Kep. Batu
Padang
Rengat
P. Siberut
Kep. Lingga
Telanaipura
Hari
Muarabungo
Muntok
Bangka
Pangkalpinang
P. Sipura
-Utara
P. Pagai-
-Selatan
Kep. Mentawai
Lubuklinggau
Musi
PALEMBANG
Bengkulu
Perabumulih
Baturadja
Manna
Tandjungpandan
Belitung
Tandjungkarang
P. Enggano
Singkawang
Pontianak
Kapuas
Kep. Karimata
Selat Karimata
Ketapang
Putussibau
Sintang
Borneo
Kalimantan
Palangkaraja
Sampit
Bandjarmasin
Tg. Sela
INDONESIA
Selat Sunda
P. Panaitan
Serang
Bogor
DJAKARTA
Tjirebon
Tegal
Pekalongan
SEMARANG
BANDUNG
Tasikmalaja
Magelang
Jogjakarta
Surakarta
Madiun
Kediri
Malang
Java
Madura
SURABAJA
Djember
Banjuwangi
Denpasar
Bali
JAVA SEA
INDIAN OCEAN
Christmas I. (Aust.)
Cocos Is. (Aust.)
1 : 25 000 000
0 200 400 600 Km
0 100 200 300 400 Mi
10°
0°
10°
100°
110°

23 b Philippines
1 : 15 000 000
P'ingtung
Taiwan
Bashi Channel
Batan Is.
Balintang Channel
Babuyan Is.
Escarpada Pt.
Laoag
Tuguegarao
Ilagan
Baguio
Dagupan
Luzon
Tarlac
Cabanatuan
QUEZON CITY
MANILA
San Pablo
Batangas
Mindoro
Catanduanes I.
Naga
Legazpi
Tablas I.
Masbate I.
Roxas
Calbayog
Samar
Ormoc
Panay
Iloilo
Cebu
Leyte
Bacolod
Dinagat I.
Negros
Bohol
Dumaguete
Surigao
Siargao I.
Cagayan de Oro
Butuan
Mindanao
Iligan
Pagadian
Cotabato
DAVAO
Zamboanga
Isabela
Basilan I.
Gen. Santos
Jolo
Jolo I.
Tinaca Pt.
Sulu Arch.
SULU SEA
PHILIPPINES
PACIFIC
Claveria
Babuyan Is.
Aparri
Vigan
Palanan
Bontoc
San Fernando
Casiguran
Lingayen
San Jose
Angeles
Olongapo
Manila B.
Lipa
Santa Cruz
Lucena
Daet
Lubang Is.
Calapan
Boac
Virac
Mamburao
Marinduque I.
Romblon
Calamian Group
Mindoro Strait
Masbate
Catbalogan
Borongan
Culion
Kalibo
Tacloban
El Nido
Agutaya
Cuyo Is.
San Jose
San Carlos
Baybay
Maasin
Cauayan
Argao
Puerto Princesa
Palawan
Tagbilaran
MINDANAO SEA
Dipolog
Oroquieta
Ozamiz
Tandag
Bislig
Agusan
SOUTH CHINA SEA
PHILIPPINE SEA
Kep. Talaud
Tahuna
Kep. Sangihe
CELEBES SEA
Tobi (U. S. A.)
OCEAN
Morotai
Galela
Tg. Torawitan
Manado
Tg. Mangkalihat
Paleleh
Gorontalo
Djailolo
Ternate
P. Ternate
Halmahera
Weda
P. Waigeo
Kep. Mapia
Kep. Schouten
Biak
Equator
Tomini
Teluk Tomini
MOLUCCA SEA
P. Batjan
Wakre
Manokwari
P. Japen
Tg. Perkam
Sarmi
Palu
Luwuk
P. Peleng
P. Obi
Salawati
Sorong
Mamberamo
Poso
P. Taliabu
Kep. Banggai
Kep. Sula
P. Mangole
P. Sanana
P. Misool
Steenkool
Teluk Sarera
Djajapura
Malili
Palopo
CERAM SEA
Wahai
Seram
Fakfak
Nabire
Kendari
Buru
Kajeli
Amahai
Ambon
P. Ambon
Kaimana
Irian (New Guinea)
Kokonau
P. Butung
Watampone
P. Kabaena
Baubau
BANDA SEA
Kep. Kai
Tual
Kep. Aru
Dobo
P. Wokam
P. Kobroor
P. Trangan
Tanahmerah
P. Selajar
PAPUA–NEW GUINEA
Maya
Digul
Kep. Tanimbar
P. Larat
P. Mdena
Saumlaki
P. Selaru
P. Lomblen
Kep. Alor
P. Wetar
P. Babar
P. Dolak
Flores
Reo
Ende
Baucau
P. Moa
Tg. Vals
Merauke
SAVU SEA
Ocussi
Dili
Timor
Waingapu
ARAFURA SEA
Kupang
Sumba
P. Sawu
P. Roti
TIMOR SEA
Melville I.
Wessel Is.
Darwin
AUSTRALIA
Nhulunbuy
Sulawesi (Celebes)
Maluku
Indonesia

Mining (1977, in 1,000 tons, metal content): crude petroleum 8,791, natural gas 0.09 billion cub.m, iron ore 185, bauxite 616, copper, gold 140 kg, manganese, tin 58,703 tons (leading world producer), antimony 250 tons, tungsten 125 tons. Electricity 6,697 mn. kWh. **Industry:** processing of tin, petroleum and agricultural products. **Production** (1977, in 1,000 tons): tin 66,305 tons (leading world producer), naphtha 192, motor spirit 60, kerosene 51, woven cotton fabrics 226 mn. sq.m, sugar 60, meat 145.

Communications: railways 2,317 km (1972), roads 25,911 km (1975), motor vehicles – passenger 572,100, commercial 196,200. Merchant shipping 552,000 GRT (1978). Civil aviation: 29.9 mn. km flown, 2,865,000 passengers carried (1977). Tourism 1,183,014 visitors. – **Exports:** natural rubber, tin, petroleum and petroleum products, wood, palm oil. Chief trading partners: Japan, U.S.A., United Kingdom, Singapore, Thailand, Fed. Rep. of Germany, U.S.S.R.

MALDIVES

Malaya Vara, area 298 sq.km, population 143,046 (1978), **republic** (President Maumoon Abdul Gayoom since 1978).

Capital: Male 15,740 inhab. (1974). – **Currency:** Maldive rupee = 100 laria. – **Economy:** coconuts 9,000 tons, copra 2,000 tons, sweet potatoes 159,000 tons, fruit; fish catch 32,300 tons (1976). – **Exports:** coconuts, copra, fish

MONGOLIA

Bügd Nayramdakh Mongol Ard Uls, area 1,565,000 sq.km, population 1,531,000 (1977), **people's democratic republic** (Chairman of the Presidium of the People's Great Khural Yumjagiin Tsedenbal since 1974).

Administrative units: 18 provinces (aimag) and 2 self-administered cities. **Capital:** Ulaanbaatar 403,000 inhab. (1979); **other towns** (1978, in 1,000 inhab.): Darchan 55, Erdenet 50. **Density** 1 person per sq.km. – **Currency:** tögrög = 100 möngö.

Economy: agricultural and industrial country with nomadic cattle raising and processing of agricultural products. Arable land only 0.5%, meadows and pastures 89.4%. **Agriculture** (1977, in 1,000 tons): wheat 300, barley 5[illegible]; livestock (1977, in 1,000 head): horses 2,205, cattle 2,417, camels 620, sheep 13,506, goats 4,548; cowhides 9,925 tons, sheepskins 12,300 tons (1977). Electricity 990 mn. kWh (1977). Food **industry** (1977, in 1,000 tons): milk 148, meat 242, butter 5.18. – **Communications:** railways 1,421 km (1975), roads 8,640 km (1970). Airport Ulaanbaatar. – **Exports:** hides and skins, wool, meat, butter. Chief trading partner: U.S.S.R. (80%).

NEPAL

Sri Nepala Sarkar, area 140,797 sq.km, population 13,421,000 (1978), **kingdom** (King Mahárájádhirája Birendra Bir Bikram Sháh Dev since 1972).

Administrative units: 14 zones. **Capital:** Kātmāndu 150,402 inhab. (1971, with agglom. 332,982). – **Currency:** Nepalese rupee = 100 paisa.

Economy: agriculture, home crafts and industrial processing of agricultural products in new plants. **Agriculture** (1977, in 1,000 tons): rice 2,285, wheat, maize (corn), millet 140, jute 56; livestock (1977, in 1,000 head); cattle 6,70[illegible], buffaloes 4,000, pigs 330, sheep 2,330, goats 2,410. Electricity 180 mn. kWh (1977). – **Communications:** railways 105 km (1970), roads 5,051 km (1973). – **Exports:** rice, jute, wood.

OMAN

Saltanat Oman, area 212,457 sq.km, population 817,000 (1977), **sultanate** (Sultan Qaboos bin Said since 1970).

Capital: Masqat (Muscat) 80,000 inhab. (1978); **other town** (1975): Matrah 20,000 inhab. – **Currency:** Omani rial = 1,000 baiza.

Economy. Agriculture (1977, in 1,000 tons): dates 50, cereals 6, fruit, lemons 13, tobacco; livestock (1977, in 1,000 head): cattle 6, sheep 77, goats 197. Fish catch 198,000 tons (1977). **Mining** (1977): crude petroleum 17,060,000 tons, natural gas. – **Communications:** roads 1,267 km (1978). – **Exports:** petroleum, dates, citrus fruit, tobacco, fish. Chief trading partners: United Kingdom, Australia.

PAKISTAN

Pakistani Islami Jumhouryat, area 803,942 sq.km, population 76,770,000 (1978) – excluding the territory of "Free Kashmir" (Azad Kashmir, 79,900 sq.km), **republic** (President Gen. Muhammad Zia ul Haq since 1978).

Administrative units: 6 provinces. **Capital:** Islāmābād 77,318 inhab. (census 1972); **other towns** (1972, in 1,000 inhab.): Karāchi 3,499, Lahore 2,165, Faīsalābād 822, Hyderābād 628, Rāwalpindi 615, Multān 542, Gujrānwāla 360, Peshāwar 268. **Population:** Punjabi, Sindhi, Urdu etc. **Density** 95 persons per sq.km. Average annual rate of population increase 3.1% (1970–75); urban population 27% (1975). 55% of the economically active inhabitants engaged in agriculture. – **Currency:** Pakistani rupee = 100 paisa.

Economy: agricultural and industrial country. Arable land 24% of the land area. **Agriculture. Crops** (1977, in 1,000 tons): wheat 9,155, maize (corn) 864, barley 127, millet 305, sorghum 400, chick-peas 649, groundnuts 64, sesame

cotton – seed 1,084, lint 532 – dates 180; **livestock** (1977, in 1,000 head): horses 440, cattle 14,361, buffaloes 10,593, sheep 19,749, goats 22,722, asses 1,926; grease wool 32,192 tons (1977), cowhides 103,982 tons. Fish catch 248,500 tons (1977). **Mining** (1977, in 1,000 tons, metal content): coal 1,112, brown coal, crude petroleum 405, natural gas 4.5 billion cub.m., chromium 3.9, antimony 93 tons, magnesite, manganese, salt. Electricity 11 billion kWh 1977). Food, textile and leather **industries. Production** (1977, in 1,000 tons): cow milk 2,036, buffalo milk 7,549, butter 245, meat 217; 28,379 mn. cigarettes, woven cotton fabrics 436 mn. m.
Communications: railways 8,811 km (1976), roads 49,926 km (1976), motor vehicles – passenger 196,100, commercial 91,700 (1975). Merchant shipping 442,000 GRT (1978). Civil aviation: 27.5 mn. km flown, 1,049,000 passengers carried (1977). Tourism 220,448 visitors (1977). – **Exports:** cotton products 25%, rice 22%, woollen and jute products, cotton. Chief trading partners: U.S.A., Fed. Rep. of Germany, Japan, United Kingdom.

PHILIPPINES

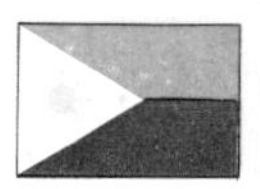

Republika ñg Pilipinas – Republic of the Philippines – República de Filipinas, area 297,413 sq.km, population 46,351,000 (1978), **republic** (President Ferdinando Edralin Marcos since 1969).

Administrative units: 67 provinces. **Capital:** Manila 1,454,352 inhab. (1975, with agglom. 5,375,000); **other towns** 1975, in 1,000 inhab.): Quezon City 995, Davao 516, Cebu 418, Caloocan 393, Iloilo 248, Pasay 241, Zamboanga 240. – **opulation:** Filipino; chief tribes: Visay, Tagal, Igorot. **Density** 156 persons per sq.km; average annual rate of population ncrease 3.3% (1970–75); urban population 32% (1975). 48% of the economically active inhabitants engaged in agriculture. – **Currency:** Philippine peso = 100 centavos.
conomy: plantational agriculture with developing industry. Arable land 17.1%, forests 41% of the land area. **Agriculture. rops:** (1977, in 1,000 tons): rice 7,150, maize (corn) 3,037, manioc 679, sweet potatoes 750, sugarcane 23,212, cocouts 11,587 (leading world producer), copra 2,400 (leading world producer), palm kernels 2.2, palm oil 12.3, abaca 46.8, bananas 1,210, pineapples 460, coffee 80, tobacco 60, vegetables; **livestock** (1977, in 1,000 head): horses 325, attle 2,373, pigs 9,700, goats 1,400; fish catch 1,510,800 tons (1977). Roundwood 32.9 mn. cub.m; natural rubber 8,200 tons (1977). **Mining** (1977, in 1,000 tons, metal content): coal 158, brown coal, uranium, iron ore, manganese 0.9, chromium 162.1, copper 267.1, mercury, nickel 30.7, silver, zinc, gold 17,363 kg, salt 213, phosphates 27. lectricity 15,800 mn. kWh, of which 4,850 mn. kWh from hydro-electric power plants (1977). **Production** (1977, in ,000 tons): cement 4,112, radio 124,000 and television receivers 107,000 units (1976), motor vehicles – passenger 4,300, commercial 24,400 units – woven cotton fabrics 198 mn. m, meat 665, sugar 2,624; 50,950 mn. cigarettes. **ommunications:** railways 1,034 km (1973), roads 99,132 km (1974), motor vehicles – passenger 386,200, commercial 31,000 (1976). Merchant shipping 1,265,000 GRT (1978). Civil aviation: 20.9 mn. km flown, 517,000 passengers arried (1977). Tourism 730,123 visitors (1977). – **Exports:** sugar, coconuts, copra, tung oil, abaca, copper concentrates, neapples, bananas, wood. Chief trading partners: Japan, U.S.A., Fed. Rep. of Germany, United Kingdom.

QATAR

nárat al-Katár, area 11,000 sq.km, population 157,000 (1977), **monarchy** (Amir Shaikh Khalifa n Hamad Al-Thani since 1972).
apital: Ad-Dawhah (Doha) 120,000 inhab. (1975). – **Currency:** Qatar riyal = 100 dirhams.
conomy: agriculture in oases (1977): dates,citrus fruit, extensive cattle raising (cattle, camels 9,000, sheep 41,000 ead), fish catch 2,700 tons, pearl fishery. **Mining** (1977, in 1,000 tons): crude petroleum 21,414, natural gas 6 billion cub.m. Electricity 900 mn. kWh (1977). **Production** (1977, in 1,000 tons): motor spirit 90, kerosene 5. – **ommunications:** roads 1,700 km, cars 31,700 (1977). Port Ad-Dawhah. – **Exports:** petroleum, dates. Chief trading artners: United Kingdom, Fed. Rep. of Germany, Japan, U.S.A.

SAUDI ARABIA

Mamlaka al-'Arabiya as-Sa'udiya, area 2,149,690 sq.km, population 9,522,000 (1977), **ngdom** (King Khalid ibn Abdul-Aziz since 1975).

dministrative units: 5 regions. **Capital:** Ar-Riyād 666,840 inhab. (1974); **other towns** (1974): Juddah 561,104, Makkah ecca) 366,801, Al-Qaṭīf 204,857, Al-Madīnah (Medina) 198,186. – **Population:** mainly Arabs, minorities: Iranians, das and others. Average annual rate of population increase 2.9% (1970–75). **Density** 4 persons per sq.km. % of the economically active inhabitants engaged in agriculture (1977). – **Currency:** Saudi rial = 100 halala.
onomy: the country with the largest resources of petroleum in the world (15,455 mn. tons in 1976) and large-scale traction. Arable land 0.5% and pastures 39.5% of the land area. Cultivation of cereals and fruit in the oases, madic raising of cattle on pastures. **Agriculture** (1977, in 1,000 tons): dates 265, wheat 135, barley 15, maize orn), millet 10, oranges 20; livestock (1977, in 1,000 head): cattle 170, camels 620, sheep 1,410, goats 775. **ning** (1977, in 1,000 tons): crude petroleum 453,160 (along the shore of the Persian Gulf and offshore wells – rd world producer), natural gas 5.9 billion cub.m. Saudi Arabia participates with Kuwait in petroleum and natural gas traction in the former Neutral Zone. Iron ore, pyrites, gold, phosphates. Electricity 2,250 mn. kWh (1976). **Production** 976, in 1,000 tons): naphtha 5,068, motor spirit 1,480, kerosene 1,098, jet fuel 559.
mmunications: railways 1,251 km, roads 16,231 km (1975), motor vehicles – passenger 59,000, commercial 52,600 974). Merchant shipping 1,019,000 GRT. Civil aviation: 38.9 mn. km flown, 3,268,000 passengers carried. – **ports:** petroleum and petroleum products (one of the largest petroleum ports in the world is Ra's At-Tannūrah), tes, oranges. Chief trading partners: Japan, Italy, United Kingdom, U.S.A.

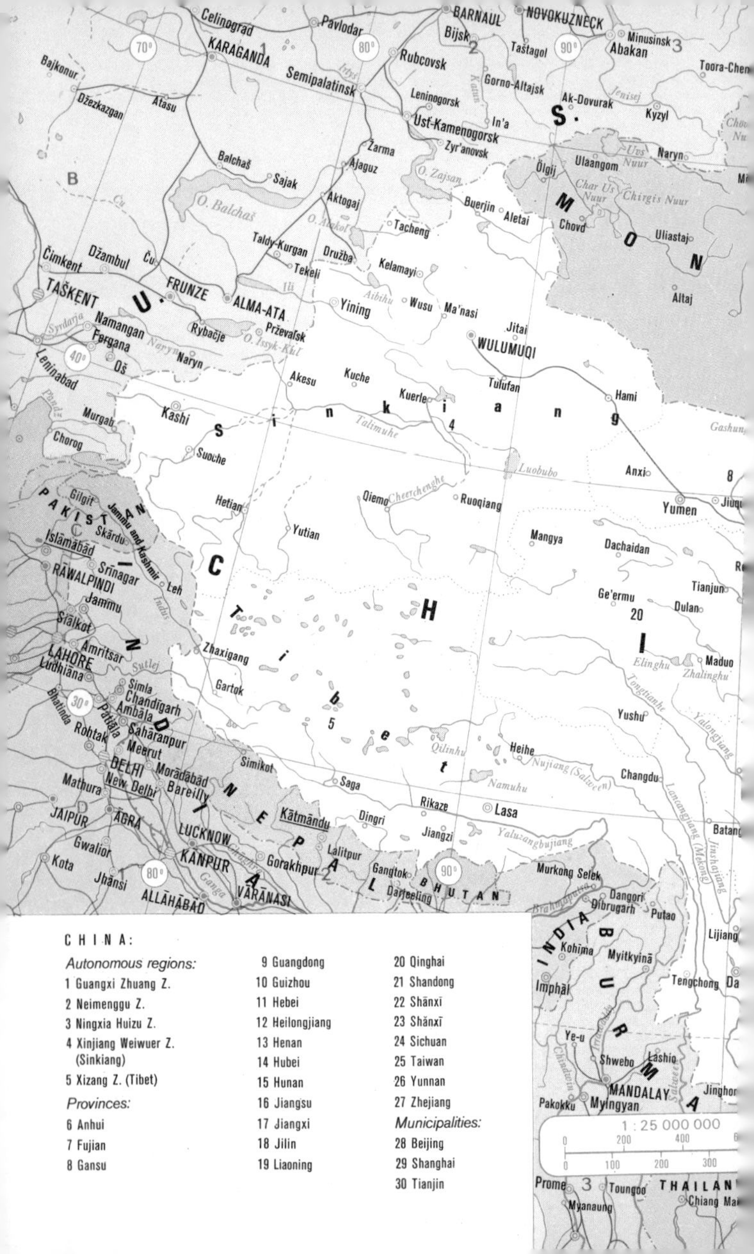

CHINA:
Autonomous regions:
1 Guangxi Zhuang Z.
2 Neimenggu Z.
3 Ningxia Huizu Z.
4 Xinjiang Weiwuer Z. (Sinkiang)
5 Xizang Z. (Tibet)
Provinces:
6 Anhui
7 Fujian
8 Gansu
9 Guangdong
10 Guizhou
11 Hebei
12 Heilongjiang
13 Henan
14 Hubei
15 Hunan
16 Jiangsu
17 Jiangxi
18 Jilin
19 Liaoning
20 Qinghai
21 Shandong
22 Shānxī
23 Shǎnxī
24 Sichuan
25 Taiwan
26 Yunnan
27 Zhejiang
Municipalities:
28 Beijing
29 Shanghai
30 Tianjin
1 : 25 000 000
Celinograd
Pavlodar
BARNAUL
NOVOKUZNECK
Bijsk
Minusinsk
Abakan
KARAGANDA
Rubcovsk
Taštagol
Bajkonur
Semipalatinsk
Gorno-Altajsk
Toora-Chem
Džezkazgan
Atasu
Leninogorsk
Ak-Dovurak
Kyzyl
Ust-Kamenogorsk
In'a
Žarma
Zyr'anovsk
Ulaangom
Naryn
Balchaš
Ajaguz
O. Zajsan
Ölgij
Sajak
O. Balchaš
Aktogaj
Buerjin
Aletai
Char Us Nuur
Chirgis Nuur
Uvs Nuur
Tacheng
Chovd
Uliastaj
Čimkent
Džambul
Taldy-Kurgan
Družba
Kelamayi
Tekeli
TAŠKENT
FRUNZE
ALMA-ATA
Yining
Wusu
Ma'nasi
Altaj
Namangan
Rybačje
Prževaľsk
O. Issyk-Kul
Jitai
Fergana
WULUMUQI
Naryn
Leninabad
Akesu
Kuche
Tulufan
Kuerle
Hami
Murgab
Kashi
Talimuhe
Sinkiang
Chorog
Suoche
Gashun
Luobubo
Anxi
Gilgit
PAKISTAN
Jammu and Kashmir
Hetian
Qiemo
Cheerchenghe
Ruoqiang
Yumen
Jiuq
Skārdu
Islāmābād
Yutian
Mangya
Dachaidan
RĀWALPINDI
Srīnagar
Leh
Jammu
Indus
Tianjun
Ge'ermu
Dulan
Siālkot
CHINA
Tibet
Amritsar
LAHORE
Ludhiāna
Sutlej
Zhaxigang
Elinghu
Maduo
Zhalinghu
Simla
Chandīgarh
Gartok
Tongtianhe
Bhatinda
Patiāla
Ambāla
Yushu
Yalongjiang
Rohtak
Sahāranpur
Meerut
Qilinhu
Heihe
Nujiang (Salween)
DELHI
New Delhi
Simikot
Morādābād
Bareilly
Changdu
Mathura
Saga
Namuhu
NEPAL
INDIA
Rikaze
Lasa
JAIPUR
ĀGRA
Kātmāndu
Dingri
Lancangjiang (Mekong)
Batang
LUCKNOW
Jiangzi
Yaluzangbujiang
Gwalior
Lalitpur
Jinshajiang
Kota
KĀNPUR
Gorakhpur
Gangtok
Murkong Selek
Jhānsi
Ganga
Darjeeling
BHUTAN
Dangori
ALLĀHĀBĀD
VĀRĀNASI
Brahmaputra
Dibrugarh
Putao
Lijiang
Kohīma
Myitkyinā
BURMA
Imphāl
Tengchong
Ye-u
Chindwin
Irrawaddy
Shwebo
Lashio
Salween
MANDALAY
Jinghon
Pakokku
Myingyan
Prome
Toungoo
THAILAND
Chiang Mai
Myanaung

S.
5
110°
Mogoča
Džalinda
120°
Gulian
6
Amur
Šimanovsk
130°
7
Urgal
Komsomol'sk-
-na-Amure
50°
A
Romanovka
Sretensk
Mangui
Huma
Svobodnyj
Belogorsk
R.
Ulan-Ude
Čita
Šilka
Jinhe
Blagoveščensk
Novoburejskij
Obluč'e
Birobidžan
Argun
Aihun
CHABAROVSK
Petrovsk-
-Zabajkal'skij
Chilok
Aginskoje
Olov'annaja
Borz'a
Nunjiang
Jiayin
Nižneleninskoje
Altanbulag
Onon
Hailaerhe
Xiguituqi
Beian
Yichun
Hegang
Jiamusi
Bikin
Darchan
Čarüngol
Bajan-Uul
Manzhouli
Hailaer
Butehaqi
QIQIHAER
12
Hailun
Shuangyashan
Dal'nerečensk
Hulunchi
Hurao
Dal'negorsk
Ulaanbaatar
Čojbalsan
Kerulen
Anda
Songhuajiang
B
Öndörchaan
A
Tamsagbulag
Aershan
HAERBIN
Jixi
Chanka
Arsenjev
Dzuunmod
Džargalant
Tailai
Ussurijsk
L
I
Baruun Urt
Wulanhaote
Baicheng
VLADIVOSTOK
Sümber
Taoan
18
JILIN
Mudanjiang
Nachodka
Mandalgov'
CHANGCHUN
Dunhua
Songhuahu
Sajnšand
Xilinhaote
Tongliao
Siping
Yanji
Ch'ŏngjin
Linxi
Shuangliao
19
Hailong
Erlian
DEM. PEOP. REP. OF KOREA
Chifeng
SHENYANG
(Mokden)
FUSHUN
BENXI
Kimch'aek
40°
Duolun
JINZHOU
Hamhŭng
Hŭngnam
Baiyunebo
2
ANSHAN
Inner Mongolia
Wuyuan
HUHEHAOTE
Jining
ZHANGJIAKOU
Chengde
Andong
Sinŭiju
Wŏnsan
28
TANGSHAN
Linhe
BAOTOU
BEIJING
30
P'YŎNGYANG
Ch'unch'ŏn
Datong
(Peking) Baoding
LÜDA
Haeju
Wuhai
Dongsheng
TIANJIN
Po Hai
SŎUL
3
Shizuishan
Ningwu
INCH'ON
REP. OF
Yantai
Weihai
TAIYUAN
SHIJIAZHUANG
Cangzhou
TAEGU
Huanghe
Yulin
11
YELLOW
TAEJŎN
Ulsan
Wuzhong
Suide
Fenyang
Yuci
Zibo
Weifang
KOREA
PUSAN
C
22
Handan
JINAN
KWANGJU
Masan
Shimonoseki
Yan'an
21
Jiaoxian
QINGDAO
Jingyuan
Changzhi
Anyang
BOSHAN
SEA
Mokp'o
KITAKYŪSHŪ
Qingyang
Linfen
Jining
Korea Strait
Tsushima-kaikyō
Hancheng
Xinxiang
Linyi
Lianyungang
FUKUOKA
Pingliang
A
Cheju
Nagasaki
Yuncheng
Kaifeng
Cheju-do
Kumamoto
Tongchuan
LUOYANG
ZHENGZHOU
XUZHOU
Weihe
Yancheng
Baoji
13
Huaiyin
Kagoshima
XI'AN
16
JAPAN
Bangbu
Nanyang
Luohe
6
NANJING
Nantong
Hanzhong
23
Huaihe
WUXI
SHANGHAI
Guanghua
HUAINAN
Wuhu
EAST
30°
Guangyuan
Ankang
Xinyang
HEFEI
29
Xiangfan
SUZHOU
Satsunan-
-shotō
Wanxian
Fengjie
14
Shaoxing
Daxian
Yichang
WUHAN
Anqing
HANGZHOU
Ningbo
CHINA
Changjiang
Nanchong
Shashi
Jinhua
Huangshi
Jiujiang
Jingdezhen
Linhai
Hechuan
Dongtinghu
Yueyang
27
SEA
Poyanghu
CHONGQING
Changde
Wenzhou
Shangrao
Nansei-shotō (Ryukyu Is.)
(Jap.)
CHANGSHA
17
Luzhou
Yuanjiang
NANCHANG
Naha
Yiyang
Fuzhou
10
Jianou
Okinawa-
-jima
Xiangtan
Zhuzhou
Zunyi
Zhijiang
Ji'an
Bijie
15
Nanping
D
Shaoyang
Hengyang
FUZHOU
GUIYANG
Sanming
T'AIPEI
Duyun
Lingling
Quanzhou
Sakishima-
-shotō
Hsinchu
Anshun
Ganzhou
7
Guilin
Chezhou
Tropic of Cancer
Dushan
Meixian
Xiamen
T'AICHUNG
Hongshuihe
Shaoguan
Yishan
Chaoan
Chiai
Taiwan
9
Liuzhou
Baise
Shantou
25
1
GUANGZHOU
(Canton)
T'AINAN
Wuzhou
KAOHSIUNG
Xijiang
NANNING
Foshan
NEW KOWLOON
Bashi Ch.
Yujiang
Yulin
Macau
(Port.)
VICTORIA
Hong Kong
(U.K.)
Dongshaqundao
(China)
20°
Pingxiang
Batan Is.
HA-NOI
Beihai
Zhanjiang
Hon-Gai
Balintang Ch.
VIETNAM
Babuyan Is.
HAI-PHONG
HAIKOU
PHILIPPINES
Gulf of Tonkin
SOUTH CHINA
E
Nam-Dinh
110°
3
5
PACIFIC OCEAN
Thanh-Hoa
120°
Tuguegarao
6
Hainandao
SEA
Nagan
Vinh
Yaxian
Luzon

SINGAPORE

Republik Singapura – Republic of Singapore, area 597 sq.km, population 2,334,000 (1978), **republic, member of the Commonwealth** (President Devan Nair since 1981).

Capital: Singapore 2,308,200 inhab. – **Population:** about 76% Chinese, 14% Malays, Indians, Pakistanis. **Density** 3,910 persons per sq.km; average annual rate of population increase 1.6% (1970–75); urban population 90.1% (1975). 2.2% of the economically active inhabitants engaged in agriculture (1977). – **Currency:** Singapore dollar = = 100 cents.

Economy: important transit port, processing of Malayan tin (the largest tin smelting plant in Asia), shipyards, petrochemical industry, processing of natural rubber, textile industry. Arable land 13.8% of the land area. **Agriculture:** sweet potatoes, manioc, coconuts, natural rubber; raising of pigs 1,250,000 head (1977); fish catch 16,400 tons (1976). Electricity 5,100 mn. kWh (1977). **Production** (1977, in 1,000 tons): vessels 86,000 GRT (1976), naphtha 2,550, motor spirit 895, kerosene 1,990.

Communications: railways 25 km, roads 2,167 km (1975), motor vehicles – passenger 142,100, commercial 51,000 (1976). Merchant shipping 6,791,000 GRT (1977). Civil aviation: 40.8 mn. km flown, 2,132,000 passengers carried (1976). Tourism 1,506,688 visitors (1977). – **Exports:** petroleum products, machinery products, transit from Malaysia (natural rubber, tin, iron ore, copra). Chief trading partners: Malaysia, Japan, U.S.A., United Kingdom, Indonesia.

SRI LANKA

Srī Lanka Janarajaya, area 65,610 sq.km, population 14,082,000 (1978), **republic, member of the Commonwealth** (President Junius Richard Jayewardene since 1978).

Administrative units: 9 provinces. **Capital:** Colombo 607,000 inhab. (1976); **other towns** (1976, in 1,000 inhab.): Dehiwala-Mt.Lavinia 166, Jaffna 117. – **Population:** about 70% Sinhalese, 11% Ceylon and Indian Tamils, 6% Moors etc. **Density** 215 persons per sq.km; average annual rate of population increase 2.22%; urban population 24.2% (1975). 53.9% of the economically active inhabitants engaged in agriculture. – **Currency:** Srī Lanka rupee = 100 cents.

Economy: tropical fruit growing on plantations and mineral mining. Arable land 13.6%, land under permanent crops 16.5%, forests 44.2% of the land area. **Agriculture** (1977, in 1,000 tons): rice 1,706, tea 213 (third world producer), coconuts 1,771, copra 160, natural rubber 148, cocoa beans; livestock (1977, in 1,000 head): cattle 1,745, buffaloes 854, pigs 41, sheep 30; fish catch 135,900 tons (1976). **Mining:** graphite 9,447 tons (1974), gem stones, salt. Electricity 1,202 mn. kWh (1976). Chemical, textile and food **industries.**

Communications: railways 1,535 km (1975), roads 26,365 km (1974), motor vehicles – passenger 93,800, commercial 49,200 (1976). Merchant shipping 93,000 GRT (1977). Civil aviation: 4.4 mn. km flown, 128,000 passengers carried (1976). Tourism 118,971 visitors (1976). – **Exports:** tea (50% of exports), natural rubber 15%, coconuts, copra, tung oil, graphite, gem stones, cocoa beans. Chief trading partners: United Kingdom, U.S.A., China, India, Japan.

SYRIA

al-Jamhouriya al-'Arabiya as Souriya, area 185,180 sq.km, population 8,103,000 (1978), **republic** (President Gen. Hafez al-Assad since 1971).

Administrative units: 14 districts (mohafazets). **Capital:** Dimashq (Damascus) 1,097,205 inhab. (1977); **other towns** (1977): Halab (Aleppo) 842,600, Hims 292,280, Hamāh 173,459. – **Population:** Syrian Arabs, minorities: Kurds 5%, Armenians, Cherkese. **Density** 44 persons per sq.km; average annual rate of population increase 3% (1970–75); urban population 45.3%. 48.6% of the economically active inhabitants engaged in agriculture (1977). – **Currency:** Syrian pound = 100 piastres.

Economy: agricultural and industrial country. Arable land 27.7%, meadows and pastures 46.6% of the land area. **Agriculture** (1977, in 1,000 tons): wheat 1,217, barley 337, cotton – seed 230, lint 160 – legumens, vegetables, Mediterranean products – chiefly olives 238, oranges 15, lemons 15; livestock (1977, in 1,000 head): cattle 584, sheep 6,817, goats 985; fish catch 2,000 tons (1976). **Mining** (1977, in 1,000 tons): crude petroleum 10,041, natural gas 0.1 billion cub.m, natural asphalt, phosphates 511, salt. Electricity 1,785 mn. kWh (1976). **Industry:** petrochemical, textile and food industries. **Production** (1976, in 1,000 tons): woven cotton fabrics 367, motor spirit 404, kerosene 365.

Communications: railways 1,333 km (1972), roads 13,513 km (1972), motor vehicles – passenger 62,800, commercial 55,900. Civil aviation: 9.8 mn. km flown, 381,000 passengers carried (1976). – **Exports:** petroleum, cotton, fruit and vegetables, cereals, hides and skins, furs, phosphates. Chief trading partners: U.S.S.R., Italy, Lebanon, Japan.

THAILAND

Prathes Thai (Muang Thai), area 514,121 sq.km, population 45,100,000 (1978), **monarchy** (King Bhumibol Adulyadej since 1946).

Administrative units: 71 provinces (changwat). **Capital:** Krung Thep (Bangkok) 4,178,000 inhab. (1975); **other towns** (1975, in 1,000 inhab.): Thon Buri 627, Chiang Mai 99. – **Population:** Thais 85% (chiefly Siamese and Lao), Chinese, Malays etc. **Density** 89 persons per sq.km; average annual rate of population increase 3.27% (1970–75); urban population 16.4% (1975). 76.8% of the economically active inhabitants engaged in agriculture. – **Currency:** baht = = 100 satang.

Economy: agricultural country with few industries. Arable land 28.8%, forests 39.9% of the land area. Rice is cultivated on 65–70% of the arable land. **Agriculture** (1977, in 1,000 tons): rice 13,590, maize (corn) 1,677, sweet potatoes 400, manioc 10,644, sugarcane 23,638, groundnuts 175, sesame 25, jute 237, cotton lint 22, bananas 1,546, pineapples 500,

oranges 52, coconuts 650, tobacco 67; livestock (1977, in 1,000 head): horses 167, cattle 4,547, buffaloes 5,685, pigs 3,020; fish catch 1,640,400 tons (1976); roundwood 21.1 mn. cub.m; natural rubber 392,500 tons (1976). **Mining** (1976, in 1,000 tons, metal content): brown coal 680, crude petroleum, iron ore 14, tungsten 2,233 tons, tin 20,453 tons, zinc, antimony 3,800 tons, manganese 13 tons, phosphates, gem stones. Electricity 10,295 mn. kWh. **Industry:** ore processing, textile and food industries. **Production** (1977, in 1,000 tons): pig iron 12, crude steel 163, tin 20.3, woven cotton fabrics 557 mn. sq.m, sugar 1,757, meat 428; 22,642 mn. cigarettes.
Communications: railways 3,855 km (1976), roads 43,900 km (1977), motor vehicles – passenger 266,100, commercial 266,700 (1975). Merchant shipping 261,000 GRT (1976). Civil aviation: 36.5 mn. km flown, 1,634,000 passengers carried (1976). Tourism 1,220,672 visitors (1977). – **Exports:** rice 15%, natural rubber 10%, maize (corn), tin, jute, wood. Chief trading partners: Japan, U.S.A., Fed. Rep. of Germany, United Kingdom.

TURKEY

Türkiye Cumhuriyeti, area 779,452 sq.km, population 43,210,000 (1978), of which the Asiatic part 755,688 sq.km and population 38,296,000 (1977), **republic** (President Gen. Kenan Evren since 1980).

Administrative units: 69 provinces (iller). **Capital:** Ankara 1,698,542 inhab. (1975); **other towns** (1975, in 1,000 inhab.): İstanbul 2,535 (with agglom. 3,865), İzmir 636 (with agglom. 1,661), Adana 467 (with agglom. 1,235), Bursa 346, Gaziantep 301, Eskişehir 258, Konya 246, Kayseri 207, Diyarbakır 170, Erzurum 163. – **Population:** over 90% Turks, about 7% Kurds, Arabs etc. **Density** 55 persons per sq.km, in the Asiatic part 51 persons per sq.km; average annual rate of population increase 2.48% (1970–75); urban population 42.9% (1975). 58.6% of the economically active inhabitants engaged in agriculture (1977). – **Currency:** Turkish Lira = 100 kuruş.
Economy: agricultural and industrial country with great regional differences. Arable land 31.9%, meadows and pastures 25.3%, forests 25.8% of the land area. **Agriculture. Crops:** (1977, in 1,000 tons): wheat 16,715, barley 4,750, maize (corn) 1,200, rye 705, rice 258, potatoes 2,900, beans 160, lentils 180, peas 245, sugar beet 8,200, sunflower 457, groundnuts 52, sesame 21, chick-peas 168, cotton – seed 980, lint 570 – olives 500, figs 190 (1975), oranges 513, tangerines 125, lemons 275, grapes 3,450, apples 835, pears 235, almonds, hazelnuts 270 (leading world producer), walnuts 120, pistachios 15, chestnuts, tomatoes 2,800, dry onions 780. **Livestock** (1977, in 1,000 head): horses 853, cattle 14,102, buffaloes 1,058, sheep 41,504, goats 18,508, asses 1,465; olive oil 81,000 tons, raisins 338,000 tons (leading world producer), cowhides 41,232 tons, sheepskins 52,296 tons, grease wool 54,500 tons, mohair (all 1977); fish catch 155,300 tons (1976), roundwood 16.9 mn. cub.m.
Mining (1976, in 1,000 tons, metal content): coal 4,632, brown coal 8,252, crude petroleum 2,595, uranium, chromium 446, iron ore 1,895, manganese, lead 9.3, zinc, mercury 156 tons, antimony 2,223 tons, cobalt, copper 30.5, magnesite 405, sulphur, bauxite 461, asbestos, marble, salt 800. Electricity 18,231 mn. kWh (1976). **Industry:** food, tobacco and textile industries. **Production** (1976, in 1,000 tons): pig iron 1,991, crude steel 1,457, cement 13,154, sulphuric acid 278, nitrogenous fertilizers 212.1, phosphate fertilizers 356.9, motor spirit 1,950, kerosene 521, naphtha 364, motor vehicles – passenger 63,000 units, commercial 47,000 – radio receivers 179,000, television receivers 619,000, woven cotton fabrics 205 mn. m, woven woollen fabrics 7 mn. m, carpets, meat 824 (1977), sugar 1,090, milk 2,950, cheese 117.7, butter 123 (1977), margarine 225.2, flour 1,362 (1975), wine 634,000 hl; 17,918 mn. cigarettes.
Communications: railways 8,138 km (1975), roads 59,069 km (1975), motor vehicles – passenger 471,500 (1976), commercial 179,000 (1974). Merchant shipping 1,288,000 GRT (1977). Civil aviation: 21.9 mn. km flown, 2,427,000 passengers carried (1976). Tourism 531,207 visitors (1977). – **Exports:** fruit and raisins (over 10%), cotton and woven cotton fabrics, tobacco and tobacco products, chromium and copper ore, animal products – chiefly hides and skins, wool; olive oil, sugar. Chief trading partners: Fed. Rep. of Germany, United Kingdom, Italy, France.

UNITED ARAB EMIRATES

Al-Imārāt al-'Arabīya al Muttahida, area 83,600 sq.km, population 862,000 (1978), **federation of emirates** (President Shaik Zayed bin Sultan al Nahyan since 1971).
Administrative units: 7 emirates. **Capital:** Abu Zaby 236,662 inhab. (1975); **other towns** (1975): Dubayy 206,861, Ash-Shāriqah 88,188. – **Population:** Arabs with high proportion of males – 100 males to 41 females. – **Currency:** Dirham = 100 fils.
The economy is based on petroleum and natural gas extraction. Cultivation of vegetables. Arable land covers only 000 ha. Fish catch 68,000 tons (1976), pearl fishery. **Mining** (1976, in 1,000 tons): crude petroleum 95,265 (of which more than two thirds emirate Abu Zaby), natural gas 0.8 billion cub.m. Electricity 600 mn. kWh. **Production** (1976, in 1,000 tons): motor spirit 65, kerosene 3. – **Exports:** petroleum, dry fish, pearls. Chief trading partners: Japan, United Kingdom, U.S.A., France.

VIETNAM

Cộng Hòa Xã Hội Chu Nghĩa Việt Nam, area 332,560 sq.km, population 49,120,000 (1978), **republic** (President of the Council of State Truong Chinh since 1981).
Administrative units: 3 self-administrated cities, 35 provinces. **Capital:** Ha-Noi 1,443,500 inhab. (1976); **other towns** (1976, in 1,000 inhab.): Thanh-pho Ho Chi Minh (Saigon) 3,460, Hai-Phong 1,191, Da-Nang 492, Nha-Trang 216, Qui-Nhon 214. – **Population:** Vietnamese 84%, other nationalities Thai, Khmer, Meo etc. **Density** 148 persons per sq.km; average annual rate of population increase 2.11% (1970–75); urban population 16.9% (1975). 72.4% of the economically active inhabitants engaged in agriculture (1977). – **Currency:** dong = 10 liao = 100 xu.

MONGOLIA
Nei Menggu Zizhiqu
Heilongjiang
Jilin
Liaoning
Hebei
Shandong
Jiangsu
Anhui
Zhejiang
HAERBIN
CHANGCHUN
JILIN
SHENYANG (Mukden)
FUSHUN
ANSHAN
BENXI
JINZHOU
BEIJING (Peking)
TIANJIN (Tientsin)
TANGSHAN
LÜDA
JINAN
QINGDAO
BOSHAN
XUZHOU
NANJING
HUAINAN
HEFEI
SHANGHAI
WUXI
SUZHOU
HANGZHOU
ZHANGJIAKOU
DEM. PEOPLE'S REP. OF KOREA
P'YŎNGYANG
REP. OF KOREA
SOUL
INCH'ŎN
TAEJŎN
TAEGU
KWANGJU
Korea Bay
Po Hai
YELLOW SEA
EAST CHINA SEA
Liaodongwan
Laizhouwan
Cheju-do
Nansei-shotō (Ryukyu Is.)
Zhoushanqundao
1 : 12 500 000
0 100 200 300 Km
0 50 100 150 200 Mi

25 a Taiwan (Formosa) 1 : 12 500 000

Economy: agricultural country with developing industrial production. Arable land 16.7%, forests 34.3% of the land area. **Agriculture. Crops** (1977, in 1,000 tons): rice 11,250, maize (corn) 260, sweet potatoes 1,330, manioc 1,500, soybeans 36, groundnuts 97, sugarcane 1,300, jute 31, cotton, oranges 53, tea 10, coffee 8, tobacco 21; **livestock** (1977, in 1,000 head): cattle 1,550, buffaloes 2,300, pigs 9,100; fish catch 1,013,500 tons (1976); roundwood 18.8 mn. cub.m (1976), natural rubber 36,000 tons (1977). **Mining** (1976, in 1,000 tons): coal 5,400, iron ore, molybdenum, tin, gold, phosphates 1,500, salt. Electricity 1,320 mn. kWh (1976). **Industry:** textile and food industries, processing of phosphates. **Production:** phosphate fertilizers 110,000 tons (1977), woven cotton fabrics 75 mn. m (1973), cement.
Communications: railways 2,000 km, roads 36,000 km (1974), motor vehicles – passenger 70,000, commercial 100,000 (1973). Civil aviation: 2.1 mn. km flown. – **Exports:** natural rubber, tea, fish, wood, minerals. Chief trading partners: U.S.S.R. and CMEA countries, Japan, EEC and South-eastern Asia countries.

YEMEN (Y.A.R.)

al Jamhuriya al Arabiya al Yamaniya, area 195,000 sq.km, population 5,642,000 (1977), **republic** (President Ali Abdullah Saleh since 1978).
Administrative units: 7 regions (liva). **Capital:** San'ā' 447,898 inhab. (1975); **other towns** (1975): Ta'izz 320,323, Al-Hudaydah 147,982. – **Population:** Arabs, minorities of other Asian and African nationals. **Density** 29 persons per sq.km; urban population 8.8%. 76.3% of the economically active inhabitants engaged in agriculture. – **Currency:** Yemen rial = 40 bugshahs.
Economy: developing agricultural country. Arable land 8.1%, meadows and pastures 35.9%. **Agriculture** (1977, in 1,000 tons): wheat 50, barley 80, sorghum 800, maize (corn) 70, coffee 5, cotton, dates 70; livestock (1977, in 1,000 head): cattle 1,050, camels 121, sheep 3,300, goats 7,600. **Mining** of salt. – **Communications:** roads 3,952 km (1975), chief port Al-Hudaydah. **Exports:** cotton, hides and skins, fish.

YEMEN, SOUTH (D.Y.)

Jumhouriya al-Yemen al Dimuqratiah al Sha'abijah, area 287,683 sq.km, population 1,853,000 (1978), **people's democratic republic** (Chairman of the Presidential Council Ali Nasser Mohamed since 1978).
Administrative units: 6 governorates (muchafaz). **Capital:** Aden 271,590 inhab. (1977); **other town:** Al-Mukallā 65,000. – **Population:** Arabs, small number of Indians and Somalis. **Density** 6 persons per sq.km; average annual rate of population increase 2.9% (1970–75); urban population 34.5% (1975). 60.5% of the economically active inhabitants engaged in agriculture. – **Currency:** Yemen dinar (YD) = 1,000 fils.
Economy: nomadic raising of cattle, petrochemical industry, major transit port. Arable land only 0.6% of the land area. **Agriculture** (1977, in 1,000 tons): wheat 27, millet 65, dates; livestock (1977, in 1,000 head): cattle 104, sheep 940, goats 1,260, camels 40; fish catch 127,300 tons (1976). Electricity 180 mn. kWh (1976). **Production** (1976, in 1,000 tons): motor spirit 98, kerosene 100, jet fuel 150.– **Communications:** roads 1,377 km (1976), motor vehicles – passenger 11,900, commercial 10,500 (1976). Aden is an important naval and air base and a major transit port. – **Exports:** petroleum products.

British territory:

HONG KONG

Crown Colony of Hong Kong, area 1,046 sq.km, population 4,514,000 (1977), **British Crown Colony** (Governor Sir Murray MacLehose).
Capital: Victoria 501,680 inhab. (1977); **other towns** (1977, in 1,000 inhab.): New Kowloon 1,594, Kowloon 725. **Density** 4,315 persons per sq.km; average annual rate of population increase 1.39% (1970–75). – **Currency:** Hong Kong dollar = 100 cents.
Economy: important industrial and trading centre, naval and air base. **Agriculture:** arable land and permanent crops 10.6% of the land area. Crop of rice 4,000 tons in 1976. Raising of pigs 461,000 head (1977). Fish catch 157,900 tons (1976). Textile, clothing and leather **industries,** shipbuilding, electronics, chemical and printing industries. Electricity 8,342 mn. kWh. **Production** (1976): radio receivers (export) 50,491,000 units, woven cotton fabrics 822 mn. sq.m, woven silk fabrics.
Communications: railways 36 km (1972), roads 1,085 km, passenger cars 120,300 (1976). Merchant shipping 610,000 GRT. Tourism 1,755,669 visitors (1977). – **Exports:** woven cotton fabrics and products, radio receivers, plastic products, chemical products, canned food. Chief trading partners: Japan, U.S.A., China, Fed. Rep. of Germany, United Kingdom.

Portuguese territory:

MACAU

Provincia de Macau, area 16 sq.km, population 279,000 (1977), **Portuguese overseas territory.**
Capital: Macau 241,252 inhab. (1970). – **Currency:** pataca = 100 avos. – **Economy:** fish catch 10,100 tons (1976). Processing of fish, cotton, tea. Transit merchant port. Tourism 2,788,139 visitors. – **Exports:** fish products, textile goods.

AFRICA

Africa lies on the both sides of Equator, with the larger part in the northern hemisphere. The name Africa is derived from a Berber tribe, the Afrigi (or Afridi), who lived in the territory of today's Tunisia. The Latin name "Africa" was applied to a Roman province extending over the area previously under the control of Carthage.

Africa covers an **area** of **30,319,000 sq.km,** i.e. 20.28% of the land surface of the Earth and is the second largest continent. It has **436 million inhabitants** (1978), and a population density 14 persons per sq.km. **Geographical position:** northernmost point: Cape Rás Ben Sekka (Tunisia) 37°21′ N.Lat.; southernmost point: Cape Agulhas (South Africa) 34°52′ S.Lat.; westernmost point: Cape Pointe des Almadies 17°38′ W.Long. (4 km northwest of Cap Vert); easternmost point: Cape Ras Hafun 51°23′ E.Long. Africa is joined to Asia by the Isthmus of Suez (120 km long), and it is separated from Europe by the Strait of Gibraltar (14 km wide). The coast of Africa, 30,500 km in length, has little articulation. The largest peninsula is the Somali Pen. (area 850,000 sq.km). The principal islands are Madagascar (area 587,041 sq.km) and the small Mascarene Is. (4,555 sq.km) in the Indian Ocean; off the northwest coast in the Atlantic Ocean lie the Canary Is. (7,273 sq.km) and the Cape Verde Is. (4,033 sq.km).

Orographically, Africa is divided into 3 main regions: the Atlas Mts., the African Tableland and the East African Highlands. The Atlas Mts. stretch over 2,000 km in north-west Africa (highest peak: Jbel Toubkal 4,165 m) adjoined by the Plateau of the Shotts and its salt lakes. To the south of the Atlas Mts. lie the extensive Sahara-Sudanese plains and plateau (average height 200–500 m). The Sahara is the world's largest desert (7,750,000 sq.km); it is a rock (hamada), gravel (reg, serir) and sand desert with dunes (ergs), with the barren Mountains of Ahaggar (3,005 m), Tibesti (Emi Koussi, 3,415 m), Aïr (2,310 m) and Dārfūr (3,088 m) in its centre. South-west of the Sahara lie the Upper Guinean Highlands (1,948 m) and the Adamaoua Highlands (2,679 m). In Central Africa a vast tectonic depression formed the Congo Basin (3 million sq.km). Its centre is 300–500 m high, and the border ridges are between 500 and 1,000 m. The Lower Guinean Highlands (2,620 m) rise at its western border. South of the Luanda-Katanga Plateau lies the synclinal Kalahari Basin, a plain (average height 950 m) and to the west, along the coast, the Namib Desert, 1,500 km in length. South Africa comprises of the Karroo Plateau, the Cape Mts. (2,326 m) and the Drakensberg Mountains, which are South Africa's highest at 3,482 m. The Ethiopian Highlands (average altitude 2,000–3,000 m) are the eastern continuation of the Saharan-Arabian Tableland (highest point: Ras Dashen, 4,620 m). The Assal Depression in the Afar Pan by the Red Sea is the lowest point in Africa, −173 m below sea level. The East African Plateau has the most varied forms: tectonic rifts (e.g. the Great Rift Valley), mountain ridges (Ruwenzori, Ngaliema – 5,119 m), volcanoes, craters (Ngorongoro, 3,648 m) and plateau. Africa's highest mountain stands here; the volcanic Kilimanjaro with its three conical peaks, Uhuru, 5,895 m being the highest.

Africa's **rivers** were formed more recently. The network of rivers and drainage is highly irregular. The average volume of water flow per year is 4,657 cubic km. Almost one third of Africa lacks any form of drainage, especially the Sahara. More than one third of Africa drains into the Atlantic Ocean. Africa's major river is the Congo (Zaïre), 4,835 km in length with a river basin of 3,822,000 sq.km, the mean discharge is 41,400 cub.m per second. The Niger reaches the Gulf of Guinea through the Niger delta; its length is 4,160 km, the river basin occupies 2,092,000 sq.km. Africa's longest river is the Nile (length: 6,671 km, river basin 2,881,000 sq.km, mean discharge 1,600 cubic m per second). It forms a vast delta (25,000 sq.km) as it flows into the Mediterranean Sea. The greatest river in South Africa is the Zambezi. Most of the great African **lakes** are of tectonic origin, the largest being Lake Victoria (68,800 sq.km) and the deepest Lake Tanganyika (1,435 m, its bottom lying 662 m below sea level).

In view of its position Africa is the warmest continent. The climatic differences between regions are conditioned by pressure systems on the mainland and the adjacent ocean. Four **climatic zones** can be distinguished: the equatorial zone (Congo Basin and the coast of the Gulf of Guinea) has a hot wet climate all year; the zone of equatorial monsoons affects one third of Africa (to 15° N.Lat. and 18° S.Lat.) with hot wet summers and warm dry winters; the zone of tropical trade-winds (Sahara and Kalahari deserts) to the north and the south of the continent suffers extreme drought; the subtropical (Mediterranean) zone has hot dry summers and temperate rainy winters. The maximum absolute temperature is found at Al-Azīzīyah (Libya) 58 °C; the highest average annual temperature, 34.4 °C was recorded in Dalol (Ethiopia) and the lowest, −15 °C, in the Atlas Mts. Maximum rainfall in Africa was measured at C. Debunja (Cameroon): 10,470 mm, while Aswān (Egypt) is the dryest place (0.5 mm).

Mean January and July temperatures in °C (annual precipitation in mm): Alger 10.3 and 24.4 (746), Al-Qāhirah 13.8 and 28.2 (25), Al-Khurtūm 22.5 and 30.8 (168), Tombouctou 22.6 and 31.5 (230), Conakry 26.5 and 25.8 (4,349), Kumasi 25.2 and 24.2 (1,530), Douala 27.1 and 24.8 (4,439), Mesewa 25.5 and 34.5 (181), Mombasa 27.8 and 23.9 (1,197), Kisangani 25.5 and 24.2 (1,530), Lusaka 20.6 and 15.5 (837), Windhoek 23.5 and 14.0 (386), Antananarivo 21.0 and 10.3 (748), Pretoria 21.0 and 10.3 (748), Cape Town 22.7 and 18.4 (644).

In terms of its **flora,** Africa is divided into two regions: the Holarctic realm in the north and the Sahara desert and the larger, the Paleotropical, south of the Sahara. Tropical evergreen rain forests in the wettest regions are bordered to the north, east and south by grass savannas (covering 35% of the land) and gallery woods in the river valleys, grass and scrub semi-deserts and deserts (xerophilous and succulent scrub). The north has Mediterranean evergreen scrub and dry forests. The flora of the Cape region is related to that of south-west Australia. The **fauna** of Africa is mostly found in the Ethiopian region. The savannas are inhabited by antelopes, elephants, giraffes, hippopotami, rhinoceroses, zebras, wildebeests, lions, leopards, hyenas, monkeys, crocodiles, ostriches, waterfowl (flamingos, pelicans, cranes, herons, etc.), vultures and insects (termites, locusts); the forests by gorillas, chimpanzees, vervets, buffaloes, parrots, beetles and butterflies. Madagascar has fauna of the Tertiary era: lemurs, running-birds, iguanas, etc. Africa has a number of extensive National Parks and wildlife reservations where the animals and the environment are protected. The best-known are: Tsavo, Serengeti, Virunga, Kruger N.P., Kalahari Gemsbok, Kafue, Selous, Southern N.P. and Gorongoza.

Africa takes up 20.3% of the area of the world, and in 1978 **436 million people** lived on the continent, i.e. 10.5% of the world's population. With 14 persons per sq.km Africa is the least densely populated continent. The unevenness of settlement is most striking in a comparison between the density of population in the Nile Valley (over 500 persons per sq.km) and that in desert areas where there is less than 1 person per sq.km. Apart from certain islands, the greatest density of population is found in Rwanda, Burundi and Nigeria, and the lowest (1 person per sq.km) in Botswana, Libya, Mauritania, Namibia and the Western Sahara. From 1970–75 the average annual

A 1
2
3
30°
20°
10°
0°
B
C
D
E
4
5
R. Ben Sekka
Tunis
•3478
Str. of Gibraltar
Tanger
•2456
Alger
Dj. Chélia
•2328
-30
Chott Melrhir
Chott Djerid
G. de Ga
Tarā
Al-Azīzīyah
Madeira
3737
Jb. Toubkal
4165
A t l a s   M t s
Gr. Erg Occidental
Gr. Erg Oriental
Tenerife
3718
Canary Is.
•836
Al-Hammadah A
Erg Iguidi
•738
Erg Chech
Ahaggar
•2254
Tropic of Cancer
Tahat
Tamanrasset
3005
S a h a r a
El Djouf
Nouadhibou
2310
Aïr
Cape Verde Is.
Tombouctou
V. de l'Azaouak
7282
Sénégal
Pte. des Almadies
São Tiago
C. Vert
Dakar
Gambie
Bamako
Niger
Ouagadougou
Bani
White Volta
Niamey
S U D
L. C
240
Black Volta
Jos
1781
Plateau
Conakry
1948
Loma Mansa
Freetown
1768
Mt. Nimba
G U I N
Komoé
L. Volta
Benue
Lagos
Niger
2679
Adamaoua
Monrovia
Abidjan
Accra
Mt. Cameroun
4070
Grain Coast
C. Palmas
Ivory Coast
Gold Coast
Slave Coast
Bight of Benin
Sanaga
Bioko
Yaoundé
Príncipe
Gulf of
Guinea
Libreville
S. Tomé
Ogooué
7859
C. Lopez
1515
Pagalu
A T L A N T I C
O C E A
Banana
Luanda
•6020
26a Nile Delta
Suez Canal
1 : 5 000 000
MEDITERRANEAN SEA
Baltīm
Rashīd
B. al-Burullus
Dumyāt
B. al-Manzilah
Būr Saʻīd (Port Saïd)
Būr Fuʼād
AL-ISKANDARĪYAH
(Alexandria)
Sīdī Sālim
Bilqās Qism
Awwāl
Al-Matarīyah
Disūq
Kafr ash-Shaykh
Damanhūr
Al-Mahallah al-Kubrā
Al-Manzilah
Al-ʻĀmirīyah
Al-Mansūrah
Al-Qantarah
Tantā
As-Sinbillāwayn
As-Sālihīyah
Suez Canal
Mīt Ghamr
Abū Kabīr
Ziftā
E G Y P T
At-Tayrīyah
Az-Zaqāzīq
Al-Ismāʻīlīyah
Shibīn al-Kawm
Bir Hooker
Minūf
Banhā
Bilbays
Al-B. al-Murrah al-Kubrā
Wādī an-Natrūn
Shibīn al-Qanātir
Fāʼid
Shubrā al-Khaymah
Imbābah
AL-QĀHIRAH
(Cairo)
As-Suways
(Suez)
Būr Tawfīq
AL-JĪZAH
Hulwān
Eastern Desert
G. of Suez
Sinai Pen.
Nile
Birkat Qārūn
Sinnūris
•1261

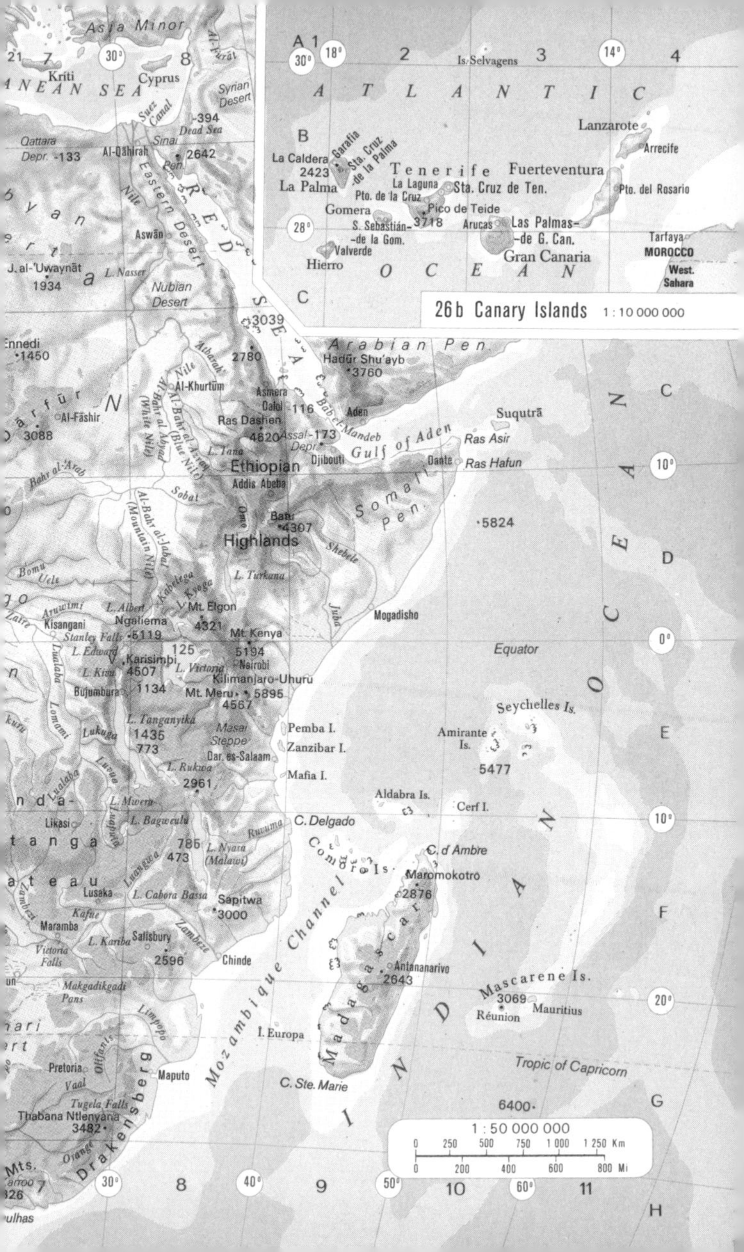
26b Canary Islands 1 : 10 000 000
ATLANTIC OCEAN
Lanzarote
Arrecife
Fuerteventura
Pto. del Rosario
Tenerife
Sta. Cruz de Ten.
La Laguna
Pto. de la Cruz
Pico de Teide 3718
La Palma
La Caldera 2423
Sta. Cruz de la Palma
Gomera
S. Sebastián-de la Gom.
Hierro
Valverde
Las Palmas-de G. Can.
Gran Canaria
Arucas
Is. Selvagens
Tarfaya
MOROCCO
West. Sahara
Asia Minor
Kriti
Cyprus
Suez Canal
Sinai
Syrian Desert
Dead Sea
Al-Qāhirah
Qattara Depr. -133
Eastern Desert
RED SEA
Aswān
L. Nasser
Nubian Desert
J. al-'Uwaynāt 1934
Arabian Pen.
Hadūr Shu'ayb 3760
Al-Khurtūm
Asmera
Dalol -116
Ras Dashen 4620
Assal-173 Depr.
Djibouti
Bab el-Mandeb
Aden
Gulf of Aden
Suqutrā
Ras Asir
Ras Hafun
Dante
Ethiopian Highlands
Addis Abeba
L. Tana
Batu 4307
Somali Pen.
Shebele
Juba
Mogadisho
L. Turkana
Equator
Mt. Elgon 4321
Mt. Kenya 5194
Nairobi
Kilimanjaro-Uhuru 5895
Mt. Meru 4567
Ngaliema 5119
Karisimbi 4507
L. Victoria
L. Albert
L. Edward
L. Kivu
Kisangani
Stanley Falls
Bujumbura
L. Tanganyika
Masai Steppe
Pemba I.
Zanzibar I.
Dar es-Salaam
Mafia I.
L. Rukwa
L. Mweru
L. Bagweulu
Likasi
L. Nyasa (Malawi)
C. Delgado
Comoro Is.
Aldabra Is.
Cerf I.
Seychelles Is.
Amirante Is.
5477
5824
C. d'Ambre
Maromokotro 2876
Madagascar
Antananarivo 2643
Mozambique Channel
Lusaka
L. Cabora Bassa
Sapitwa 3000
Maramba
L. Kariba
Salisbury 2596
Victoria Falls
Chinde
Makgadikgadi Pans
Limpopo
Pretoria
Maputo
Drakensberg
Thabana Ntlenyana 3482
Tugela Falls
I. Europa
C. Ste. Marie
Mascarene Is.
Réunion 3069
Mauritius
Tropic of Capricorn
6400
INDIAN OCEAN
1 : 50 000 000
0 250 500 750 1 000 1 250 Km
0 200 400 600 800 Mi

LONGEST RIVERS

Name	Length in km	River Basin in sq.km
Nile – Kagera	6,671	2,881,000
Congo (Zaïre) – Lualaba	4,835	3,822,000
Niger	4,160	2,092,000
Zambeze (Zambezi)	2,660	1,450,000
Ubangi – Uele	2,280	770,000
Kasai (Kwa, Cassai)	2,200	875,000
Shebele	1,950	
Al-Bahr al-Azraq /Blue Nile/ – Abay	1,900	324,500
Volta – Volta Noire	1,900	440,000
Orange	1,860	1,020,000
Okavango (Cubango)	1,800	785,000
Luvua – Luapula	1,800	
Juba	1,650	200,000
Limpopo – Krokodil	1,600	440,000
Lomani	1,500	110,000
Benue (Benoué)	1,450	319,000
Chari – Ouham	1,450	880,000
Sénégal – Bafing	1,430	450,000
Cuando	1,400	
Kwango (Cuango)	1,400	
Aruwimi – Ituri	1,300	116,100

LARGEST LAKES

Name	Area in sq.km	Greatest Depth in m	Altitude in m
L. Victoria (Ukerewe)	68,800	125	1,134
L. Tanganyika	32,880	1,435	773
L. Nyasa (Malawi)	28,480	785	473
L. Chad	20,700	4–7	240
L. Turkana+ /L. Rudolf/	8,500	73	375
Chott Melrhir+	6,700	.	–30
Chott Djerid+	5,700	.	16
L. Albert	5,345	48	619
L. Mweru	4,920	18	917
L. Tana	3,630	72	1,830
L. Bangweulu	2,850	4	1,067
L. Kivu	2,650	80	1,455
L. Rukwa	2,640	4	793
L. Kyoga	2,600	5	1,033
B. al-Manzilah	2,600	.	2
L. Mai-Ndombe	2,320	15	340
L. Edward	2,150	117	914
Chott ech Chergui+	2,000	.	940
B. al-Burullus	1,930	.	2
L. Chilwa	1,240	.	600
L. Abaya	1,162	13	1,285

+ salt lake

LARGEST ISLANDS

Name	Area in sq.km	Name	Area in sq.km	Name	Area in sq.km
Madagascar	587,041	Mauritius	1,865	Île de Djerba	1,050
Suqutrā /Socotra/	3,579	Fuerteventura	1,722	São Tiago	991
Réunion	2,510	Zanzibar I.	1,658	Pemba	984
Bioko	2,017	Gran Canaria	1,376	Dahlak Kebir I.	900
Tenerife	1,946	Grande Comore	1,148	São Tomé	836

HIGHEST MOUNTAINS

Name (Country)	Height in m	Name (Country)	Height in m	Name (Country)	Height in m
Kilimanjaro-Uhuru (Tanz.)	5,895	Mt. Elgon (Kenya-Ugan.)	4,321	Lesatima (Kenya)	3,994
Mt. Kenya (Kenya)	5,194	Batu (Eth.)	4,307	Amba Ferit (Eth.)	3,975
Kilimanjaro-Mawenzi (Tanz.)	5,149	Abuye Meda (Eth.)	4,305	Mt. Kinangop (Kenya)	3,906
Ngaliema (Margherita) (Ugan.-Zaïre)	5,119	Guna (Eth.)	4,231	Jbel Tignousti (Mor.)	3,825
Ras Deshen (Eth.)	4,620	Guge (Eth.)	4,200	Ari n'Ayachi (Mor.)	3,737
Mt. Meru (Tanz.)	4,567	Abune Yosef (Eth.)	4,190	Gurag (Eth.)	3,719
Buahit (Eth.)	4,510	Jbel Toubkal (Mor.)	4,165	Loolmalassin (Tanz.)	3,648
V. Karisimbi (Rwanda-Zaïre)	4,507	Birhan (Eth.)	4,154	Thabana-Ntlenyana (Les.)	3,482
Talo (Eth.)	4,413	Muhavura (Ugan.)	4,113	Emi Koussi (Chad)	3,415
		Irhil M'goun (Mor.)	4,071		

ACTIVE VOLCANOES

Name (Country)	Altitude in m	Latest eruption
Mt. Cameroun (Cameroon)	4,070	1910
Nyiragongo (Zaïre)	3,470	1977
Pico de Teide (Canary Is.)	3,718	1909
Nyamulagira (Zaïre)	3,056	1977
Ol Doinyo Lengai (Tanzania)	2,878	1960
Pico (Cape Verde Is.).	2,829	1951
Piton de la Foumaise (Réunion)	2,631	1977
La Caldera (Canary Is.)	2,423	1971

LARGEST NATIONAL PARKS

Name (Country)	Area in sq.km
Kafue (Zambia)	22,400
Salonga (Zaïre)	22,300
Tsavo (Kenya)	20,800
Kruger (South Africa)	18,170
Southern N.P. (Sudan)	16,000
Serengeti (Tanzania)	14,500
Wankie (Zimbabwe)	13,353
Komoé (Ivory Coast)	11,500
Ruaha (Tanzania)	11,500

AFRICA

Country	Area in sq.km	Population	Year	Density per sq.km
Algeria	2,381,740	18,515,000	1978	8
Angola	1,246,700	6,761,000	1977	5
Benin	112,622	3,377,000	1978	30
Botswana	600,372	728,000	1978	1.2
Burundi	27,834	3,966,000	1977	142
Cameroon	475,442	7,914,000	1977	17
Canary Islands (provinces of Spain)	7,273	1,299,300	1976	179
Cape Verde	4,033	315,000	1977	78
Central African Republic	622,984	2,640,000	1977	4
Chad	1,284,000	4,324,000	1978	3
Comoros	2,171	370,000	1977	170
Congo	342,000	1,454,000	1978	4
Djibouti	23,000	220,000	1977	10
Egypt	1,001,449	40,287,000	1979	40
Equatorial Guinea	28,051	326,000	1977	12
Ethiopia	1,221,900	29,339,000	1978	24
Gabon	267,667	534,000	1977	2
Gambia	11,295	569,000	1978	50
Ghana	238,537	10,475,000	1977	44
Guinea	245,857	4,646,000	1977	19
Guinea-Bissau	36,125	544,000	1977	15
Ivory Coast	322,464	7,062,000	1977	22
Kenya	582,646	14,856,000	1978	25
Lesotho	30,355	1,248,000	1977	41
Liberia	111,370	1,684,000	1977	15
Libya	1,759,540	2,608,000	1978	1.5
Madagascar	587,041	8,881,000	1978	15
Madeira (province of Portugal)	796	265,600	1976	334
Malawi	118,484	5,571,567	1977	47
Mali	1,239,710	6,035,272	1977	5
Mauritania	1,030,700	1,481,000	1977	1.4
Mauritius and dependencies	2,045	909,000	1977	445
Morocco	458,730	18,592,000	1978	41
Mozambique	784,961	9,678,000	1977	12
Namibia	824,295	936,000	1977	1.1
Niger	1,266,995	4,994,000	1978	4
Nigeria	923,768	80,627,000	1978	87
Réunion (Fr.)	2,510	496,000	1978	198
Rwanda	26,338	4,450,000	1978	169
Saint Helena and dependencies (U.K.)	419	5,147	1976	12
São Tomé and Principe	964	82,000	1977	85
Senegal	196,722	5,198,000	1977	26
Seychelles and dependencies	444	62,000	1977	140
Sierra Leone	71,740	3,470,000	1977	48
Somalia	637,657	3,354,000	1977	5
South Africa	1,221,037	26,764,000	1977	22
Spanish North Africa	33	121,600	1976	3,685
Sudan	2,505,813	16,953,000	1977	7
Suqutrā /Socotra (South Yemen)	3,626	16,000	1977	4
Swaziland	17,363	507,000	1977	29
Tanzania	945,087	16,560,000	1978	18
Togo	56,600	2,409,000	1978	43
Tunisia	164,150	6,216,000	1978	38
Uganda	236,036	12,780,000	1978	54
Upper Volta	274,200	6,319,000	1977	23
Western Sahara	266,000	152,000	1978	0.6
Zaïre	2,345,409	27,080,000	1978	12
Zambia	752,614	5,472,000	1978	7
Zimbabwe	390,580	6,930,000	1978	18

population increase in Africa was **2.6%,** the birth rate was 46.3, and the death rate 19.8 per 1,000. Relatively few persons live in towns, in 1975 the figure was only 24.2%. Africa has 10 cities with over one million inhabitants.

Africa's **economy** is typical of that of developing countries. With the exception of the Republic of South Africa, which is an advanced industrial and agricultural country, **agriculture** predominates in the economy of most African countries with the stress on the production of plantation crops. Plant production takes precedence over livestock breeding. Africa produces a major share in the world production of oil crops (groundnuts, palm kernels, palm oil), cocoa, sisal,

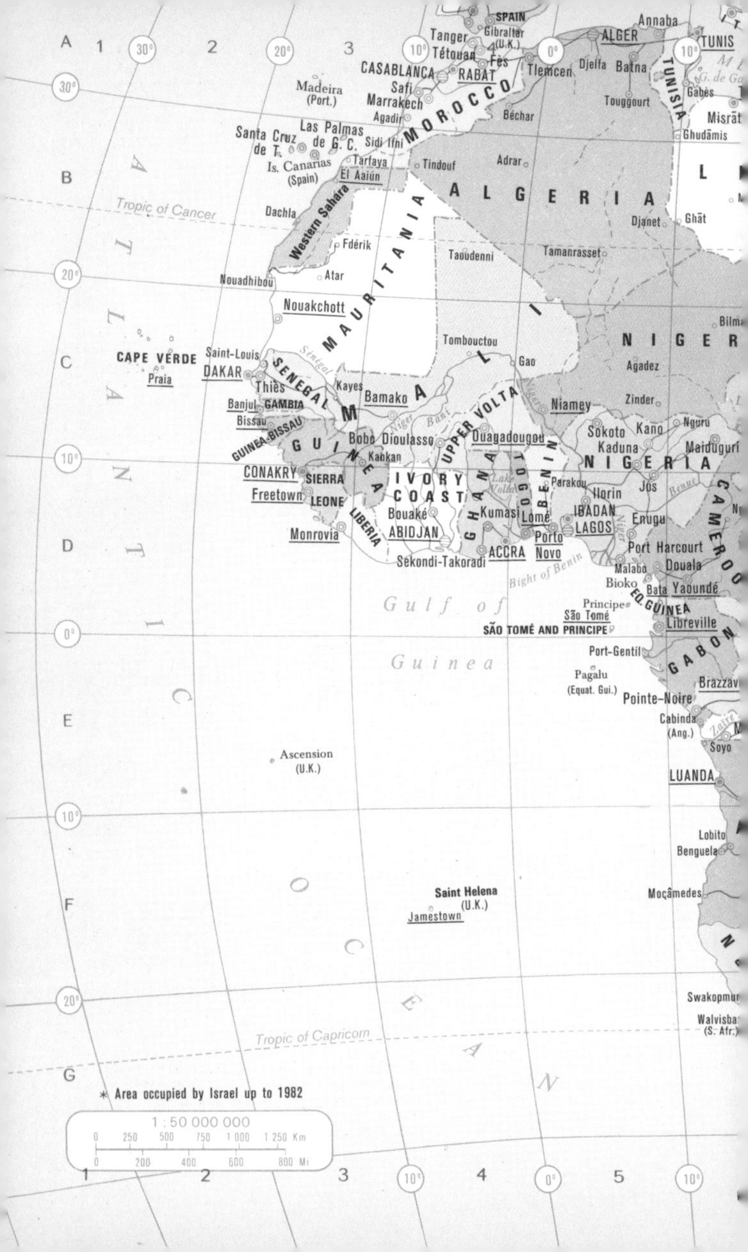

SPAIN
Gibraltar (U.K.)
Tanger
Tétouan
Fès
CASABLANCA
RABAT
Safi
Marrakech
Agadir
MOROCCO
Madeira (Port.)
Las Palmas de G. C.
Santa Cruz de T.
Is. Canarias (Spain)
Sidi Ifni
Tarfaya
El Aaiún
Tindouf
Béchar
Tlemcen
ALGER
Annaba
Djelfa
Batna
Touggourt
TUNIS
TUNISIA
Gabès
Misrāt
Ghudāmis
Adrar
ALGERIA
Djanet
Ghāt
Tamanrasset
Tropic of Cancer
Dachla
Western Sahara
Fdérik
Atar
Nouadhibou
Nouakchott
MAURITANIA
Taoudenni
Tombouctou
Gao
MALI
NIGER
Agadez
Zinder
Bilma
CAPE VERDE
Praia
Saint-Louis
DAKAR
Thiès
SENEGAL
Sénégal
Kayes
Bamako
Banjul
GAMBIA
Bissau
GUINEA-BISSAU
GUINEA
Niger
Bani
Bobo Dioulasso
UPPER VOLTA
Ouagadougou
Niamey
Sokoto
Kano
Nguru
Kaduna
Maiduguri
NIGERIA
Kankan
CONAKRY
SIERRA LEONE
Freetown
IVORY COAST
GHANA
TOGO
BENIN
Lake Volta
Parakou
Ilorin
Jos
Benue
CAMEROON
LIBERIA
Monrovia
Bouaké
ABIDJAN
Kumasi
Lomé
Porto Novo
IBADAN
LAGOS
Enugu
Port Harcourt
Douala
Sekondi-Takoradi
ACCRA
Bight of Benin
Malabo
Bioko
Bata
Yaoundé
EQ. GUINEA
Principe
São Tomé
SÃO TOMÉ AND PRINCIPE
Libreville
GABON
Gulf of Guinea
Port-Gentil
Pagalu (Equat. Gui.)
Pointe-Noire
Brazzav
Cabinda (Ang.)
Zaire
Soyo
LUANDA
Ascension (U.K.)
Lobito
Benguela
Saint Helena (U.K.)
Jamestown
Moçâmedes
Swakopmu
Walvisba (S. Afr.)
Tropic of Capricorn
ATLANTIC OCEAN
* Area occupied by Israel up to 1982
1 : 50 000 000
0 250 500 750 1 000 1 250 Km
0 200 400 600 800 Mi

27 a Tunisia and East Algeria
1 : 15 000 000
MEDITERRANEAN SEA
ALGER
TUNIS
ALGERIA
TUNISIA
EGYPT
SAUDI ARABIA
SUDAN
ETHIOPIA
SOMALIA
KENYA
UGANDA
TANZANIA
ZAMBIA
MALAWI
MOZAMBIQUE
MADAGASCAR
ZIMBABWE
SOUTH AFRICA
SEYCHELLES
YEMEN
SOUTH YEMEN
DJIBOUTI
RWANDA
BURUNDI
COMOROS
MAURITIUS
SWAZILAND
LESOTHO
INDIAN OCEAN
AL-QAHIRAH
ADDIS ABEBA
NAIROBI
DAR ES-SALAAM
Mogadisho
Antananarivo
Lusaka
SALISBURY
PRETORIA
Maputo
Equator
Gulf of Aden
Mozambique Channel
RED SEA

dates and spices. It also possesses great wealth in its forests, especially in the wet equatorial parts of the continent and has rich sources of water power, which are still underdeveloped. The enormous wealth of mineral resources has only been partly prospected. **Mining** is fairly widespread. Africa leads world production in diamonds, gold, platinum, and contributes an important share in uranium, copper, manganese, chromium, cobalt, vanadium, bauxite, antimony and phosphates. The only **industry** that is to be found practically in all African countries is the food industry. **Transport and foreign trade** are of immense importance for the economic development of the continent.

ALGERIA

El Djemhouria El Djazaïria Demokratia Echaabia – République Algérienne Démocratique et Populaire, area 2,381,740 sq.km, population 18,515,000 (1978), **democratic people's republic** (President Bendzhedid Shadlí since 1979).
Administrative units: 31 departements. **Capital:** Alger 1,503,720 inhab. (1974); **other towns** (1974, in 1,000 inhab.): Oran (Ouahran) 485, Constantine 350, Annaba 313, Tizi-Ouzou 224, Blida 159, Sétif 157, Sidi bel Abbès 151, Skikda 128, Batna 115, Tlemcen 115, Mostaganem 102. – **Currency:** Algerian dinar = 100 centimes. **Economy:** agricultural country with developing industry, especially mining. **Agriculture**: **crops** (1977, in 1,000 tons): wheat 1,200, barley, potatoes 500, grapes 500, oranges 360, tangerines 148, lemons, olives 113, olive oil 15, dates 140, figs 61 (1975), fruit, tomatoes 136, tobacco, sugar beet 85; **livestock** (1977, in 1,000 head): cattle 1,300, sheep 9,540, goats 2,220, camels 135, horses, mules, asses; poultry 16,900, eggs 19,000 tons; fish catch 35,100 tons (1976); cork oak 14,999 tons (1972), alfalfa **Mining** (1977, in 1,000 tons, metal content): crude petroleum 51,444 (Hassi Messaoud, Edjeleh, Ohanet); natural gas 10 billion cub.m (Hassi R'Mel), iron ore 1,490, zinc, lead, mercury 1,049 tons (Ras el Ma, third world producer), silver, salt 150, phosphates 818, pyrites. Electricity 4,414 mn. kWh (1977). **Industry:** food processing (canning plants, oil processing plants, mills), construction of metallurgy (Annaba), new plants for machinery and chemical industries, petroleum processing. **Production** (1977, in 1,000 tons): meat 134, milk 663, wine 360, cement 2,350.– **Communications:** railways 4,074 km, roads 78,408 km. Merchant shipping 1,056,000 GRT. – **Exports:** petroleum, natural gas, wine, fruit.

ANGOLA

República Popular de Angola, area 1,246,700 sq.km, population 6,761,000 (1977), **republic** (President José Eduardo Dos Santos since 1979).
Administrative units: 16 districts. **Capital:** Luanda 540,000 inhab. (1972); **other towns** (1970, in 1,000 inhab.): Huambo 62, Lobito 60, Benguela 41, Malanje 32, Cabinda 21. – **Currency:** kwanza = 100 lwei.
Economy: **agriculture** (1977, in 1,000 tons): coffee 72, sisal 20, maize (corn) 450, millet, manioc, cotton, groundnuts, palm oil, sugarcane, bananas, pineapples, tobacco; **livestock** (1977, in 1,000 head): cattle 3,050, goats 920, sheep; fish catch 153,600 tons (1976); roundwood 7.8 mn. cub.m (1976). **Mining** (1976, in 1,000 tons, metal content): crude petroleum 8,640 (1977), iron ore 1,664 (1975), manganese, diamonds 660,000 carats, salt 100. – **Communications:** railways 3,049 km, roads 72,323 km. – **Exports:** petroleum, coffee, diamonds, iron ore etc.

BENIN

République populaire du Benin, area 112,622 sq.km, population 3,377,000 (1978), **republic** (President Lieut.-Col. Mathieu Kérékou since 1972).
Administrative units: 6 departments. **Capital:** Porto-Novo 104,000 inhab. (1975); **other towns** (in 1,000 inhab.) Cotonou 178 (1975), Abomey 38 (1973). – **Currency:** CFA franc = 100 centimes.
Economy: developing agricultural country. **Agriculture** (1977, in 1,000 tons): maize (corn), manioc 610, coffee, bananas palm kernels 70, palm oil 35, groundnuts 48, cotton; **livestock** (1977, in 1,000 head): cattle 833, sheep 886, goats 856 fish catch 25,500 tons. – **Communications:** railways 579 km, roads 6,937 km. – **Exports:** palm kernels and oil.

BOTSWANA

Republic of Botswana, area 600,372 sq.km, population 728,000 (1978), **republic, member of the Commonwealth** (President Dr Quett Ketumile Jonny Masire since 1980).
Administrative units: 16 districts. **Capital:** Gaborone 33,142 inhab. (1976); **other towns** (1971, in 1,000 inhab.) Serowe 43, Kanye 39. – **Currency:** pula = 100 thebe. **Economy:** livestock rearing (1977, in 1,000 head): cattle 2,400 sheep 450, goats 1,100; cultivation of millet, sorghum and maize (corn). **Mining** (1976, in 1,000 tons, metal content) nickel 12,581 tons, copper 12.5, diamonds 2.4 mn. carats, coal 288. – **Communications:** railways 634 km, roads 8,019 km – **Exports:** meat, diamonds, nickel.

BURUNDI

République du Burundi – Republica y'u Burundi, area 27,834 sq.km, population 3,966,000 (1977), **republic** (President Col. Jean-Baptiste Bagaza since 1976).
Administrative units: 8 provinces. **Capital:** Bujumbura 157,000 inhab. (1976). **Density** 142 persons per sq.km. – **Currency:** Burundi franc = 100 centimes. **Economy:** developing agricultural country with livestock rearing. **Agriculture** (1977, in 1,000 tons): coffee 22, maize (corn), sorghum, manioc 902, sweet potatoes 873, bananas 932, groundnuts cotton; **livestock** (1977, in 1,000 head): cattle 775, sheep 312, goats 571; fish catch 20,333 tons in Lake Tanganyika **Mining:** tin, deposits of nickel. – **Exports:** coffee, cotton, hides etc.

AMEROON

ėpublique Unie du Cameroun, area 475,442 sq.km, population 7,914,000 (1977), **republic**
resident Hadji Ahmadou Ahidjo Babatoura since 1960).
dministrative units: 7 provinces. **Capital:** Yaoundé 274,399 inhab. (1975); **other towns** (1975, in 1,000 inhab.): ›uala 486 (1976), Nkongsamba 71 (1970), Bamenda 49 (1970). – **Currency:** CFA franc = 100 centimes.
conomy: developing agricultural country. **Agriculture** (1977, in 1,000 tons): cocoa beans 90, coffee 90, bananas 105, oundnuts 150, palm oil 82, cotton; **livestock** (1977, in 1,000 head): cattle 2,917, sheep 2,100, goats 1,553, pigs 700; h catch 71,600 tons; roundwood 8.3 mn. cub.m, rubber 18,000 tons. **Mining:** tin, gold, titanium. Electricity 1.3 billion kWh 976). – **Communications:** railways 1,172 km, roads 56,673 km. – **Exports:** cocoa, coffee, wood, bananas.

APE VERDE

epública do Cabo Verde, area 4,033 sq.km, population 315,000 (1977), **republic** (President stides Pereira since 1975).
apital: Praia 21,494 inhab. (1970). – **Currency:** Cape Verde escudo = 100 centavos. **Economy:** maize (corn), sweet tatoes, manioc, sugarcane, groundnuts, bananas, coffee; fishing. Important naval station.

ENTRAL AFRICAN REPUBLIC

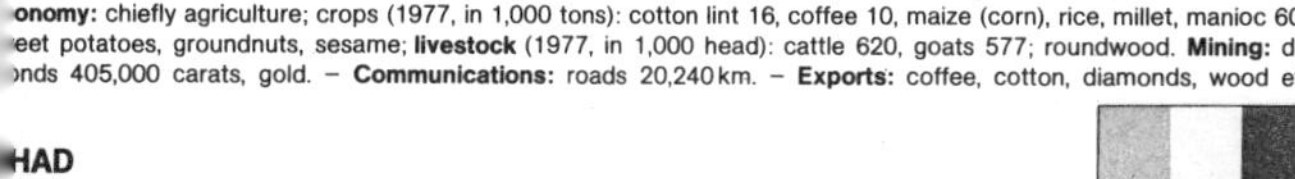

ea 622,984 sq.km, population 2,640,000, republic (Headed by Gen. A. Kolingba since 1981).
ministrative units: 14 prefectures. **Capital:** Bangui 187,000 inhab. (1971). – **Currency:** CFA franc = 100 centimes.
onomy: chiefly agriculture; crops (1977, in 1,000 tons): cotton lint 16, coffee 10, maize (corn), rice, millet, manioc 600, eet potatoes, groundnuts, sesame; **livestock** (1977, in 1,000 head): cattle 620, goats 577; roundwood. **Mining:** dia›nds 405,000 carats, gold. – **Communications:** roads 20,240 km. – **Exports:** coffee, cotton, diamonds, wood etc.

HAD

publique du Tchad, area 1,284,000 sq.km, population 4,324,000 (1978), **republic** (President ukouni Oueddei since 1979).
ministrative units: 14 prefectures. **Capital:** Ndjamena 150,000 inhab. (1974). – **Currency:** CFA franc = 100 centimes.
onomy: backward agricultural country. **Agriculture:** cotton lint, maize (corn), rice, millet, manioc, sweet potatoes, undnuts, sugarcane, dates, sesame, tobacco; **livestock** (1977, in 1,000 head): cattle 3,716, sheep 2,448, goats 2,448, nels; roundwood; fish catch 115,000 tons (1976). – **Communications:** roads 10,760 km.- **Exports:** cotton, cattle, meat.

OMOROS

t Comorien, area 2,171 sq.km, population 370,000 (1977), **republic** (President Ahmed Abdal- Abderemane since 1978).
pital: Moroni 16,300 inhab. (1976). **Density** 170 persons per sq.km. **Currency:** CFA franc = 100 centimes. **Production export:** vanilla, ylang-ylang, cloves, sisal, coconuts, copra, cinnamon, essential oils.

ONGO

publique populaire du Congo, area 342,000 sq.km, population 1,454,000 (1978), **republic** esident Maj. Denis Sassou-Nguesso since 1979).
ninistrative units: 9 regions. **Capital:** Brazzaville 298,967 inhab. (1974, with suburbs); **other towns:** Pointe- re 142,000 inhab. (1974). – **Currency:** CFA franc = 100 centimes.
nomy: agricultural country. **Agriculture** (1977, in 1,000 tons): manioc 769, maize (corn), sweet potatoes, coffee, oa beans, palm oil 5.4, groundnuts 24, sugarcane, bananas; fish catch 19,400 tons; roundwood 2.4 mn. cub.m (1976). **ing** (1976, in 1,000 tons, metal content): crude petroleum 2,002, lead 2.5, gold, potassium salts 445. – **Communica- s:** railways 798 km, roads 10,992 km, navigable waterways. – **Exports:** wood, petroleum, palm kernels, sugar.

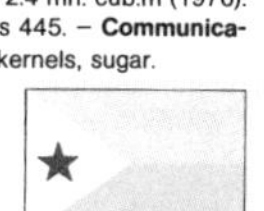

IBOUTI

ublique de Djibouti, area 23,000 sq.km, population 220,000 (1977), **republic** (President san Gouled Aptidon since 1977).
ital: Djibouti 130,000 (1976). – **Currency:** Djibouti franc = 100 centimes. **Economy:** coffee grown for export; **stock** (1977, in 1,000 head): goats 585, sheep, camels 25. – **Communications:** railways 90 km, roads 1,875 km.

YPT

umhūriya Misr al-'Arabīya, area 1,001,449 sq.km, population 40,287,000 (1979), **republic** esident Muhammad Hosni Mubarak since 1981).
ninistrative units: 25 governorates. **Capital:** Al-Qāhirah (Cairo) 5,414,000 inhab. (1978, with agglomeration

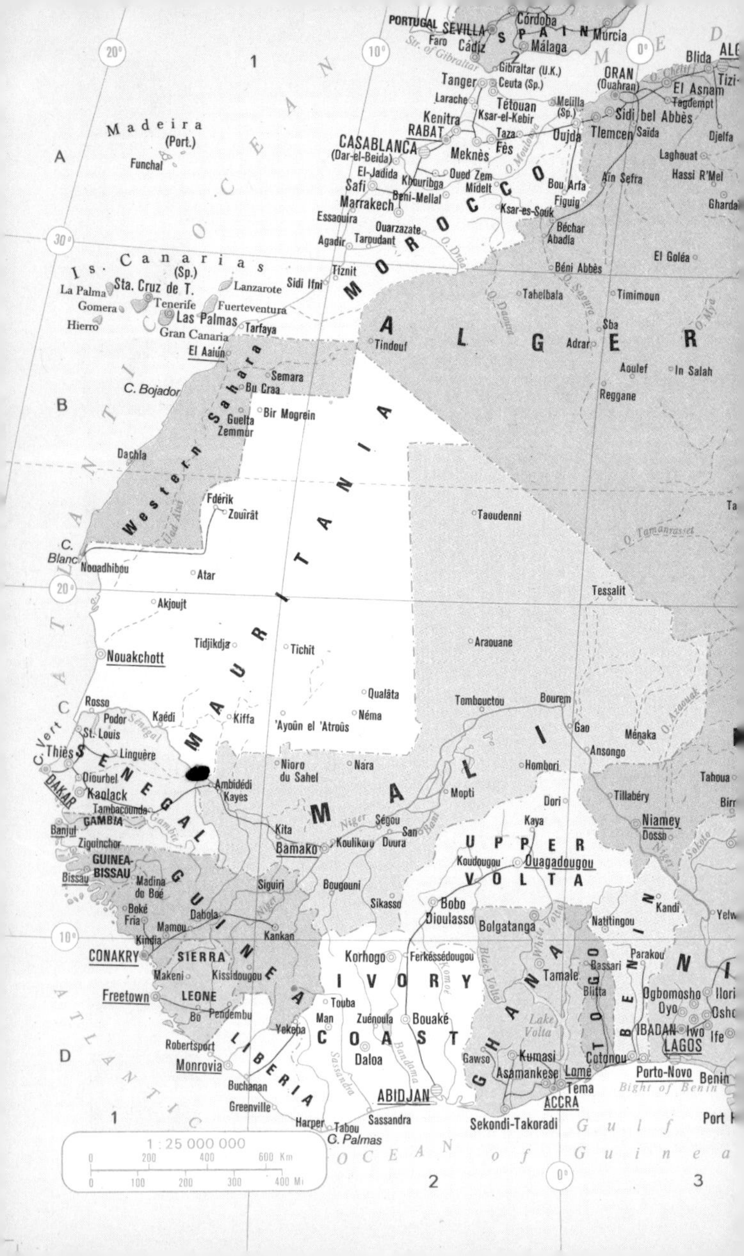
PORTUGAL
SEVILLA
Córdoba
SPAIN
Murcia
Faro
Cádiz
Málaga
Str. of Gibraltar
Gibraltar (U.K.)
ORAN
(Ouahran)
Blida
Tanger
Ceuta (Sp.)
El Asnam
Tagdempt
Larache
Tétouan
Melilla (Sp.)
Kenitra
Ksar-el-Kebir
Sidi bel Abbès
RABAT
Taza
Oujda
Tlemcen
Saïda
Djelfa
CASABLANCA
(Dar-el-Beida)
Meknès
Fès
Laghouat
El-Jadida
Khouribga
Oued Zem
Midelt
Bou Arfa
Aïn Sefra
Hassi R'Mel
Safi
Beni-Mellal
Figuig
Marrakech
Ksar-es-Souk
Essaouira
Ouarzazate
Béchar
Abadla
Agadir
Taroudant
MOROCCO
El Goléa
Madeira
(Port.)
Funchal
ATLANTIC OCEAN
Is. Canarias
(Sp.)
La Palma
Sta. Cruz de T.
Lanzarote
Gomera
Tenerife
Fuerteventura
Hierro
Las Palmas
Gran Canaria
Tiznit
Sidi Ifni
Tarfaya
Béni Abbès
Tabelbala
Timimoun
O. Saoura
O. Daoura
O. Mya
Sba
ALGERIA
Tindouf
Adrar
Aoulef
In Salah
El Aaiún
Semara
Bu Craa
Reggane
C. Bojador
Western Sahara
Guelta Zemmur
Bir Mogrein
Dachla
MAURITANIA
Fdérik
Zouîrât
Taoudenni
O. Tamanrasset
C. Blanc
Nouadhibou
Atar
Tessalit
Akjoujt
Tidjikdja
Tichît
Araouane
Nouakchott
Qualâta
Rosso
Tombouctou
Bourem
Podor
Kaédi
Kiffa
Néma
'Ayoûn el 'Atroûs
C. Vert
St. Louis
Sénégal
Gao
Ménaka
O. Azaouak
Thiès
Linguère
Ansongo
SENEGAL
Nioro du Sahel
Nara
Hombori
DAKAR
Diourbel
Ambidédi
Kayes
Mopti
Tahoua
Kaolack
MALI
Dori
Tillabéry
Tambacounda
GAMBIA
Kita
Ségou
Kaya
Niamey
Dosso
Banjul
Ziguinchor
Bamako
Koulikoro
San
Doura
UPPER VOLTA
Niger
GUINEA-BISSAU
Koudougou
Ouagadougou
Bissau
Madina do Boé
Siguiri
Bougouni
Sikasso
Bobo Dioulasso
Kandi
Boké
Fria
Dabola
GUINEA
Bolgatanga
Natitingou
Kindia
Mamou
Kankan
CONAKRY
SIERRA LEONE
Korhogo
Ferkéssédougou
Parakou
Makeni
Kissidougou
IVORY COAST
Tamale
Bassari
Freetown
Bo
Pendembu
Touba
Komoe
Black Volta
White Volta
Bafilo
Ogbomosho
Ilori
Oyo
Oshogbo
Man
Zuénoula
Bouaké
GHANA
TOGO
BENIN
Lake Volta
Yekepa
LIBERIA
Robertsport
Daloa
IBADAN
Iwo
Ife
LAGOS
Monrovia
Gawso
Kumasi
Cotonou
Porto-Novo
Benin
Asamankese
Lomé
Tema
Buchanan
Sassandra
Bandama
ABIDJAN
ACCRA
Bight of Benin
Greenville
Harper
Tabou
C. Palmas
Sekondi-Takoradi
Gulf of Guinea
Port
1 : 25 000 000
0 200 400 600 Km
0 100 200 300 400 Mi
20°
10°
0°
30°
A
B
C
D
1
2
3

Asamankese 101, Tamale 99, Bolgatanga 93. – **Currency:** cedi = 100 pesewas.
Economy: based on agriculture specialized in products for export. Rich resources of raw materials. **Agriculture** (1977, in 1,000 tons): cocoa beans 310 (leading world producer, Ashanti region), maize (corn) 366, millet, rice, manioc 2,500, sugarcane, groundnuts, palm kernels 30, palm oil 15, coconuts 304, copra 17, coffee, oranges 140, lemons, pineapples, bananas, tomatoes 100; **livestock** (1977, in 1,000 head): cattle 850, sheep 1,500, goats 1,800, pigs 380; poultry 10,500; fish catch 237,700 tons (1976); roundwood 12.6 mn. cub.m (1976), natural rubber 3,600 tons (1977). **Mining** (1976, in 1,000 tons, metal content): manganese 128, gold 16,619 kg, diamonds 2,283,000 carats, bauxite 243.6 (1977), salt. Electricity 2,226 mn. kWh (1976), of this 98.8% hydro-electric power stations (on the R. Volta at Akosombo). – **Communications:** railways 1,300 km, roads 35,015 km (1974). Merchant shipping 183,000 GRT (1977). River transport on the R. Volta. – **Exports:** cocoa, gold, wood, diamonds, manganese.

GUINEA

République Revolutionnaire Populaire de Guinée, area 245,857 sq.km, population 4,646,000 (1977), **republic** (President Ahmed Sékou Touré since 1958).
Administrative units: 29 regions. **Capital:** Conakry 525,675 inhab. (1973); **other towns** (1973, in 1,000 inhab.): Kindia 58, Fria 54, Kankan 52 (1970). **Currency:** syli = 100 cauris.
Economy: agricultural products for export, mineral mining. **Agriculture** (1977, in 1,000 tons): rice 320, maize (corn), manioc 484, sisal, groundnuts, palm oil 37.8, coffee, bananas 70, pineapples, tobacco; **livestock** (1977, in 1,000 head): cattle 1,600, sheep 420, goats 385; roundwood 3.1 mn. cub.m (1976). **Mining** (1976, in 1,000 tons, metal content): iron ore 1,040 (1970), bauxite 11,316 (second world producer, deposits Boké, Fria), diamonds 80,000 carats. Electricity 500 mn. kWh. **Industry:** production of aluminium, food processing. – **Communications:** railways 818 km, roads 28,400 km. – **Exports:** aluminium, coffee, pineapples, bananas, diamonds, groundnuts etc.

GUINEA-BISSAU

República da Guiné-Bissau, area 36,125 sq.km, population 544,000 (1977), **republic** (Head of State Chairman of the Revolutionary Council Maj. João Bernardo Vieira since 1980).
Capital: Bissau 71,169 inhab. (1970). – **Currency:** Guinean peso = 100 centavos. **Economy:** agricultural country. **Agriculture** (1977, in 1,000 tons): rice, groundnuts 30, palm kernels 7, palm oil 5, coconuts 23; roundwood 530,000 cub.m (1976).

IVORY COAST

République de Côte d'Ivoire, area 322,464 sq.km, population 7,062,000 (1977), **republic** (President Félix Houphouet-Boigny since 1960).
Administrative units: 24 departments. **Capital:** Abidjan 1,150,000 inhab. (with agglomeration, 1977); **other towns** (1975, in 1,000 inhab.): Bouaké 220, Korhogo 100 (1971), Daloa 59.5. – **Currency:** CFA franc = 100 centimes.
Economy: agricultural country, export products grown on plantations. **Agriculture** (1977, in 1,000 tons): millet, sorghum, rice 440, manioc 689, cotton lint 28, sesame, palm oil 185, coconuts 105, copra 14, groundnuts, coffee 291 (third world producer), cocoa beans 240 (second world producer), bananas 160, pineapples 285; **livestock** (1977, in 1,000 head): cattle 650, sheep 1,050; fish catch 77,000 tons; roundwood 10.1 mn. cub.m, natural rubber 20,000 tons (1977). **Mining:** diamonds 60,000 carats (1976), manganese. Electricity 960 mn. kWh (1976). **Processing** of agricultural products. **Communications:** railways 1,173 km, roads 45,170 km.- **Exports:** coffee, wood, cocoa, bananas.

KENYA

Jamhuri ya Kenya, area 582,646 sq.km, population 14,856,000 (1978), **republic, member of the Commonwealth** (President Daniel Arap Moi since 1978).
Administrative units: 8 provinces. **Capital:** Nairobi 776,000 inhab. (1977); **other towns:** Mombasa 371,000 inhab. (1977). Average annual rate of population increase 3.6%. – **Currency:** Kenyan shilling = 100 cents.
Economy: agricultural country with developing livestock and industrial production. **Agriculture** (1977, in 1,000 tons): maize (corn) 1,700, wheat, barley, rice, millet, manioc, sweet potatoes; on plantations: cotton, sisal 34, sugarcane, pyrethrum 14.4 (1971, leading world producer), groundnuts, sesame, palm oil, copra, coffee 96, tea 86, pineapples 110; **livestock** (1977, in 1,000 head): cattle 7,350, sheep 3,900, goats 4,300, camels 540; poultry; fish catch 40,900 tons (1976);

add page 126

8,539,000); **other towns** (1975, in 1,000 inhab.): Al-Iskandarīyah (Alexandria) 2,320, Al-Jīzah 893, Shubrā al Khaymah 373, Būr Sa'īd (Port Said) 349, As-Suways (Suez) 315 (1970), Al-Mahallah al Kubrā 296, Tantā 284, Aswān 259, Al-Mansūrah 238, Asyūt 203, Az-Zaqāzīq 201. – **Currency:** Egyptian pound = 100 piastres. **Economy:** agricultural and industrial country; cultivated land only 2.8% in the Nile delta and valley. **Agriculture: crops** (1977, in 1,000 tons): wheat 1,872, maize (corn) 2,900, barley, rice 2,270, potatoes 970, sweet potatoes, sugarcane 8,000, cottonseed 710, lint 435, groundnuts, sesame, fruit-oranges 990, tangerines, lemons, grapes 265, bananas 117, dates 417 (leading world producer), figs-vegetables, tomatoes 2,400, onions 670, watermelons 1,250; **livestock** (1977, in 1,000 head): cattle 2,148, buffaloes 2,294, camels 101, asses 1,574, sheep 1,938, goats 1,393, poultry 26,681; fish catch 106,574 tons (1976). **Mining** (1976, in 1,000 tons, metal content): crude petroleum 21,036 (1977, Gulf of Suez region), natural gas 1.2 billion cub.m, iron ore 621, manganese, phosphates 486, titanium, salt 508. Electricity 11 billion kWh (1976), of this 93% hydro-electric power stations; chief Aswān dam on the R. Nile. Most developed **industries:** textiles and food processing. **Production** (1976, in 1,000 tons): pig iron 569, crude steel 457, cement 3,290, nitrogenous fertilizers 200, cotton yarn 193, meat 414 (1977), milk 682 (1977), sugar 576; 20.8 billion cigarettes (1975). **Communications:** mainly along the Nile valley and Suez Canal zone; railways 7,224 km, roads 25,976 km, navigable waterways 3,400 km. Merchant shipping 408,000 GRT. **Suez Canal:** opened in 1869, length Port Said – Suez 161 km, width on level 70–125 m, depth 11–12 m, passage permitted to vessels up to 10.36 m draught. Civil aviation (1976): 20.9 mn. km flown and 1,113,000 passengers carried. – Tourism: 984,000 visitors (1976). – **Exports:** cotton, rice, petroleum, fruit.

EQUATORIAL GUINEA

República de Guinea Ecuatorial, area 28,051 sq.km, population 326,000 (1977), **republic** (Head of the State and Prime Minister Theodoro Obiango Nguema Mbasogo since 1979).
Administrative units: 2 provinces. **Capital:** Malabo 52,000 inhab- **Currency:** ekuele = 100 centimos. **Agriculture:** manioc, sweet potatoes, palm oil, coffee, cocoa beans; roundwood- **Exports:** coffee, cocoa, wood.

ETHIOPIA

Area 1,221,900 sq.km, population 29,339,000 (1978), **republic** (Head of State Chairman of the Provisional Military Administration Council Lieut.-Col. Menghistu Haile Mariam since 1975).
Administrative units: 14 provinces. **Capital:** Addis Abeba 1,133,200 inhab. (1977); **other towns** (1971, in 1000 inhab.): Asmera 340 (1976), Dire Dawa 64, Dese 47, Harer 46, Jima 45. – **Currency:** birr = 100 cents. **Economy:** agricultural country with predominantly livestock production. **Agriculture** (1977, in 1,000 tons): wheat 592, barley 830, maize (corn) 1,150, millet, sorghum 671, potatoes, sugarcane 1,249, cotton lint 24, flax 50, groundnuts, sesame 70, castor beans 14 (1976), coffee 175, bananas, legumes, tobacco; **livestock** (1977, in 1,000 head): cattle 26,119, sheep 23,149, goats 17,064, camels 966, horses 1,520, mules 1,431, asses 3,875; poultry 52,156; roundwood 24.2 mn. cub.m (1976). **Mining** (1976): gold 933 kg, platinum 10 kg, salt 88,000 tons, potassium salt. – **Communications:** railways 694 km, roads 23,158 km. – **Exports:** coffee, hides, skins, oil seeds, meat.

GABON

République Gabonaise, area 267,667 sq.km, population 534,000 (1977), **republic** (President Omar Bongo since 1967).
Administrative units: 9 regions. **Capital:** Libreville 186,154 inhab. (1976); **other towns** (1976, in 1,000 inhab.): Port-Gentil 85, Lambaréné 24. **Density** 2 persons per sq.km. – **Currency:** CFA franc = 100 centimes. **Economy:** agricultural country with mining and wood processing industries. Forests cover 75% of the land area. **Crops:** rice, manioc, coffee, cocoa beans, bananas; roundwood 2.6 mn. cub.m (1976). **Mining** (1977, in 1,000 tons, metal content): crude petroleum 11,268, natural gas 60 mn. cub.m, manganese 1,094 (1976, third world producer), uranium 1900 tons, gold. – **Communications:** railways 185 km, roads 6,848 km. – **Exports:** petroleum (over 80%), wood, manganese, uranium, cocoa, coffee.

GAMBIA

Republic of Gambia, area 11,295 sq.km, population 569,000 (1978), **republic, member of the Commonwealth** (President Al Hadji Sir Dawda Kairaba Jawara since 1970).
Capital: Banjul 48,333 inhab. (1977). **Currency:** dalasi = 100 bututs. **Economy:** agriculture specialized mainly in growing groundnuts (145,000 tons), rice, millet, cotton, palm kernels and oil, cattle. – **Communications:** roads 2,990 km, goods transport on the R. Gambia. – **Exports:** groundnuts and groundnut oil.

GHANA

Republic of Ghana, area 238,537 sq.km, population 10,475,000 (1977), **republic, member of the Commonwealth** (Chairman of the National Defence Council Flight Lieutenant Jerry Rawlings since 1982).
Administrative units: 8 regions and 1 capital district. **Capital:** Accra 564,194 inhab. (1970, with agglomeration 848,825); **other towns** (1970, in 1,000 inhab.): Kumasi 260 (with agglomeration 343), Sekondi-Takoradi 161,

roundwood 12.4 mn. cub.m (1976). **Mining**: magnesite 17,000 tons (1976), gold, salt, soda. Electricity 1,119 mn. kWh (1976). **Industries:** processing of agricultural products, canning plants, processing of petroleum, fertilizers and textiles. **Production** (1976, in 1,000 tons): cement 983, sugar 182. – **Communications:** railways 4,125 km, roads 50,092 km. – **Exports:** coffee, tea, petroleum products, sisal etc.

LESOTHO

Kingdom of Lesotho – 'Muso oa Lesotho, area 30,355 sq.km, population 1,248,000 (1977), **kingdom, member of the Commonwealth** (King Moshoeshoe II since 1966).
Administrative units: 9 districts. **Capital:** Maseru 29,049 inhab. (1972). – **Currency:** maluti = 100 licente. **Economy:** developing agriculture and livestock rearing. **Agriculture:** wheat, maize (corn), sorghum; **livestock** (1977, in 1,000 head): cattle 600, sheep 1,680, goats 920. **Mining** of diamonds and precious stones. – **Exports:** wool, livestock, diamonds.

LIBERIA

Republic of Liberia, area 111,370 sq.km, population 1,684,000 (1977), **republic** (headed by Samuel Kanyon Doe since 1980).
Administrative units: 9 provinces. **Capital:** Monrovia 171,580 inhab. (1974). – **Currency:** Liberian dollar = 100 cents. **Economy:** agricultural country with a relatively developed mining industry. **Agriculture** (1977, in 1,000 tons): rice 230, manioc 270, palm kernels 13, palm oil 27, coffee 5, bananas 67, pineapples; roundwood 2.1 mn. cub.m (1976), natural rubber 80,000 tons. **Mining** (1976, in 1,000 tons): iron ore 14,010 (metal content), gold 140 kg, diamonds 400,000 carats. – **Communications:** railways 560 km, roads 4,749 km. Merchant shipping 79,983,000 GRT (1977, first place in the world). – **Exports:** iron ore, natural rubber, diamonds, wood, coffee.

LIBYA

Al-Jamahiriyah Al-Arabiya Al-Libya Al-Shabiya Al-Ishtirakiya, area 1,759,540 sq.km, population 2,608,000 (1978), **republic** (Secretary General – Gen. Muammar al-Qadhafi since 1969).
Administrative units: 10 provinces. **Capital:** Tarābulus (Tripoli) 551,477 inhab. (1973); **other towns** (1973, in 1,000 inhab.): Banghāzī 282, Misrātah 103, Zāwiyat al-Baydā' 59 (1970). – **Currency:** Libyan dinar = 1000 dirhams.
Economy: developing agricultural country with major petroleum mining. Arable land 1.4% only on the coast and in the oases, unproductive land 94.4%. **Agriculture** (1977, in 1,000 tons): barley 200, olives 100, olive oil 20, dates 70, figs, agrums, tobacco, alfalfa; raising of livestock: sheep 3,000,000, goats, camels. **Mining** (1977, in 1,000 tons): crude petroleum 100,140, deposits Sarīr, Zaltan, Jālū etc.; pipelines to the coast; natural gas 3.8 billion cub.m (1976), salt. – **Communications:** roads 3,850 km. Merchant shipping 674,000 GRT (1977). – **Exports:** petroleum 95%.

MADAGASCAR

République démocratique de Madagascar – Repoblika demokratika Malagasy, area 587,041 sq.km, population 8,881,000 (1978), **republic** (President Capt. Didier Ratsiraka since 1975).
Administrative units: 6 provinces. **Capital:** Antananarivo 366,530 inhab. (1972); **other towns** (1972, in 1,000 inhab.): Mahajanga 67, Toamasina 60, Fianarantsoa 59, Antseranana 45. – **Currency:** Malagasy franc = 100 centimes. **Economy:** developing agricultural country. **Agriculture** (1977, in 1,000 tons): rice 2,200, maize (corn), sweet potatoes 287, manioc 1,300, cotton, sisal 21, sugarcane 1,200, groundnuts, coconuts, oranges 85, bananas 449, pineapples, pepper 3,500 tons, vanilla 850 tons, cloves, coffee 95, tobacco; **livestock** (1977, in 1,000 head): cattle 9,800, pigs 655, sheep 744, goats 1,376; poultry 13,628; fish catch 54,950 tons; roundwood 6.4 mn. cub.m. **Mining** (1975, in 1,000 tons, metal content): chromium 87.7 (1976), diamonds 1,740,000 carats, precious stones, ilmenite, graphite 17.8, salt 27, mica. – **Communications:** railways 884 km, roads 8,600 km. – **Exports:** coffee, cloves, vanilla, sugar, chromium.

MALAWI

Republic of Malawi, area 118,484 sq.km, population 5,571,567 (1977), **member of the Commonwealth** (President Dr Hastings Kamuzu Banda since 1966).
Administrative units: 3 regions. **Capital:** Lilongwe 102,924 inhab. (1977); **other towns** (1977, in 1,000 inhab.): Blantyre 229, Zomba 16. – **Currency:** Malawi-kwacha = 100 tambals. **Economy:** developing agricultural country. **Agriculture** (1977, in 1,000 tons): maize (corn) 1,250, rice, sorghum, manioc, sweet potatoes, cotton, sugarcane, groundnuts 100, tung oil, tea 32, bananas; **livestock** (1977, in 1,000 head): cattle 729, goats 763; inland fishing 74,900 tons (1976). **Deposits** of coal, iron ore, gold, uranium, bauxite and asbestos. – **Communications:** railways 566 km, roads 10,980 km. – – **Exports:** tobacco, tea, groundnuts, cotton.

MALI

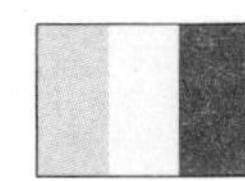

République du Mali, area 1,239,710 sq.km, population 6,035,272 (1977), **republic** (President Gen. Moussa Traoré since 1968).
Administrative units: 7 regions and 1 capital district. **Capital:** Bamako 404,022 inhab. (1977, with agglomeration);

1
Nouamrhar
16°
Akjoujt
2
12°
3
8°
4
Talorza
A
Nouakchott
MAURITANIA
Tidjikdja
Tichit
Moudjéria
Boutilimit
Kroufa
Mederdra
Aleg
Tamchaket
Qualâta
Rosso
Podor
Bogué
Dagana
Kaédi
Kiffa
'Ayoûn el 'Atroûs
Néma
Timbédra
Mbout
Sénégal
St.-Louis
Louga
Kébémer
Linguère
Matam
SENEGAL
Sélibaby
DAKAR
Thiès
Touba
Rufisque
Diourbel
C. Vert
Bakel
Nioro du Sahel
Ballé
Nara
Ambidédi
Sandaré
Sokolo
Kidira
Guinguinéo
Kayes
Diéma
Mourdiah
Kaolack
Koungheul
Niono
Banjul
Maka
Tambacounda
Sadiola
Bafoulabé
Didiéni
B
GAMBIA
Banamba
Brikama
Basse Santa Su
Toukoto
M
Ségou
Kolda
Vélingara
Gambie
Kita
Niger
Doura
Ziguinchor
Farim
Kédougou
Satadougou
Bafing
Kati
Koulikoro
Canchungo
GUINEA-
Koundara
Bamako
12°
Bissau
Geba
Bafatá
Niagassola
Korodougou
BISSAU
Gaoual
Bolama
Madina do Boé
Sélingué
Siguiri
Arq. dos Bijagós
Labé
Dinguiraye
Bougouni
Sikasso
Boké
Télimélé
Pita
GUINEA
Kamsar
Dalaba
Dabola
Kouroussa
Boffa
Fria
Mamou
Kankan
Baoulé
C
Kindia
Niger
Faranah
Milo
Tingréla
Dubréka
CONAKRY
Kabala
Boundiali
SIERRA
Odienné
Korhog
Port Loko
Makeni
Kérouané
Pepel
Lunsar
Kissidougou
Freetown
Rokel
Sewa
Sefadu
LEONE
Macenta
Beyla
IV
8°
Moyamba
Bo
Pendembu
Voinjama
Touba
Mankono
Kenema
Nzérékoré
Séguéla
Bonthe
Yekepa
Man
Zuénoula
Zimi
Bendaja
St. Paul
Sanniquellie
Bomi Hills
Gbarnga
CO
Ganta
Daloa
Robertsport
Bong
LIBERIA
D
Monrovia
Harbel
Toulépleu
Guiglo
Nzi
Marshall
Cestos
Buchanan
Tchien
Taï
Soubré
Gagno
Sassandra
Cavalla
Greenville
Sassandr
Tabou
San-Pédro
Grand Cess
4°
Harper
C. Palmas
ATLANTIC OCEAN
E
1 : 12 500 000
0 100 200 300 Km
0 50 100 150 200 Mi
16°
2
12°
3
8°
4

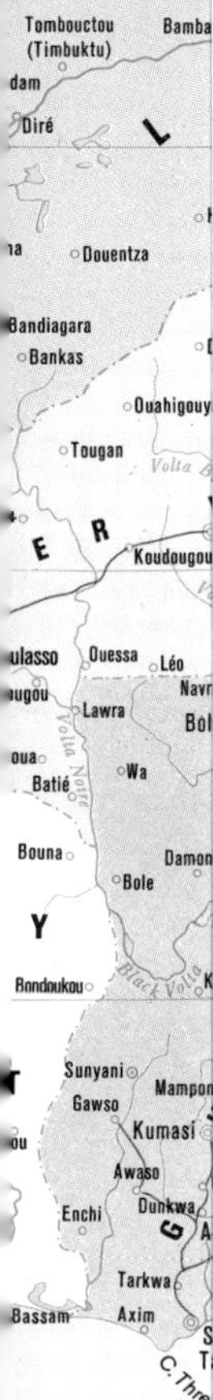

NAMIBIA

Suidwes-Afrika – South-West Africa, area 824,295 sq.km, population 936,000 (1977), **trust territory of the United Nations administered by the Republic of South Africa. Administrative units:** 5 bantustans. **Capital:** Windhoek 64,700 inhab. **Currency:** South African rand = 100 cents. **Economy:** livestock (1977, in 1,000 head): cattle 2,875, sheep 5,085 (3 mn. karakul sheep), goats 2,026; fish catch 574,200 tons (1976). **Mining** (1977, in 1,000 tons, metal content): diamonds 1,694,000 carats, tin 696 tons, zinc 45.5, lead 41, copper 46.6, tungsten, vanadium 709 tons, silver, uranium, salt 210. – **Communications:** railways 2,354 km, roads 34,915 km.

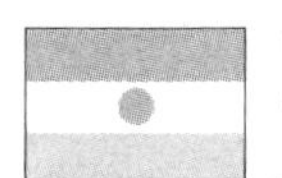

NIGER

République du Niger, area 1,266,995 sq.km, population 4,994,000 (1978), **republic** (President of the Supreme Military Council Col. Seyni Kountché since 1974).
Administrative units: 7 departments. **Capital:** Niamey 130,299 inhab. (1975).- **Currency:** CFA franc = 100 centimes. **Economy:** developing agricultural country with animal production predominating. **Agriculture** (1977, in 1,000 tons): millet 1,000, sorghum, rice, manioc, cotton, sugarcane, groundnuts 60, dates; **livestock** (1977, in 1,000 head): cattle 2,900, sheep 2,560, goats 6,200, horses 210, asses 370, camels 265. **Mining** (1976): uranium 1,460 tons, tin. – **Communications:** roads 7,468 km, river transport on the R. Niger. – **Exports:** uranium (over 60%), groundnuts, cattle.

NIGERIA

Federal Republic of Nigeria, area 923,768 sq.km, population 80,627,000 (1978), **federal republic, member of the Commonwealth** (President Alhaji Shehu Shagari since 1979).
Administrative units: 19 states. **Capital:** Lagos 1,060,848 inhab. (1975); **other towns** (1975, in 1,000 inhab.): Ibadan 847, Ogbomosho 432, Kano 399, Oshogbo 282, Ilorin 282, Abeokuta 253, Port Harcourt 242, Ilesha 224, Zaria 224, Onitsha 220, Iwo 214, Ado-Ekiti 213, Kaduna 202, Mushin 197, Maiduguri 189. **Density** 87 persons per sq.km; average annual rate of population increase 3%. – **Currency:** naira = 100 kobo.
Economy: agricultural country with rapidly developing industry. **Agriculture** (1977, in 1,000 tons): maize (corn) 1,395, rice 600, millet 2,600 (third world producer), sorghum 3,750, manioc 10,600, sweet potatoes 200, cotton -seed 146, -lint 73, sugarcane 750, coconuts 90, palm kernels 340, palm oil 660 (second world producer), groundnuts 330, sesame 70, coffee, cocoa beans 210, tropical fruit, tobacco 17; **livestock** (1977, in 1,000 head): cattle 11,500, sheep 8,100, goats 23,600 (third world population), pigs 950, poultry 95,000, eggs 119,700 tons; fish catch 494,800 tons (1976); roundwood 68.9 mn. cub.m (1976), natural rubber 90,000 tons (1977). **Mining** (1977, in 1,000 tons): crude petroleum 104,304 (deposits in the Niger delta), natural gas 632 mn. cub.m (1976), coal 288, tin 3,264 tons, columbite 1,312 tons (1975). Electricity 3,400 mn. kWh (1976), of this 74% hydro-electric power stations. **Production** (1976, in 1,000 tons): food processing of domestic agricultural production; beer 3.2 mn. hl, meat 488, milk 316; 12.6 billion cigarettes, cement 1,274, woven cotton fabrics 367 mn. sq.m.- **Communications:** railways 3,524 km, roads 96,932 km. Merchant shipping 336,000 GRT (1977), river transport. – **Exports:** petroleum 90%, cocoa, tin, palm kernels, groundnut oil.

RWANDA

République rwandaise – Republica y'u Rwanda, area 26,338 sq.km, population 4,450,000 inhab. (1978), **republic** (President Gen. Juvénal Habyalimana since 1973).
Administrative units: 10 prefectures. **Capital:** Kigali 90,000 inhab. (1977).- **Currency:** Rwanda franc = 100 centimes. **Economy:** developing agricultural country. **Agriculture:** coffee (20,000 tons in 1977), sorghum, potatoes, manioc, sweet potatoes, groundnuts, tea, tobacco; **livestock:** cattle 668,000 head (1977), sheep and goats; roundwood 3.9 mn. cub.m (1976). – **Exports:** coffee, tin, tea, tungsten.

SÃO TOMÉ AND PRINCIPE

República democrática de São Tomé e Príncipe, area 964 sq.km, population 82,000 (1977), **republic** (President Dr Manuel Pinto da Costa since 1975).
Capital: São Tomé 17,380 inhab. (1970). – **Currency:** dobra = 100 centavos. – **Economy:** agriculture (1977): coffee, cocoa beans 9,000 tons, coconuts 43,000 tons, copra, palm oil, spice.

add page 130

other towns (1972, in 1,000 inhab.): Mopti 43, Ségou 40, Kayes 37. – **Currency:** Mali franc = 100 centim
Economy: agricultural country with predominating livestock production. **Agriculture** (1977, in 1,000 tons): r
182, maize (corn), millet, manioc, sweet potatoes, cotton -seed 78,-lint 43, groundnuts 230; **livest**
(1977, in 1,000 head): cattle 4,076, sheep 4,437, goats 4,057, horses 160, asses 429, camels 188, pou
10,284; inland fishing 100,000 tons (1976). Considerable **resources** of iron ore, manganese, lithium and baux
– **Communications:** railways 646 km, roads 14,704 km, river transport on the R. Niger. – **Exports:** livest
cotton, groundnuts, fish.

MAURITANIA

République Islamique de Mauretanie, area 1,030,700 sq.km, population 1,481,0 (1977), **republic** (President Lt.-Col. Mohamed Khouna Ould Haidalla since 1980).

Administrative units: 12 regions and 1 capital district. **Capital:** Nouakchott 134,986 inhab. (1976). – **Curren** ouguiya = 5 khoums. **Economy:** agricultural country with predominating livestock production. **Agricultu** growing of millet, maize (corn), rice, gum arabic 5,646 tons (1972); **livestock** (1977, in 1,000 head): ca 1,400, sheep 4,700, goats 3,100, camels 700. **Mining:** iron ore 6,233,000 tons (1976, metal content), copp ilmenite, salt, phosphates. – **Communications:** railways 652 km, roads 6,186 km. – **Exports:** iron and cop ore, gum arabic, fish, cattle.

MAURITIUS AND DEPENDENCIES

Area 2,045 sq.km, population 909,000 (1977), **independent state, member of Commonwealth** (Prime Minister Dr Sir Seewoosagur Ramgoolam since 1967).
Dependencies: Agalega, Cargados-Carajos and other islands. Rodrigues Island is now part of Mauri
Administrative units: 9 districts. **Capital:** Port Louis 139,399 inhab. (1976, with agglomeration); **other tow** (in 1,000 inhab.): Beau-Bassin 82, Curepipe 53. **Density** 445 persons per sq.km. – **Currency:** Mauri rupee = 100 cents.
Economy: agricultural country with monocultural cultivation of sugarcane – 6,900,000 tons in 1977; cocon copra, coffee, tea, bananas, vanilla, tobacco. Mining of salt. Food industry (sugar 715,000 tons in 19 petroleum refinery. – **Communications:** roads 1,308 km. – **Exports:** sugar more than 80%, tea etc.

MOROCCO

al-Mamlaka al-Maghrebia, area 458,730 sq.km, population 18,592,000 (1978), **k dom** (King Hassan II since 1961).
Administrative units: 28 provinces and 2 urban prefectures. **Capital:** Rabat 367,620 inhab. (1971, with ag meration 530,366 inhab.); **other towns** (1971, in 1,000 inhab.): Casablanca (Dar-el-Beida) 1,506, Marrak 333, Fès 325, Meknès 248, Tanger 188, Oujda 176, Salé 156, Kenitra 139, Tétouan 139, Safi 129, Khouribga Mohammedia 70, Agadir 61, El-Jadida 56, Beni-Mellal 54.– **Currency:** dirham = 100 centimes.
Economy: agricultural country with a relatively developed mining industry. **Agriculture** (1977, in 1,000 to wheat 1,288, barley 1,347, maize (corn), rice, potatoes, cotton, sugar beet, groundnuts, olives 252, olive oil oranges 610, tangerines 130, lemons, grapes 215, dates 70, figs, tomatoes 470, tobacco; **livestock** (197 1,000 head): cattle 3,650, sheep 14,300, goats 4,940, asses 1,200, horses 312, camels 210; poultry 21,0 fish catch 281,434 tons (1976). **Mining** (1977, in 1,000 tons, metal content): coal 708, natural gas, iron 202, manganese 60, copper, lead 105, zinc 15.4, nickel, cobalt 2,454 tons (1971), molybdenum, antim 1,713 tons, silver 70 tons, phosphates 17,600 (third world producer, deposits Khouribga and Youssou pyrites, fluorite. Electricity 3,329 mn. kWh (1976). **Industry:** developed food processing (mills, oil presses, su refineries; fish, fruit and vegetable canning plants). Traditional textile industry; developing chemical construction industries; processing of petroleum. **Production** (1977, in 1,000 tons): meat 178, milk 518, w 70, cement 2,800. – **Communications:** railways 1,756 km, roads 26,382 km. Tourism 1,107,700 visit (1976). – **Exports:** phosphates (55%), citrus fruit, canned fish etc.

MOZAMBIQUE

República Popular de Moçambique, area 784,961 sq.km, population 9,678 (1977), **people's republic** (President Samora Moises Machel since 1975).
Administrative units: 9 districts. **Capital:** Maputo 354,684 inhab. (1970); **other towns** (1970, in 1, inhab.): Nampula 126, Sofala (Beira) 114, Quelimane 72.– **Currency:** Mozambique limpad = 100 centa
Economy: developing agricultural country. **Agriculture** (1977, in 1,000 tons): maize (corn) 350, rice sugarcane 2,700, cotton seed 60, cotton lint 30, sisal 18, groundnuts 100, coconuts 420, copra 85, tea citrus fruit, bananas; **livestock** (1977, in 1,000 head): cattle 1,350, goats; roundwood 9.1 mn. cub.m (19
Mining: coal 384,000 tons, copper, bauxite, salt, beryllium, fluorite. Electricity 1,915 mn. kWh, of this from hydro-electric power stations. – **Communications:** railways 3,793 km, roads 39,173 km. – **Exp** cashew nuts, woven fabrics, sugar, cotton, wood.

SENEGAL

République du Sénégal, area 196,722 sq.km, population 5,198,000 (1977), **republic** (President Abdou Diouf since 1981).

Administrative units: 7 regions. **Capital:** Dakar 798,792 inhab. (1976); **other towns** (1976, in 1,000 inhab.): Thiès 117, Kaolack 107, Saint-Louis 81 (1970), Rufisque 58 (1970), Ziguinchor 46 (1970). – **Currency:** CFA franc = 100 centimes. **Economy:** agricultural country, cultivates chiefly groundnuts. **Agriculture** (1977, in 1,000 tons): groundnuts 700, millet 432, rice 62, oranges, bananas; **livestock** (1977, in 1,000 head): cattle 2,440, sheep 1,760, goats 895; fish catch 360,900 tons (1976); roundwood 2.7 mn. cub.m (1976). **Mining** (1976, in 1,000 tons): phosphates 1,552, salt 142. – **Communications:** railways 1,186 km, roads 15,422 km, river transport. – **Exports:** groundnut oil, phosphates.

SEYCHELLES AND DEPENDENCIES

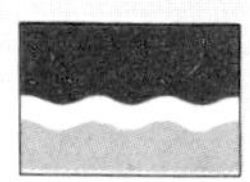

Republic of Seychelles, area 444 sq.km, population 62,000 (1977), **republic, member of the Commonwealth** (President F. Albert René since 1977).

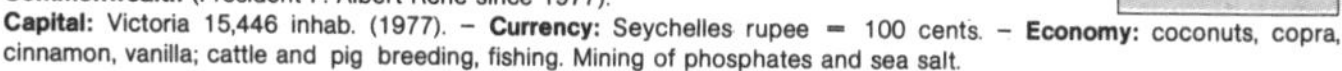

Capital: Victoria 15,446 inhab. (1977). – **Currency:** Seychelles rupee = 100 cents. – **Economy:** coconuts, copra, cinnamon, vanilla; cattle and pig breeding, fishing. Mining of phosphates and sea salt.

SIERRA LEONE

Republic of Sierra Leone, area 71,740 sq.km, population 3,470,000 (1977), **republic, member of the Commonwealth** (President Dr Siaka Probyn Stevens since 1971).

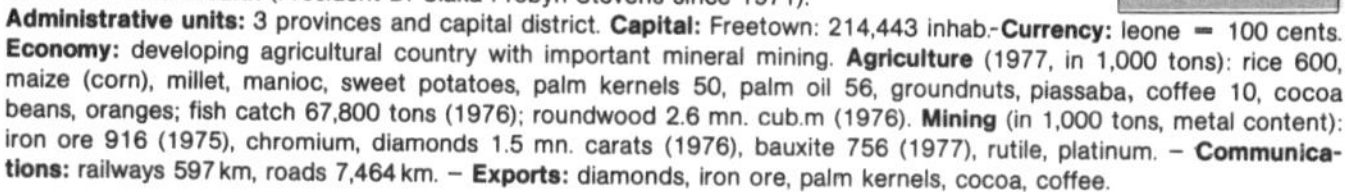

Administrative units: 3 provinces and capital district. **Capital:** Freetown: 214,443 inhab.-**Currency:** leone = 100 cents. **Economy:** developing agricultural country with important mineral mining. **Agriculture** (1977, in 1,000 tons): rice 600, maize (corn), millet, manioc, sweet potatoes, palm kernels 50, palm oil 56, groundnuts, piassaba, coffee 10, cocoa beans, oranges; fish catch 67,800 tons (1976); roundwood 2.6 mn. cub.m (1976). **Mining** (in 1,000 tons, metal content): iron ore 916 (1975), chromium, diamonds 1.5 mn. carats (1976), bauxite 756 (1977), rutile, platinum. – **Communications:** railways 597 km, roads 7,464 km. – **Exports:** diamonds, iron ore, palm kernels, cocoa, coffee.

SOMALIA

Al-Jumhouriya As-Somaliya Al-Domocradia, area 637,657 sq.km, population 3,354,000 (1977), **republic** (President of the Supreme Revolutionary Council Marshal Mohammed Siyad Barre since 1969).

Administrative units: 8 regions. **Capital:** Mogadisho 230,000 inhab. (1972); **other towns** (1973, in 1,000 inhab.): Marka 62, Hargeysa 60, Kismayu 60, Berbera 50 (1967). – **Currency:** Somalian shilling = 100 centimes.

Economy: agricultural country. **Agriculture** (1977, in 1,000 tons): maize (corn), sorghum, cotton, sugarcane 400, groundnuts, sesame, bananas 150; fish catch 32,600 tons (1976); collection of gum arabic; **livestock** (1977, in 1,000 head): cattle 2,654, sheep 7,212, goats 8,212, camels 2,000 (second world population). – **Communications:** roads 17,223 km. – **Exports:** cattle, meat, bananas.

SOUTH AFRICA

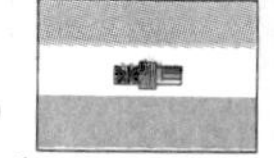

Republiek van Suid-Afrika – Republic of South Africa, area 1,221,037 sq.km, population 26,764,000 (1977), **republic** (President Marais Viljoen since 1979).

Administrative units: 4 provinces, 8 bantustans. **Capital:** Pretoria 543,950 inhab. (1970); **other towns** (1970, in 1,000 inhab.): Johannesburg 655 (1,432 with agglomeration), Cape Town 691 (1,097 with agglom.), Durban 730, Port Elizabeth 387 (469 with agglom.), Vereeniging 170, Alberton 150, Benoni 149, Bloemfontein 148, Welkom 132, Germiston 126, East London 118, Pietermaritzburg 114, Kimberley 104. – **Currency:** rand = 100 cents. **Economy:** highly developed industrial and agricultural country with enormous mineral wealth, economically the most important country of Africa. **Agriculture** (1977, in 1,000 tons): wheat 1,815, maize (corn) 9,714, sorghum 386, potatoes 740, cotton -seed 77, -lint 39, sugarcane 19,770, sunflower seeds 476, groundnuts 241, soybeans, oranges 630, lemons 20, grapefruits 90, bananas 100, pineapples 182, grapes 1,100, fruit, tobacco 39; **livestock** (1977, in 1,000 head): cattle 12,800, goats 5,250, sheep 31,200, pigs 1,400, horses, eggs 170,400 tons; fish catch 638,000 tons (1976), whaling; roundwood 10.5 mn. cub.m (1976). **Mining:** in a number of branches leading world output (1976, in 1,000 tons): coal 85,572 (1977), iron ore 16,395 (1977), manganese 2,409 (second world producer), chromium 1,087 (leading world producer), copper 216 (1977), tin 2,880 tons (1977), zinc 73 (1977), nickel 23,371 tons, antimony 12,599 tons (1977, second world producer), vanadium 9,892 tons (leading world producer), gold 711,500 kg (leading world producer), silver 138 tons (1977), platinum 58,000 kg (1974 – leading world producer), magnesite 63, uranium 3,412 tons (third world producer), diamonds 7,022,000 carats (third world producer), salt 224, phosphates 12,362, pyrites, sulphur, mica, asbestos 370 (third world producer). Electricity 80,196 mn. kWh (1977). **Principal industries:** iron metallurgy, machinery, shipbuilding, chemical, construction, textiles and foodstuffs. **Production** (1976, in 1,000 tons): pig iron 6,611, coke 4,608, crude steel 6,926, naphtha 22,000, copper 198, aluminium 78, nitrogenous fertilizers 332, phosphate fertilizers 385, synthetic rubber 34.6, cement 7,048, paper 904, cotton yarn 48.3, woven cotton fabrics 105 mn. m, meat 908, butter 26.3, milk 2,529, beer 5.4 mn. hl, wine 6.3 mn. hl; 23.4 billion tobacco products. – **Communications:** railways 22,432 km, roads 320,000 km. – **Exports:** diamonds, food products, metals, minerals, textiles, machines.

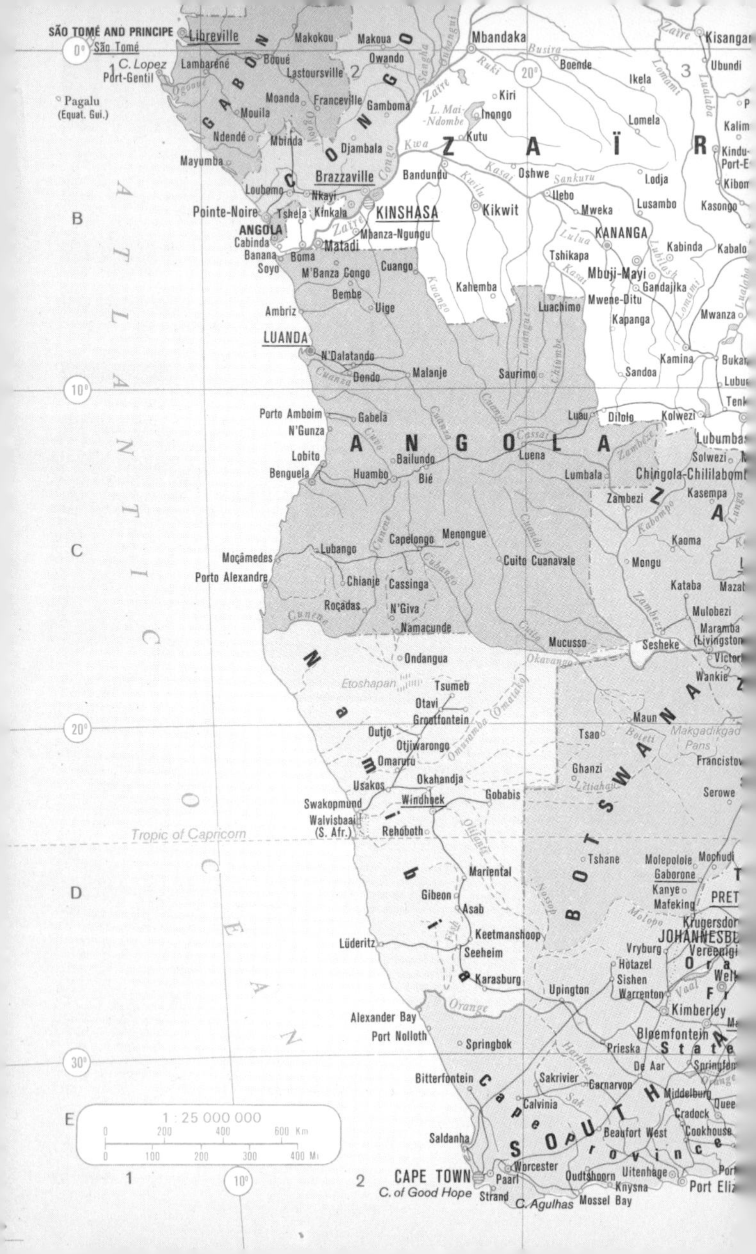
SÃO TOMÉ AND PRINCIPE
Libreville
São Tomé
C. Lopez
Port-Gentil
Pagalu
(Equat. Gui.)
GABON
CONGO
Lambaréné
Makokou
Makoua
Owando
Booué
Lastoursville
Moanda
Franceville
Mouila
Gamboma
Ndendé
Mbinda
Djambala
Mayumba
Brazzaville
Loubomo
Nkayi
Pointe-Noire
Tshela
Kinkala
KINSHASA
ANGOLA
Cabinda
Banana
Soyo
Boma
Matadi
Mbanza-Ngungu
Mbandaka
Boende
Ikela
Kisangani
Ubundi
Kiri
Inongo
L. Mai-Ndombe
Kutu
Lomela
ZAÏRE
Bandundu
Oshwe
Lodja
Kikwit
Ilebo
Mweka
Lusambo
Kasongo
KANANGA
Tshikapa
Kabinda
Kabalo
Mbuji-Mayi
Gandajika
Mwene-Ditu
Kahemba
Luachimo
Kapanga
Mwanza
Kamina
Sandoa
Cuango
M'Banza Congo
Bembe
Ambriz
Uige
LUANDA
N'Dalatando
Dondo
Malanje
Saurimo
Porto Amboim
Gabela
N'Gunza
Luau
Dilolo
Kolwezi
Lubumbashi
ANGOLA
Lobito
Benguela
Bailundo
Huambo
Bié
Luena
Lumbala
Solwezi
Chingola-Chililabombwe
Zambezi
Kasempa
Capelongo
Menongue
Lubango
Moçâmedes
Porto Alexandre
Chianje
Cassinga
Cuito Cuanavale
Mongu
Kaoma
Roçadas
N'Giva
Namacunde
Kataba
Mulobezi
Maramba (Livingstone)
Mucusso
Sesheke
Ondangua
Etoshapan
Tsumeb
Otavi
Grootfontein
Outjo
Otjiwarongo
Omaruru
Usakos
Okahandja
Swakopmund
Windhoek
Walvisbaai (S. Afr.)
Rehoboth
Gobabis
Wankie
Maun
Tsao
Makgadikgadi Pans
Francistown
Ghanzi
Serowe
BOTSWANA
Namibia
Tropic of Capricorn
Tshane
Molepolole
Mochudi
Gaborone
Kanye
Mafeking
Mariental
Gibeon
Asab
Keetmanshoop
Lüderitz
Seeheim
Karasburg
Krugersdorp
JOHANNESBURG
Vryburg
Vereeniging
Hotazel
Sishen
Warrenton
Upington
Kimberley
Alexander Bay
Port Nolloth
Springbok
Bloemfontein
Prieska
De Aar
Springfontein
Bitterfontein
Sakrivier
Carnarvon
Middelburg
Calvinia
Cradock
Cape Province
SOUTH AFRICA
Saldanha
Beaufort West
Cookhouse
Worcester
Paarl
Oudtshoorn
Uitenhage
CAPE TOWN
C. of Good Hope
Strand
C. Agulhas
Knysna
Mossel Bay
Port Elizabeth
ATLANTIC OCEAN
Ogooué
Sangha
Ubangui
Zaire
Congo
Busira
Ruki
Kwa
Kasai
Kwilu
Sankuru
Lulua
Lubilash
Lomami
Lualaba
Kwango
Luangue
Chiumbe
Cuanza
Cuvo
Cuango
Cassai
Zambeze
Kabompo
Cunene
Cubango
Cuando
Cuito
Okavango
Omuramba (Omatako)
Boteti
Letiahau
Olifants
Nossop
Molopo
Fish
Orange
Vaal
Hartbees
Sak
0°
10°
20°
30°
B
C
D
E
1
2
3
1 : 25 000 000
0 200 400 600 Km
0 100 200 300 400 Mi

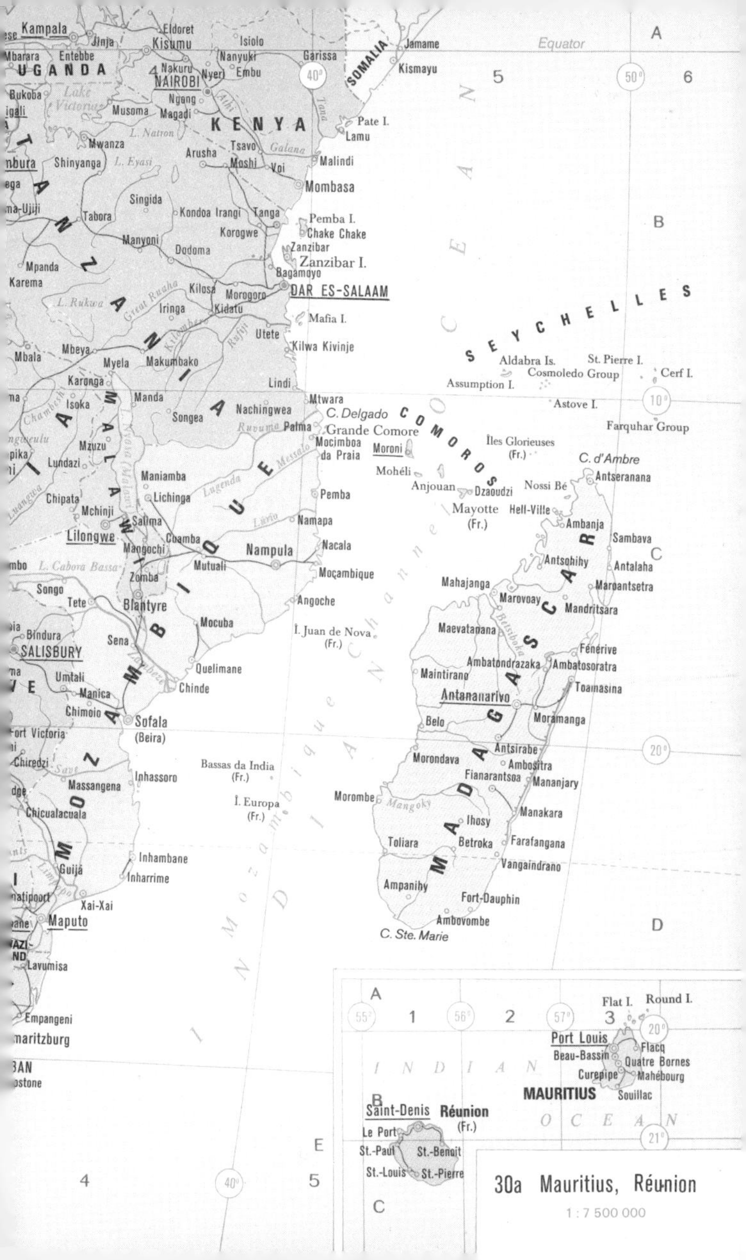

30a Mauritius, Réunion
1 : 7 500 000

SUDAN

Jamhuryat es-Sudan Al Democratia, area 2,505,813 sq.km, population 16,953,000 (1977), **republic** (President Marshal Jafaar Mohammed an Nimeiry since 1969).
Administrative units: 6 provinces and 1 autonomous region. **Capital:** Al-Khurtūm (Khartoum) 333,906 inhab. (1977); **other towns** (1973, in 1,000 inhab.): Umm Durmān (Omdurman) 299, Al-Khurtūm Bahrī 151, Būr Sūdān (Port Sudan) 133, Wad Madanī 107, Al-Ubayyid 90, An-Nuhūd 80 (1972). – **Currency:** Sudanese pound = 100 piastres.
Economy: developing agricultural country. **Agriculture** (1977, in 1,000 tons): millet 410, sorghum 1,600, wheat, rice, cotton -seed 350, -lint 208, sugarcane 1,640, groundnuts 850, sesame 220 (third world producer), dates 106, bananas 83, grapes 53, tomatoes 143; **livestock** (1977, in 1,000 head): cattle 15,892, sheep 15,248, goats 11,592, camels 2,813 (leading world population), asses 675; roundwood 22.4 mn. cub.m, gum arabic 43,918 tons (1976, 90% of world production). **Mining:** iron ore, copper, chromium, manganese, salt. – **Communications:** railways 4,787 km, roads 73,577 km. – **Exports:** cotton 45%, groundnuts, sesame, gum arabic, livestock.

SWAZILAND

Kingdom of Swaziland – Ka Ngwane, area 17,363 sq.km, population 507,000 (1977), **constitutional monarchy, member of the Commonwealth** (King Sobhuza II since 1921).
Administrative units: 4 districts. **Capital:** Mbabane 24,000 inhab. (1975). – **Currency:** lilangeni = 100 cents. **Economy:** developing agricultural country. **Agriculture** (1977, in 1,000 tons): maize (corn), rice, cotton, sugarcane 203, citrus fruit 75, cattle and goat breeding. **Mining:** iron ore 1,229,000 tons (1976), coal, tin, asbestos.– **Exports:** iron ore, sugar etc.

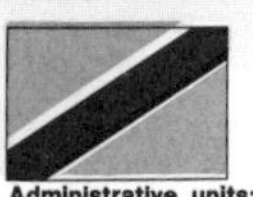

TANZANIA

United Republic of Tanzania, area 945,087 sq.km, population 16,560,000 (1978), **republic, union of Tanganyika and Zanzibar, member of the Commonwealth** (President Dr Julius Kambarage Nyerere since 1964).
Administrative units: 20 provinces. **Capital:** Dar-es-Salaam 517,000 inhab. (1975); **other towns** (in 1,000 inhab.): Zanzibar 70 (1970), Tanga 65 (1970), Dodoma 54 (1973). – **Currency:** Tanzanian shilling = 100 cents.
Economy: agricultural country. **Agriculture** (1977, in 1,000 tons): maize (corn) 968, rice 194, millet, sorghum, manioc 4,000, sweet potatoes 335, potatoes, cotton -seed 113, -lint 59, sisal 110 (second world producer), sugarcane 1,297, coconuts 314, copra 27, palm oil, cashew nuts 96, groundnuts, pyrethrum 3,700 tons, coffee 59, tea 17, bananas 790, pineapples 46, tobacco 21, cloves 4,000 tons (1974, leading world producer, Zanzibar and Pemba islands); **livestock** (1977, in 1,000 head): cattle 14,817, sheep 3,000, goats 4,700, asses; fish catch 180,700 tons; roundwood 37.4 mn. cub.m (1976). **Mining** (1976): gold, tin, diamonds 450,000 carats, salt, mica. – **Communications:** railways 3,500 km, roads 47,866 km. – **Exports:** coffee, sisal, cotton, diamonds etc.

TOGO

République Togolaise, area 56,600 sq.km, population 2,409,000 (1978), **republic** (President Gen. Gnassingbe Eyadema since 1967).
Administrative units: 5 regions. **Capital:** Lomé 214,200 inhab. (1976).- **Currency:** CFA franc = 100 centimes. **Economy:** developing agricultural country. **Agriculture** (1977, in 1,000 tons): millet, maize (corn), manioc 468, cotton, coconuts, copra, palm oil 6.2, groundnuts, coffee 12, cocoa beans 17, bananas; raising of cattle, sheep, goats, pigs; fish catch 14,400 tons (1976). **Mining** (in 1,000 tons): phosphates 2,067 (1976), deposits of iron ore and bauxite. – **Communications:** railways 498 km, roads 7,175 km. – **Exports:** phosphates, cocoa, coffee.

TUNISIA

Al-Djoumhouria Attunisia, area 164,150 sq.km, population 6,216,000 (1978), **republic** (President Habib Bourguiba since 1957).
Administrative units: 18 governorates. **Capital:** Tunis 550,404 inhab. (1975, with agglomeration 647,640); **other towns** (1975, in 1,000 inhab.): Sfax 171, Djerba 70, Sousse 69.5.- **Currency:** Tunisian dinar = 1,000 millimes.
Economy: agricultural country with developed mining industry. **Agriculture** (1977, in 1,000 tons): wheat 570, barley, potatoes, olives 615, olive oil 132, oranges 72, tangerines 38, lemons 25, dates 50, fruit, almonds, grapes 170, tomatoes 270, tobacco; **livestock** (1977, in 1,000 head): cattle 890, sheep 3,600, goats 950, camels 190, asses; fish catch 42,700 tons; cork 8,800 tons, alfalfa 79,000 tons. **Mining** (1977, in 1,000 tons): crude petroleum 4,260, iron ore 255, lead 10.2, zinc 5.8, silver, salt 288, phosphates 3,600. – **Communications:** railways 2,025 km, roads 21,309 km. – **Exports:** petroleum, olive oil, phosphates, wine, fruit.

UGANDA

Republic of Uganda, area 236,036 sq.km, population 12,780,000 (1978), **republic, member of the Commonwealth** (President Milton Obote since 1980).
Admininistrative units: 4 regions. **Capital:** Kampala 330,700 inhab. (1969); **other towns** (1969, in 1,000 inhab.): Jinja 53, Bugembe 47. **Currency:** Uganda shilling = 100 cents. **Economy:** developing agricultural country. **Agriculture** (1977, in 1,000 tons): maize (corn) 661, millet 471, sorghum 516, manioc 1,100, sweet potatoes 660, cottonseed 91, groundnuts 208, sesame 40, tea 15, bananas 350; **livestock** (1977, in 1,000 head): cattle 4,900, sheep 1,100, goats; fish

catch 152,400 tons; roundwood 14.6 mn. cub.m (1976). **Mining** (1976, tons, metal content): copper 7,000, tin, tungsten 139, phosphates, salt. – **Communications:** railways 1,301 km, roads 28,140 km.- **Exports:** coffee, cotton, tea, copper.

UPPER VOLTA

République de Haute-Volta, area 274,200 sq.km, population 6,319,000 (1977), **republic** (Head of State Chairman of the Military Committee Col. S. Zebro since 1980).
Administrative units: 10 departments. **Capital:** Ouagadougou 168,607 inhab. (1975). – **Currency:** CFA franc = 100 centimes. **Economy:** developing agricultural country. **Agriculture** (1977, in 1,000 tons): maize (corn), millet 330, sorghum 600, rice 23, manioc, cotton -seed 35, -lint 20, sugarcane 300, groundnuts 89, sesame, tobacco; **livestock** (1977, in 1,000 head): cattle 1,900, sheep 1,300, goats 2,377. – **Communications:** railways 517 km, roads 16,662 km.

WESTERN SAHARA

Area 266,000 sq.km, population 152,000 (1978). **The territory occupied by Morocco. Capital:** El Aaiún 28,010 inhab. (1974). – **Economy:** large supplies of phosphates (mining 173,000 tons in 1976) and deposits of gold and uranium.

ZAÏRE

République du Zaïre, area 2,345,409 sq.km, population 27,080,000 (1978), **republic** (President Gen. Mobutu Sese Seko since 1965).
Administrative units: 8 regions and capital district. **Capital:** Kinshasa 2,008,352 inhab. (1974); **other towns** (1974, in 1,000 inhab.): Kananga 601, Lubumbashi 403, Mbuji-Mayi 337, Kisangani 311, Bukavu 182, Kikwit 150, Likasi 146 (1970), Matadi 144, Mbandaka 134. – **Currency:** zaïre = 100 makuta.
Economy: agricultural country with developed mining industry and metallurgy. **Agriculture** (1977, in 1,000 tons): maize (corn) 515, millet, rice, manioc 12,300 (second world producer), sweet potatoes, cottonseed 36, sisal, sugarcane 810, palm oil 145, groundnuts 330, coffee 93, tea 10, cocoa beans, pineapples 372, tobacco; **livestock** (1977, in 1,000 head): cattle 1,144, goats 2,679, pigs 705; fish catch 117,900 tons; roundwood 13.7 mn. cub.m (1976), natural rubber 27,000 tons. **Mining** (1977, in 1,000 tons, metal content): crude petroleum 1,128, manganese 94.7, copper 427, zinc 73, tin 3,612 tons, cobalt 13,644 tons (1975, leading world producer), tungsten 288 tons, cadmium, gold 2,686 kg, silver, uranium, diamonds 11,820,000 carats (leading world producer). Electricity 3,502 mn. kWh (1976). **Production** (1976, in 1,000 tons): copper 274, zinc 60.6, cement 770.- **Communications:** railways 5,280 km, roads 145,000 km, navigable waterways 13,500 km. – **Exports:** copper, cobalt, coffee, diamonds, palm oil, wood etc.

ZAMBIA

Republic of Zambia, area 752,614 sq.km, population 5,472,000 (1978), **republic, member of the Commonwealth** (President Dr Kenneth David Kaunda since 1964).
Administrative units: 8 provinces. **Capital:** Lusaka 415,000 inhab. (1974); **other towns** (1974, in 1,000 inhab.): Kitwe-Kalulushi 314, Ndola 222, Chingola-Chililabombwe 202, Mufulira 136. – **Currency:** kwacha = 100 ngwee.
Economy: chiefly mining and metallurgy of non-ferrous metals. **Agriculture:** maize (corn) 980,000 tons (1977), manioc, groundnuts, tobacco; **livestock** (1977, in 1,000 head): cattle 1,860, goats; poultry 15,376; fish catch 54,300 tons (1976); roundwood 4 mn. cub.m (1976). **Mining** (1977, in 1,000 tons, metal content): coal 780, copper 774, lead 13.6, zinc 45, manganese, tin, cobalt 1,700 tons, gold 341 kg (1976), silver 33 tons (1976). Electricity 7,034 mn. kWh (1976). – **Communications:** railways 2,187 km, roads 34,963 km. – **Exports:** copper 92%, zinc, cobalt, lead, maize etc.

ZIMBABWE

Republic of Zimbabwe, area 390,580 sq.km, population 6,930,000 (1978), **republic, member of the Commonwealth** (President Canaan Banana since 1980).
Administrative units: 7 provinces. **Capital:** Salisbury 564,000 inhab. (1977, with agglomeration); **other towns** (1977, in 1,000 inhab.): Bulawayo 339, Gwelo 66, Umtali 58, Que Que 52, Wankie 28. – **Currency:** dollar = 100 cents.
Economy: developed agriculture and mining industry. **Agriculture** (1977, in 1,000 tons): maize (corn) 1,300, wheat, millet, cottonseed 66, sugarcane 2,500, groundnuts 120, tea, citrus fruit, bananas, tobacco 88; **livestock** (1977, in 1,000 head): cattle 6,247, sheep 818, goats 2,167; roundwood 5.9 mn. cub.m (1976). **Mining** (1976, in 1,000 tons, metal content): coal 2,820, iron ore 384, chromium 305, copper, tin, nickel, tungsten, gold 25,000 kg, silver, antimony, magnesite, emeralds, phosphates, pyrites, asbestos 165. Electricity 5,653 mn. kWh (1976). – **Communications:** railways 3,278 km, roads 45,856 km. – **Exports:** tobacco, asbestos, copper, chromium, sugar.

BRITISH TERRITORIES:

SAINT HELENA AND DEPENDENCIES – British colony and 5 dependencies. Saint Helena – area 122 sq.km, population 5,147 (1976). **Capital:** Jamestown 1,516 inhab. **Currency:** English pound = 100 pence. **Dependencies:** Ascension, Tristan da Cunha and volcanic islands 297 sq.km.

FRENCH TERRITORY:

RÉUNION – Area 2,510 sq.km, population 496,000 (1978), **French overseas department. Capital:** Saint-Denis 103,513 inhab. (1974). – **Currency:** CFA franc = 100 centimes.

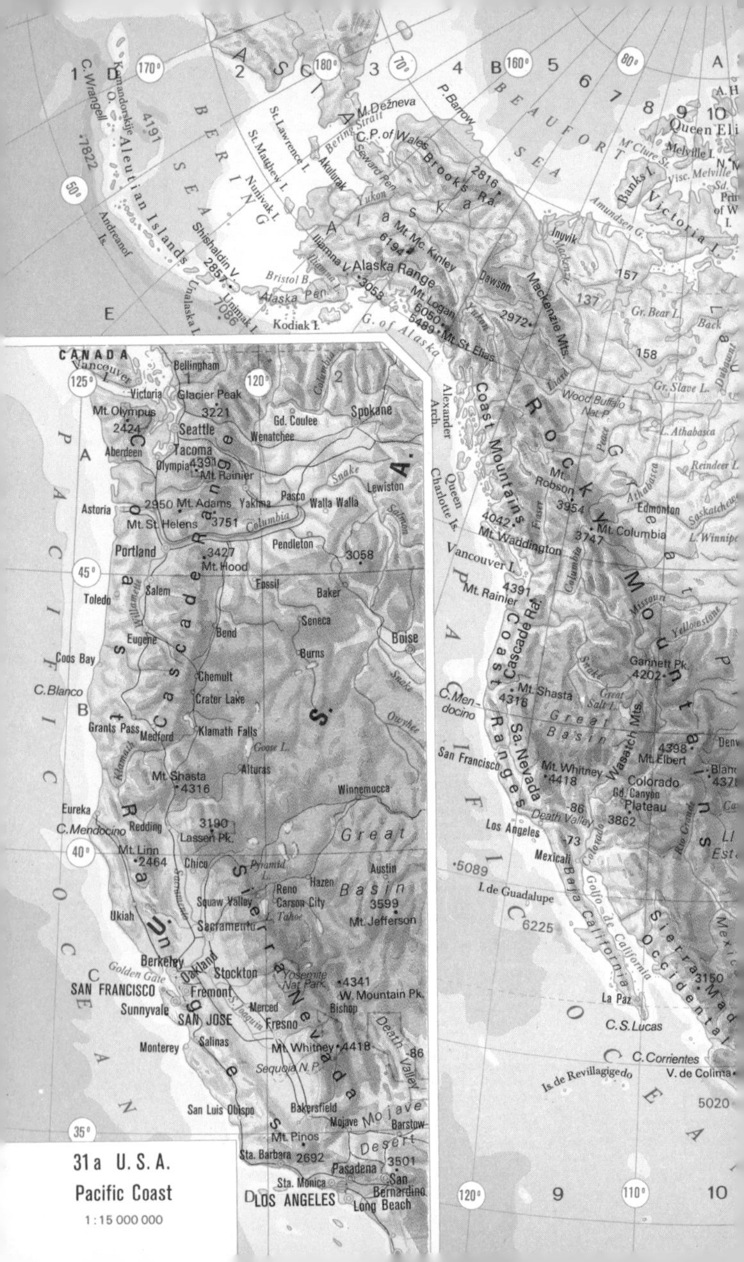

31a U. S. A.
Pacific Coast
1 : 15 000 000

Greenland
Iceland
ATLANTIC OCEAN
LABRADOR SEA
Labrador Peninsula
Plateau
Davis Strait
Hudson Strait
Baffin Island
Newfoundland
St. John's
C. Race
C. Charles
Nova Scotia
C. Sable
G. of St. Lawrence
Cabot Str.
C. Breton I.
Québec
Ottawa
St. Lawrence
L. Superior
L. Huron
L. Michigan
L. Ontario
L. Erie
Chicago
New York
Washington
Appalachian Mountains
Mt. Mitchell
2037
C. Hatteras
Bermuda
Atlantic Lowlands
Mississippi Lowlands
Mexico Plain
New Orleans
Florida
C. Canaveral
Miami
Bahamas
Gulf of Mexico
La Habana
Cuba
Jamaica
Hispaniola
Puerto Rico
Greater Antilles
Lesser Antilles
CARIBBEAN SEA
Yucatán Str.
C. Catoche
Yucatan Pen.
Guadeloupe
Dominica
Martinique
St. Lucia
Barbados
Grenada
Tobago
Trinidad
Caracas
Georgetown
Orinoco
Pta. Gallinas
L. de Maracaibo
P. Cristóbal Colon
5775
Cord. de Mérida
5002
P. Bolivar
5493
Magdalena
Panama Canal
Isthmus of Panama
Pta. Mariato
Chirripó
3837
Managua
L. de Nicaragua
Guatemala
V. Tajumulco
4210
Tropic of Cancer
Equator
Azores
EUROPE
Iberian Pen.
1 : 50 000 000
0 250 500 750 1 000 1 250 Km
0 200 400 600 800 Mi

NORTH AMERICA

The physical geography of the North American continent includes Central America, the islands of the Caribbean Sea and Greenland. It is the largest continent in the western hemisphere. The original term "the New World" (Mundus Novus) was replaced, in the first half of the 16th century, by the name America, after the Florentine Amerigo Vespucci, an Italian seafarer who took part in Columbus's expeditions in the late 15th century.

North America covers an **area** of **24,247,000 sq.km**, i.e. 16.22% of the land area of the world; it has **360 mn. inhabitants** (1978) and a population density of 15 persons per sq.km. **Geographical position:** northernmost point: on the mainland – Cape Murchison on the Boothia Peninsula 71°50' N.Lat., of the entire continent – cape K. Morris Jesup 83°40' N.Lat. in Greenland; southernmost point: cape Punta Mariato on the Península de Azuero in Panama 7°12' N.Lat.; easternmost point: on the mainland – Cape Charles 55°40' W.Long. in Labrador, in Greenland cape Nordostrundingen 11°39' W.Long.; westernmost point: on the mainland – Cape Prince of Wales on the Seward Peninsula in Alaska 168°05' W.Long. (here America comes within 75 km of Asia), of the entire continent – Cape Wrangell on the Aleutian island of Attu 172°27' E.Long. The American continent reaches its maximum length between Point Barrow and Punta Mariato, a total length of roughly 8,700 km; its width from the most westerly point of Alaska as far as Canso on the peninsula of Nova Scotia measures 5,950 km; and the narrowest point is on the Isthmus of Panama, a mere 48 km.

Geological evolution gave North America highly varied contours, an uneven coastline and large numbers of islands (mostly of continental origin). North America has the longest coastline of all continents, reaching a length of 75,600 km. 1 km of shore corresponds to 320 sq.km on the continent. **Islands and peninsulas** makes up 25.4% of the area of the continent. The islands – 4,160,000 sq.km in area – are concentrated chiefly in the Arctic North. There lies Greenland, the largest island in the world (2,175,600 sq.km), and also the vast Canadian Arctic Archipelago (area: 1,405,000 sq.km) with the two largest islands Baffin I. (507,414 sq.km) and Victoria I. (217,274 sq.km). The second island region is that of the Caribbean Sea with the Greater (207,700 sq.km) and the Lesser Antilles (14,200 sq.km). The Aleutian Is. (37,850 sq.km) stretch northwest in the Pacific Ocean, together with the Alexander Archipelago (36,780 sq.km) and Queen Charlotte Is. (9,600 sq.km). The largest peninsula is Labrador (1,320,000 sq.km).

The surface of North America is divided in a north-south direction by the mountains of the Cordilleras with high plateaux between the ranges in the west of the continent, by the vast Great Plains and the Central Lowlands, the alluvial Mississippi Lowlands (1,000 km long and 80–100 km wide) in the interior of the continent, and by the eastern ranges (the Laurentin Plateau and the Appalachian Mts.), which line the coastal plain on the Atlantic Ocean and the Gulf of Mexico. The highest mountain of the continent is Mt. McKinley (6,194 m) in the Alaska Range in the Pacific Mountain System, which includes the Cascade Range (4,391 m) and the Sierra Nevada (Mt. Whitney, 4,418 m). The lowest point is Death Valley (–86 m) on the western edge of the Great Basin in California. The Rocky Mts. (4,398 m) enclose the Great Basin without outlet to the sea. The geologically younger Cordilleras in Central America have a number of volcanoes (V. Citlaltépetl 5,699 m), many of them active, and this is also a region of frequent earthquakes. The Laurentin Plateau, known as the Canadian Shield (comprising half of Canada) is geologically the oldest part of the continent; it is considerably worn down by glaciers, with a vast number of shallow valleys, rivers, streams and lakes. In the east of the continent the Paleozoic Appalachian Mts. reach the height of 2,037 m.

The river system, especially in the North of the continent, was greatly affected by Quaternary glaciation. The annual mean discharge is 7,130 cub.km. Most of the river basins drain into the Atlantic Ocean through the mouth of the third longest river in the world, the Mississippi-Missouri (length 6,212 km, draining an area of 3,250,000 sq.km, with an average annual flow 19,800 cub.m per second) and the St. Lawrence (3,058 km), which flows from the Great Lakes. Both of these are very important shipping routes. The Mackenzie, the longest river of Canada (4,240 km, drainage area 1,813,000 sq.km) drains into the Arctic Ocean, and in the West, the Colorado (with the famous Grand Canyon), the Columbia and the Fraser flow into the Pacific Ocean. The longest river in Alaska is the Yukon (3,185 km), which drains into the Bering Sea. The northern rivers have an ample supply of water from glaciers and snow-fall throughout the year, but they freeze up in winter. On the other hand, the rivers in the Central Basins of the South-West of the United States and Mexico often dry up in summer. The Great Lakes, the largest fresh water **lakes** in the world, covering an area of 246,515 sq.km, are of glacial and tectonic origin.

North America stretches across 3 **climatic zones** in the northern hemisphere. In the far North, on the Arctic islands and in Greenland there is the polar-Arctic-climate (mean annual temperature between 0 °C and –20 °C, precipitation c. 200 mm). The central part of the continent has an extreme climate (with great differences in temperature in summer and winter). A transitional zone of subtropical climate is found around the Gulf of Mexico (sufficient rainfall), in northern Mexico and southern California – U.S.A. (little rainfall). The climate of the Central American mainland and the islands in the Caribbean Sea is tropical, oceanic. In winter, blizzards blow from the North across the open continent, the temperature drops by about 20°C and in summer tornadoes and hurricanes strike the South-east of the continent and especially the islands. The highest temperature, 57°C, was measured in Death Valley and the lowest, at Fort Good Hope, was –78.2°C. Mean January and July temperatures in °C (and annual precipitation in mm): Barrow –26.4 and 4.5 (112), Dawson –29.4 and 15.6 (321), Cambridge Bay –30.7 and 9.6 (170), Ûpernavik –21.8 and 4.9 (230), Prince Rupert 1.7 and 13.4 (2,417), Churchill –28.3 and 12.1 (405), Winnipeg –19.4 and 20.2 (501), St. John's –4.8 and 15.5 (1,348), Montréal –9.4 and 21.3 (1,061), New York 0.1 and 24.6 (1,076), Chicago –3.3 and 23.9 (843), Salt Lake City –2.6 and 24.7 (353), San Francisco 10.2 and 14.9 (528), Miami 19.2 and 27.4 (1,518), México 11.6 and 15.6 (765), Kingston 24.7 and 27.6 (802), Colón 26.2 and 26.7 (3,680).

There are two **natural vegetation** regions in North America, the holarctic in the North and the neotropical in the South. From North to South we can find arctic deserts, tundra, North Canadian coniferous forest, a zone of Pacific forests, Laurentinian and South Atlantic mixed forest, grassland, steppe, shrub with cacti and thorn bushes, Central American coniferous forest and tropical swamp forest, savanna and tropical rain forest. There are also two zones of **animal life** : the neo-arctic (e.g. elk, "wapiti" stag, grizzly bear, racoon, beaver, puma, wild turkey, waterfowl, etc.) and the neotropical (e.g. jaguar, armadillo, tapir, porcupine, alligator, parrots, humming birds, etc.). Buffaloes exist only in reservations. Great attention is being paid to conservation; national parks, monuments and reservations have been established, the largest being Wood Buffalo N.P. in Canada (44,807 sq.km) and Yellowstone N.P. in the U.S.A.

LONGEST RIVERS

Name	Length in km	River Basin in sq.km
Mississippi-Missouri	6,212	3,250,000
Mackenzie-Athabasca	4,240	1,813,000
Yukon	3,185	848,400
St. Lawrence	3,058	1,260,000
Rio Grande (Bravo del Norte)	3,023	580,000
Nelson-Saskatchewan	2,575	1,250,000
Arkansas	2,334	416,000
Colorado	2,334	590,000
Ohio-Allegheny	2,102	525,700
Columbia	1,954	668,000
Saskatchewan	1,940	336,700
Peace	1,923	303,000
Snake	1,671	382,300
Red	1,638	214,500
Churchill	1,609	410,000
Canadian	1,458	.
Tennessee	1,387	105,950
Fraser	1,368	232,800
Kuskokwim	1,287	126,900
Ottawa	1,271	.

LARGEST ISLANDS

Name	Area in sq.km
Greenland	2,175,600
Baffin I.	507,414
Victoria I.	217,274
Ellesmere I.	196,221
Cuba	110,922
Newfoundland	108,852
Hispaniola	77,218
Banks I.	70,023
Devon I.	55,243
Axel Heiberg I.	43,175
Melville I.	42,146
Southampton I.	41,211
Prince of Wales I.	33,336
Vancouver I.	31,282
Somerset I.	24,784
Bathurst I.	16,041
Prince Patrick I.	15,847
King William I.	13,110
Ellef Ringnes I.	11,294
Jamaica	11,424

LARGEST LAKES

Name	Area in sq.km	Greatest Depth in m	Altitude in m
L. Superior	82,414	393	183
L. Huron	59,596	226	177
L. Michigan	58,016	281	177
Great Bear Lake	31,328	137	157
Great Slave Lake	28,570	140	158
L. Erie	25,745	64	174
L. Winnipeg	24,388	21	217
L. Ontario	19,553	237	75
Lago de Nicaragua	8,430	70	34
L. Athabasca	7,935	91	213
Reindeer L.	6,651	.	337
Nettilling L.	5,542	.	30
L. Winnipegosis	5,374	12	253
L. Nipigon	4,848	123	261
L. Manitoba	4,660	7	248
Great Salt Lake	4,365	16	1,283
Lake of the Woods	4,349	21	323
Dubawnt L.	3,833	.	236
L. Melville	3,069	.	tidal
Wollaston L.	2,681	.	398

HIGHEST MOUNTAINS

Name (Country)	Height in m
Mount McKinley (U.S.A.)	6,194
Mt. Logan (Can.)	6,050
V. Citlaltépetl (Mex.)	5,699
Mt. St. Elias (Can.-U.S.A.)	5,489
Mt. Foraker (U.S.A.)	5,303
Ixtacihuatl (Mex.)	5,286
Mt. Lucania (Can.)	5,227
Mt. Blackburn (U.S.A.)	5,036
Mt. Bona (U.S.A.)	5,029
Mt. Sanford (U.S.A.)	4,940
Mt. Wood (Can.)	4,842
Mt. Vancouver (Can.-U.S.A.)	4,785
Toluca (Zinantecatl, Mex.)	4,577
Mt. Whitney (U.S.A.)	4,418
Mt. Elbert (U.S.A.)	4,398
Mt. Rainier (U.S.A.)	4,391
Mt. Shasta (U.S.A.)	4,316
Mt. Kennedy (Can.)	4,238
V. Tajumulco (Guat.)	4,210
Mt. Waddington (Can.)	4,042
Mt. Robson (Can.)	3,954
Chirripó (Costa Rica)	3,837
Mt. Columbia (Can.)	3,747
Gunnbjørns Fjeld (Green.)	3,700

ACTIVE VOLCANOES

Name (Country)	Altitude in m	Latest eruption
V. Popocatépetl (Mex.)	5,452	1932
Mt. Wrangell (Alaska, U.S.A.)	4,268	1907
V. Acatenango (Guat.)	3,976	1972
V. de Colima (Mex.)	3,885	1975
V. Fuego (Guat.)	3,835	1977
V. Irazú (Costa Rica)	3,432	1967
Lassen Pk. (U.S.A.)	3,190	1915
Parícutin (Mex.)	3,170	1952
Mt. St. Helens (U.S.A.)	2,950	1980
Shishaldin V. (Alaska, U.S.A.)	2,857	1977

FAMOUS NATIONAL PARKS

Name (Country)	Area in sq.km
Wood Buffalo (Can.)	44,807
Katmai N.M. (U.S.A., Alaska)	10,917
Yellowstone (U.S.A.)	8,985
Mt. McKinley (U.S.A., Alaska)	7,850
Jasper (Can.)	6,760
Everglades (U.S.A.)	5,660
Grand Canyon (U.S.A.)	4,930
Banff (Can.)	4,126
Olympic (U.S.A.)	3,647

32 a New York 1 : 900 000

ICELAND
Reykjavík
16
17
C
18
D
19
E
SPAIN
PORTUGAL
(Denmark)
Angmagssalik
Arctic Circle
Disko
Qeqertarssuaq (Godhavn)
Nûk (Godthåb)
Qaqortoq (Julianehåb)
Kap Farvel
Davis Strait
Frobisher Bay
Port Burwell
Hudson Strait
LABRADOR SEA
ATLANTIC OCEAN
F
Azores (Port.)
Labrador
Ft. Chimo
Schefferville
Goose Bay
Newfoundland
Channel-Port-aux-Basques
St. John's
C. Race
Ft. George
Sept-Iles
G. of St. Lawrence
St. Pierre-et Miquelon (Fr.)
Sydney
Chicoutimi
Cochrane
Québec
St. John
Halifax
Sault Ste. Marie
MONTRÉAL
Ottawa
Augusta
TORONTO
BOSTON
G
Buffalo
Albany
DETROIT
NEW YORK
CLEVELAND
PHILADELPHIA
BALTIMORE
WASHINGTON
Tropic of Cancer
Richmond
Norfolk
Bermuda (U.K.)
C. Hatteras
Nashville
Charlotte
Birmingham
Atlanta
Charleston
Savannah
Montgomery
JACKSONVILLE
Mobile
H
NEW ORLEANS
Tampa
BAHAMAS
Turks and Caicos Is. (U.K.)
Miami
Nassau
Strs. of Florida
Gd. Turk
DOMINICAN REP.
STO. DOMINGO
Virgin Is. (U.K.)
St. Kitts (U.K.)
ANTIGUA
Guadeloupe (Fr.)
SAN JUAN
DOMINICA
Martinique (Fr.)
LA HABANA
CUBA
Santiago de C.
Virgin Is. (U.S.A.)
Puerto Rico (U.S.A.)
ST. LUCIA
BARBADOS
PORT-AU-PRINCE
HAITI
ST. VINCENT
GRENADA
Greater Antilles
Yucatán Str.
Mérida
Cayman Is. (U.K.)
JAMAICA
Kingston
Port of Spain
TRINIDAD AND TOBAGO
CARIBBEAN SEA
Lesser Antilles
Campeche
BELIZE
Belmopan
Neth. Antilles
Barquisimeto
CARACAS
Georgetown
Paramaribo
I
Pto. Barrios
GUATEMALA
HONDURAS
Tegucigalpa
Panama Canal Zone (U.S.A.)
Sta. Marta
MARACAIBO
VENEZUELA
Orinoco
GUYANA
SURINAM
Cayenne
French Guiana
EL SALVADOR
NICARAGUA
Cartagena
COLOMBIA
San Cristóbal
Macapá
Managua
COSTA RICA
Colón
Panamá
PANAMA
MEDELLÍN
BRAZIL
San José
12
13
14
15
1 : 50 000 000
0 250 500 750 1 000 1 250 Km
0 200 400 600 800 Mi

NORTH AMERICA

Country	Area in sq.km	Population	Year	Density per sq.km
Antigua (U.K.)	422	72,000	1978	163
Bahamas	13,864	226,000	1978	16
Barbados	431	254,000	1977	589
Belize	22,965	149,000	1977	6
Bermuda (U.K.)	53	62,000	1978	1,170
Canada	9,976,139	23,644,800	1979	2
Cayman Islands (U.K.)	260	13,200	1977	51
Costa Rica	50,900	2,111,000	1978	41
Cuba	114,524	9,889,000	1977	86
Dominica	751	80,000	1977	106
Dominican Republic	48,442	5,124,000	1978	106
El Salvador	21,041	4,255,000	1977	202
Greenland (Denmark)	2,175,600	49,338	1979	0.02
Grenada	344	108,000	1977	314
Guadeloupe and Depend. (Fr.)	1,780	327,000	1978	184
Guatemala	108,889	6,621,000	1978	61
Haiti	27,750	4,833,000	1978	174
Honduras	112,088	2,918,000	1978	26
Jamaica	10,991	2,140,000	1978	195
Martinique (Fr.)	1,102	316,000	1978	287
Mexico	1,972,546	66,944,000	1978	34
Montserrat (U.K.)	98	12,162	1976	124
Nicaragua	148,000	2,395,000	1978	16
Panama	75,650	1,826,000	1978	24
Panama Canal Zone (U.S.A.)	1,676	38,000	1978	23
Puerto Rico (U.S.A.)	8,897	3,319,000	1977	373
St. Kitts – Nevis – Anguilla (U.K.)	357	71,000	1978	199
Saint Lucia	616	117,000	1978	190
Saint Pierre et Miquelon (Fr.)	242	6,200	1977	26
Saint Vincent	389	106,000	1978	272
Turks and Caicos Islands (U.K.)	430	6,300	1978	15
United States	9,363,166	220,000,000	1979	23
Virgin Islands (U.K.)	153	12,700	1978	83
Virgin Islands (U.S.A.)	344	98,000	1978	285

Today's **racial and ethnic composition** of the inhabitants of North America (incl. Central America) is the cor sequence of an intermingling of races and nationalities that began after the discovery of the continent. The populatio can be divided into five groups: 1. the original inhabitants – Indians and Eskimos, 2. inhabitants of European origi (immigrants and their descendants), 3. inhabitants of African origin (Negroes and their descendants), 4. mixed race between Indians, Whites and Negroes, 5. a small number of immigrants from Asia. In 1978 the total **number c inhabitants** was **360 million,** not quite 67% living in the U.S.A. and Canada (of this: 94,000 Eskimos, over 3/4 millio Indians and more than 20 million Negroes and mixed races) and 33% in Mexico and the countries and islands c Central America (where Indians, Negroes and mixed races predominate). Population density per sq.km is highes with 114 persons, on the islands of the Caribbean Sea (Barbados 569), 32 persons in Central America and lowe in North America, with 11 persons per sq.km. Between 1960 and 1975 **the population** of the continent **increase at the** high **rate of 2.1%** (Honduras 3.48%, the U.S.A. only 0.86%), resulting from a high birth rate of 30.5 pe thousand (e.g. Honduras 49.3 per thousand in 1970–75) and a decreasing death rate of 9.3 per thousand (e. Guadeloupe 6.4 per thousand).The annual migration inflow amounts to 600,000 persons. There has been a remarkab growth in the number of inhabitants of towns and conurbations. Urban populations are highest in North America 76.4 (Canada 78.3%, the United States 76.2% in 1975), lower in Central America 57.1% (Mexico 63%), and least on the islanc in the Greater Antilles 48% (Cuba 61.4%, Haiti only 20.6%). North America had 11 cities with a population of ove a million (without agglomeration) and 25 cities with more than 500 000 inhabitants in 1977.

Economy: The United States is economically the strongest and technologically the most advanced count in the western world. It produces more than 35% of the entire industrial production and roughly 20% of th agricultural production of the world, excluding the socialist countries, and handles almost one fifth of worl trade. Canada, too, is a highly advanced industrial and agricultural country with an important ore mining indust and very considerable agricultural output. The most developed country of Latin America is Mexico with it important mineral resources. Monoculture on plantations is the predominant form of agriculture in Central Americ

CANADA

area 9,976,139 sq.km, population 23,644,800 (1 April 1979), **independent federal stat member of the Commonwealth** (Prime Minister Pierre Elliott Trudeau since 1980, Governo -General Edward Richard Schreyer since 1978).

Admininistrative units: 10 provinces and 2 territories (see table). **Capital:** Ottawa 304,462 inhab. (census 1976, wit agglomeration 693,288 inhab.); **other towns** (1976, in 1,000 inhab. and with agglom. in brackets): Montréal 1,08

02), Toronto 633 (2,803⁺), Winnipeg 561 (578), North York⁺ 558, Calgary 470, Edmonton 461 (554), Vancouver 410 66), Scarborough⁺ 387, Hamilton 312 (529), Etobicoke⁺ 297, Mississauga 250, Laval 246, London 240 (270), dsor 197, Québec 177 (542), Regina 150 (151), York⁺ 141, Saskatoon 134, Kitchener 132 (272), Burnaby 132, nt Catharines 123, Longueuil 122, Halifax 118 (268), Thunder Bay 111 (119), Oshawa 107 (135), East York⁺ 107, lington 104, Brampton 103, Sudbury 98 (157), Saint John's 87 (143), Saint John 86 (113), Sherbrooke 77 (105), toria 63 (218). **Population** (1971, in 1,000): English Canadians 7,966, French Canadians 6,180, Indians 276 (1974), imos 18; Irish 1,582, Germans 1,317, Italians 731, Ukrainians 581, Dutch 426, Poles 316 and others. **Density** re than 2 persons per sq.km; average annual rate of population increase 1.4%, birth rate 18.6 per 1,000 and death 7.7 per 1,000 (1970–76). Urban population 78.3% (1975). Economically active inhabitants 9,869,000, of which 5.9% worked in advanced agriculture, 30.1% in industry and 64% in services (1977). – **Currency:** Canadian ar = 100 cents.

Province	Area in sq.km	Population (in 1,000, 1978)	Density per sq.km	Capital (population in 1,000, 1976)
Alberta	661,185	1,889	2.9	Edmonton (461)
British Columbia	948,596	2,534	2.7	Victoria (63)
Manitoba	650,086	1,050	1.6	Winnipeg (561)
New Brunswick	73,437	696	9.5	Fredericton (45)
Newfoundland	404,517	573	1.4	St. John's (87)
Nova Scotia	55,491	851	15	Halifax (118)
Ontario	1,068,582	8,492	7.9	Toronto (633)
Prince Edward Island	5,657	121	21	Charlottetown (21)
Québec	1,540,680	6,406	4.2	Québec (177)
Saskatchewan	651,902	947	1.5	Regina (150)
Northwest Territories	3,379,682	46.4	0.01	Yellowknife (10)
Yukon Territory	536,324	22.4	0.04	Whitehorse (15)
CANADA	9,976,139⁺	23,627.8	2.4	Ottawa (304)

⁺ including interior water areas

nomy: highly advanced industrial country. Canada's contribution to industrial production of non-socialist ntries is roughly 3%; it is the ninth largest in volume of industrial production, third in agricultural production, eighth in the total turnover of foreign trade. In 1976/77 Canada had the highest output of nickel (30% of world ut), zinc (17.5%) and newsprint; it held second place in the output of molybdenum, asbestos, gold, uranium, num, cobalt, titanium, potassium salt, in the production of motor spirit and wood pulp; third place in the output lver, production of zinc, sawnwood, electric energy and commercial vehicles; fourth place in the output of copper, and natural gas.

cipal industries: metallurgy, mining of fuels and ores, electricity, processing of fuels, engineering, food processing, d-working, paper and textile industry.

ng (1976, in 1,000 tons, metal content): coal 25,572 (1978, Alberta, Nova Scotia), brown coal 5,040 (1978, atchewan), crude petroleum 73,000 (1977, Alberta, Saskatchewan), natural gas 79,058 mn. cub.m (1977, Alberta, e all, Medicine Hat), iron ore 33,218 (1977, Québec – Schefferville, Gagnon; Newfoundland – Labrador City ush), nickel 235 (1977, Ontario – Sudbury, Falconbridge; Manitoba – Thompson, Lynn Lake), asbestos 1,536 bec – Thetford Mines, Asbestos, Amos), uranium 4,850 tons (Ontario – Elliott Lake; Saskatchewan – Uranium City), 52,456 kg (Ontario – Kirkland L., Porcupine, Red Lake; Québec – Val-d'Or, Malartic; Northwest Territories – wknife; Yukon – region Klondike), silver 1,330 tons (1977, British Columbia – Kimberley; Ontario – Timmins, ganda; Yukon – Faro), platinum 11,973 kg (1974, Ontario – Sudbury; Manitoba – Thompson), cobalt 1,561 tons 4, Ontario – Cobalt, Lac Preissac), lead 366 (1978, British Columbia – Kimberley, Slocan; Yukon – Faro, Mayo), 1,248 (1978, Ontario – Timmins; Québec – Mattagami; Manitoba – Flin Flon; British Columbia – Kimberley, on, Remac; Yukon – Faro, Elsa; Northwest Ter. – Pine Point), copper 782 (1977, Manitoba – Sherridon; rio – Manitouwadge, Copper Cliff; Québec – Noranda, Chibougamau, Murdochville; British Columbia – Highland y, Merritt), molybdenum 14,415 tons (British Columbia – Peachland, Revelstoke), tungsten 2,160 tons (Yukon – sten), antimony 3,700 tons (1977), columbite 1,917 tons (1975, Québec – Oka), salt 5,994 (largest Pugwash – Scotia; Goderich, Sarnia – Ontario), potassium salt 5,125 (1977, Saskatchewan), gypsum 6,240.

strial production – some branches belong to the largest in the world: production of paper and chemical pulp, motor spirit and oils, electric energy, aluminium, lead, zinc, sawnwood. In the South-east of Canada engineering, metallurgy, textile, chemical and food industries predominate; in the South-west of the country er and paper industry, metallurgy of non-ferrous metals (especially aluminium), and engineering. Ontario is nost industrialized province – providing 65% of industrial production value.

llurgy (1977, in 1,000 tons): pig iron 10,584 and crude steel 14,904 (1978, largest steel plants Hamilton), nium 1,021 (5.8% of world production; Arvida, Kitimat, Baie-Comeau, Shawinigan), zinc 496 (7.3% of world uction; Trail, Flin Flon, Timmins), lead 187 (Trail, Belldune Point); copper – smelted 496 (6.4% of world pro- on, Noranda, Montréal, Flin Flon), – refined 509 (5.6% of world production, Montréal, Copper Cliff, Murdoch- cadmium 1,265 tons (1978, third world producer, Flin Flon, Trail). Processing of cobalt – Lynn Lake, Sudbury, Saskatchewan; nickel – Thompson, Bécancour, Sudbury; uranium – Port Hope in Ontario.

neering: (1977): production of motor vehicles – passenger 1,162,800 units, – commercial 613,200 (Windsor, wa, Toronto, Hamilton, St. Thomas)–aircraft (Toronto, Milton, Montréal, Vancouver); railway carriages and loco- es (Trenton, Montréal, London, Rivière-du-Loup, Winnipeg); vessels (Montréal, Lauzon, Sorel, Halifax, Victoria); otechnical industry (Toronto, Montréal, Ottawa, Trois-Rivières, Hamilton) – 712,000 radio and 439,000 television vers (1976); machine tools, mining, textiles, agricultural machines (Toronto, Montréal, Vancouver, Winnipeg, Québec).

U. S. S. R.
Uelen
Providenija
Gambell
Bering Strait
St. Lawrence I. (U.S.A.)
Wales
Nome
Kotzebue
Point Hope
Noatak
P. Barrow
Barrow
Wainwright
Umiat
Colville
Beechey Point
Prudhoe Bay
BEAUFORT SEA
Mackenzie Bay
BERING SEA
St. Matthew I.
Mekoryuk
Nunivak I.
Norton S.
Akulurak
Holy Cross
Bethel
Quinhagak
Aniak
Kuskokwim
Nulato
Unalakleet
Yukon
Kobuk
Koyukuk
Bettles Field
Wiseman
Alaska
(UNITED STATES)
Tanana
Nenana
Mc Grath
Fort Yukon
Porcupine
Circle
Fairbanks
Eagle
Big Delta
Tanacross
Bristol Bay
Naknek
Alaska Pen
Chignik
Anchorage
Willow
Palmer
Kenai
Seward
Susitna
Copper
Valdez
Cordova
Kodiak
Kodiak I.
Gulf of Alaska
PACIFIC OCEAN
Yakutat
Aklavik
Inuvik
Tuktoyaktuk
Arctic Red River
Peel
Mackenzie
Yukon
Dawson
Mayo
Stewart
Snag
Fort Selkirk
Faro
Ross River
Pelly
Whitehorse
Teslin
Skagway
Juneau
Sitka
Alexander Archip.
Petersburg
Wrangell
Stewart
Ketchikan
Watson Lake
Liard
Ft. Good Hope
Norman Wells
Fort Norman
Wrigley
Fort Simpson
Fort Liard
Fort Nelson
High Level
C
British Columbia
Stikine
Skeena
Finlay
Ware
Fort St. John
Hines Creek
Dawson Creek
Grande Prairie
Peace
Prince Rupert
Hazelton
Smithers
Kitimat
Masset
Queen Charlotte Is.
Ocean Falls
Prince George
Fraser
Quesnel
Jasper
Williams L.
Columbia
Pt. Hardy
Vancouver I.
Courtenay
Squamish
Kamloops
Lytton
Revelstoke
Hope
Kelowna
Vancouver
Victoria
Trail
1
2
3
4
5
B
C
60°
170°
160°
150°
140°
33 a Vancouver Area
1 : 10 000 000
Ocean Falls
Bella Coola
Anahim Lake
Redstone
Williams Lake
Fitzhugh Sound
Tatlayoko Lake
Hanceville
Dog Creek
Klinaklini
Allison Harbour
Kingcome Inlet
Knight Inlet
Clinton
Port Hardy
Englewood
CANADA
Bralorne
Port Alice
Kelsey Bay
Lillooet
Ashcroft
Spences Bridge
Rock Bay
Pemberton
Lytton
Kyuquot
Tahsis
Campbell River
Powell River
Merritt
Nootka
Courtenay
Texada I.
Squamish
Nootka Sound
Sechelt
Princeton
Vancouver I.
Parksville
North Vancouver
Nanaimo
New Westminster
Hope
Osoyoos
Grand Forks
Trail
Tofino
Port Alberni
Vancouver
Metaline Falls
Duncan
Burnaby
Barkley Sound
Bellingham
Republic
Strait of Juan de Fuca
Port Renfrew
Esquimalt
Anacortes
Newhalem
Kettle Falls
Addy
Neah Bay
Burlington
Omak
Victoria
Port Townsend
Arlington
Stehekin
Grand Coulee
Port Angeles
Everett
Monroe
Chelan
Miles
Spokane
Seattle
Bellevue
Waterville
Davenport
U.S.A.
Bremerton
Snoqualmie
Adrian
Cheney
Moclips
Auburn
Wenatchee
Rosalia
Hoquiam
Tacoma
Enumclaw
Lind
Marengo
Aberdeen
Eatonville
Ashford
Ellensburg
Hooper
PACIFIC OCEAN
Idaho
Twin Falls
A
B
C
1
2
3
4
8
51°
48°
126°
123°
120°

E. Ringnes I.
Isachsen
Queen
Patrick I.
Mackenzie King I.
Elizabeth I
Melville I.
Viscount Melville S
Victoria Island
Cambridge Bay
Perry Island
Bathurst Inlet
Contwoyto L.
Clinton-Colden L.
Reliance
Dubawnt L.
Uranium City
L. Athabasca
Chipewyan
Wollaston L.
Churchill L.
Flin Flon
The Pas
Prince Albert
L. Winnipegosis
Saskatoon
Regina
Yorkton
Swift Current
Moose Jaw
Glendive
Yellowstone
Sheridan

add page 146

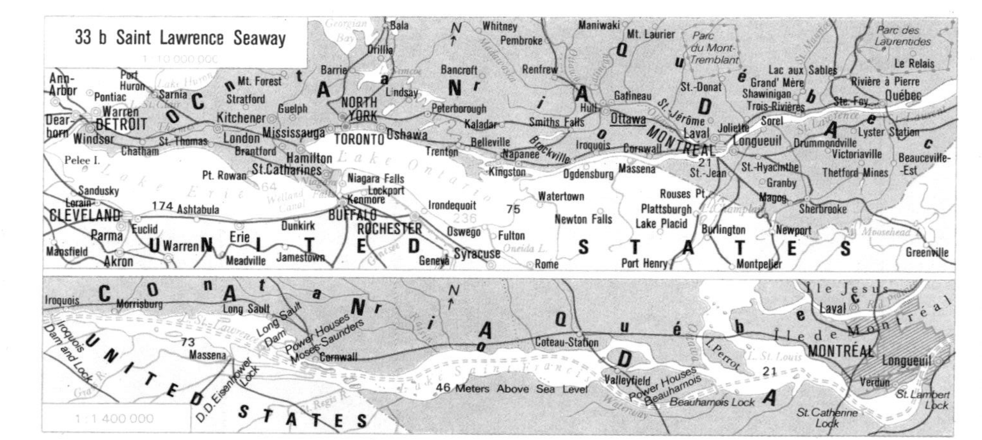

33 b Saint Lawrence Seaway
1:10 000 000
Ann-Arbor
Dear-born
DETROIT
Windsor
Pontiac
Warren
Port Huron
Sarnia
Chatham
St. Thomas
London
Kitchener
Stratford
Mt. Forest
Guelph
Brantford
Hamilton
St. Catharines
Mississauga
TORONTO
NORTH YORK
Barrie
Orillia
Bala
Lindsay
Oshawa
Peterborough
Bancroft
Whitney
Pembroke
Renfrew
Kaladar
Belleville
Trenton
Napanee
Kingston
Brockville
Smiths Falls
Ottawa
Hull
Gatineau
Maniwaki
Mt. Laurier
Iroquois
Cornwall
MONTRÉAL
Laval
St.-Jérôme
St.-Donat
Parc du Mont-Tremblant
Joliette
Longueuil
Sorel
Trois-Rivières
Shawinigan
Grand' Mère
Lac aux Sables
Rivière à Pierre
Ste.-Foy
Québec
Le Relais
Parc des Laurentides
Lyster Station
Drummondville
Victoriaville
Thetford Mines
Beauceville-Est
St.-Hyacinthe
Granby
Magog
Sherbrooke
Newport
Greenville
St.-Jean
Rouses Pt.
Plattsburgh
Lake Placid
Burlington
Montpelier
Port Henry
Massena
Ogdensburg
Watertown
Newton Falls
Rome
Fulton
Oswego
Syracuse
Geneva
Irondequoit
ROCHESTER
Lockport
Niagara Falls
Kenmore
BUFFALO
Dunkirk
Jamestown
Erie
Meadville
Warren
Ashtabula
Pt. Rowan
Pelee I.
Sandusky
Lorain
CLEVELAND
Euclid
Parma
Akron
Mansfield
Lake Ontario
Lake Erie
Welland Canal
21
75
236
174
64
CANADA
Ontario
Québec
UNITED STATES
1:1 400 000
Iroquois
Iroquois Dam and Lock
Morrisburg
Long Sault
Long Sault Dam
Moses Saunders Power Houses
Cornwall
Massena
D.D. Eisenhower Lock
73
Coteau-Station
Valleyfield
Beauharnois Power Houses
Beauharnois Lock
46 Meters Above Sea Level
I. Perrot
Île de Montréal
Île Jésus
Laval
MONTRÉAL
Verdun
Longueuil
St. Lambert Lock
St. Catherine Lock
L. St. Louis
Lake Saint Francis
21

Chemical industry – chief production centres: Sarnia, Welland, Trail, Ft. Saskatchewan, Calgary, Brandon, Redwater and others. Production (1976, in 1,000 tons): sulphur 4,559, sulphuric acid 2,842; fertilizers – nitrogenous 960, – phosphoric 670, – potash 5,656, organic chemicals 841, synthetic rubber 210, plastics 722 (1974). Capacity of petroleum refineries 104 mn. tons (largest Montréal, Sarnia, Québec, Edmonton, Vancouver, Regina, Calgary, Ft. McMurray, Winnipeg, Toronto, Halifax, Point Tupper, St. John and Come-by--Chance), production: motor spirit 26,730 (second world producer), aviation spirit 6,656, light oils 23,573, heavy oils 18,972, asphalt 2,877 (all 1977).
Wood and paper industry (1976, in 1,000 tons): roundwood 104.7 mn. cub.m, sawnwood 36.9 mn. cub.m, newsprint 8,160 (1977, leading world producer), mechanical wood pulp 6,847, chemical wood pulp 10,725 (second world producer), other paper 3,455. Largest paper mills include Thunder Bay, Port Alberni, Kapuskasing, Campbell River, Hull, Corner Brook, St. John; wood processing plants: Vancouver, Prince George, Edmonton, Merritt, La Sarre and Ottawa. **The textile industry** (Québec, Ontario – production 1976): cotton fabrics 272 mn. sq.m, woollen fabrics 26 mn. sq.m; synthetic fibres 101,900 tons.
The food industry is concentrated in large cities and ports. Production (1977, in 1,000 tons): meat 2,311, milk 7,751, cheese 160, butter 119, flour 1,933 (1976), sugar 163, beer 20.5 mn. hl. **Production** of cement 9.9 mn. tons (1977, Ontario, Québec, Nova Scotia), tyres 20 mn. units; 61.6 billion cigarettes.
Electricity: installed capacity of electric power stations 65.6 mn. kW (of which 61% were hydro-electric power stations in 1976). Produced 316,548 mn. kWh in 1977; capacity of nuclear power stations 3.9 mn. kW and production 20,800 mn. kWh. Largest power stations: hydro-electric – La Grande 5,328 MW (Québec), Churchill Falls 5,225 MW (Labrador), Manicouagane – Riv. aux Outardes, Péribonca, Beauharnois (Québec), Niagara-Queenston (Ontario) and Peace R., Kemano (British Columbia); nuclear – Bruce-Tiverton 2,984 MW and Pickering-Toronto 2,572 MW (Ontario) and others.
Agriculture: arable land 4.8%, meadows and pastures 2.5%, forests 35.4%, other land 57.3% of the land area. In 1976 634,481 tractors and 163,557 combined harvesters were in use. Only 579,000 people work in the advanced, highly mechanized system of agriculture. **Animal production** predominates (1977, in 1,000 head): cattle 13,717 (dairy cows+ 2,081, Alberta, Saskatchewan, Québec+, Ontario+), pigs 6,170 (Ontario, Québec), sheep 418 (Alberta, Ontario), horses 345, poultry 96,323 (Ontario, Québec, British Columbia); eggs 312,085 tons, honey 27,747 tons, fresh hides 134,350 tons, fish catch 1,135,700 tons.
Vegetable production: mostly cultivation of cereals and fodder (Edmonton – Regina – Winnipeg). **Crops** (1977, in 1,000 tons): wheat 19,651 (Saskatchewan 63%, fifth world producer), barley 11,515 (Alberta 52%, third world producer), oats 4,303 (Alberta 33%, third world producer), maize (corn) 3,675 (1976), potatoes 2,498, linseed 650 (Manitoba 61%), soybeans 517 (Ontario), vegetables 1,477 and fruit 640 (Ontario, British Columbia, Québec), sugar beet 1,008 (Alberta), tobacco 103 (Ontario). Hunting of fur animals.
Communications (1976); railways 70,715 km mostly electrified, freight 196 billion ton-km; roads 834,152 km, of which 636,393 km were surfaced and 7,820 km of the Trans-Canada Highway: St. John's (Newfoundland) Victoria (Vancouver I.); motor vehicles in use 11.7 mn., of which 9.4 mn. passenger cars. Civil aviation: 297.6 mn. km flown and 16.8 mn. passengers carried (1976). Largest airports: Montréal, Toronto and Vancouver. Naval transport: merchant shipping 2,823,000 GRT (1977, of which ore vessels 1.6 mn. GRT) including vessels for inland navigation with a gross tonnage of 4.4 mn., annual freight turnover in ports 171 mn. tons (1976). Almost 65% of merchant shipping operates on the 3,650 km long "great inland waterway" from the St. Lawrence to Great Lakes. Largest ports: Vancouver, Sept-Îles, Montréal, Thunder Bay, Hamilton, Port-Cartier, Halifax etc. Length of pipelines 32,862 km (1977). Tourism 13 mn. visitors.
Foreign trade (1977)- **Exports** (Canadian share in world export is 3.9%): transport equipment 23%, ores and other raw materials 21% (iron ore 6.2%, wood 5.7%), industrial products 17.4% (paper 6.5%), fuels 12.5% (crude petroleum 4.6%), cereals 5.5%; above all to the U.S.A. 69.5%, Japan 5.7%, the United Kingdom 4.3%, Fed. Rep. of Germany, the Netherlands, Italy, Belgium, China. **Imports** (Canadian share in world import is 3.7%): machines and apparatus 49.5% (transport equipment 25.8%), industrial products 12.3%, crude petroleum 8.3%, chemical 6% and others; above all from the U.S.A. 70.4%, Japan 4.3%, the United Kingdom 3%, Venezuela, Fed. Rep. of Germany, Saudi Arabia, France and others.

UNITED STATES OF AMERICA

United States of America (U.S.A.), **area 9,363,166 sq.km, population 220,000,000** (1979), **census population** 1 April 1970 **203,211,926** (excluding armed forces stationed abroad), **federal republic** (President Ronald Wilson Reagan since 1981).

Administrative units: 50 federal states and 1 federal district (the capital). Under the sovereignty of the U.S.A. belong overseas territories: 1. one self-governing federal state: the Commonwealth of Puerto Rico; 2. territories: American Samoa, Guam, the Commonwealth Islands, Northern Mariana Is.; 3. the United Nations Trust Territory of the Pacific Islands; 4. rented territory: Panama Canal Zone.
Capital: Washington 689,000 inhab. (1978, with agglomeration 3.2 mn.); **other towns** (in 1,000 inhab., 1978, with agglom. in brackets): New York 7,420 (17,040), Chicago 2,980 (7,635), Los Angeles 2,765 (9,200), Philadelphia 1,805 (5,320), Houston 1,460 (2,345), Detroit 1,245 (4,660), San Diego 820 (2,100), Baltimore 814 (1,895), San Antonio 804 (960), Dallas 800 (2,495), Indianapolis 711 (1,078), Phoenix 669 (1,270), San Francisco 658 (4,500), Memphis 655 (825), Milwaukee 641 (1,382), Boston 624 (3,825), Cleveland 595 (2,255), San Jose 590, New Orleans 560 (1,125), Jacksonville 547 (619), Columbus 527 (922), St. Louis 495 (2,250), Seattle 481 (1,855), Denver 474 (1,305), Nashville-Davidson 459 (560), Kansas City (Mo.) 454 (1,235), Pittsburgh 430 (2,190), Atlanta 406 (1,760), El Paso 402 (1,020), Cincinnati 390 (1,425), Buffalo 387 (1,225), Omaha 383 (576), Oklahoma City 371 (691), Toledo 366 (573), Minneapolis 359 (1,950), Miami 348 (2,420), Portland 348 (1,109), Long Beach 341, Fort Worth 340, Tulsa 335 (505),

State (abbv.)	Area in sq.km	Population (in 1,000, 1978)	Density per sq.km	Capital (population in 1,000, 1978)
Alabama (Ala.)	133,667	3,725	28	Montgomery (161)
Alaska	1,518,775	408	0.3	Juneau (18.4)
Arizona (Ariz.)	295,024	2,360	8	Phoenix (669)
Arkansas (Ark.)	137,539	2,144	16	Little Rock (143)
California (Calif.)	411,015	22,018	54	Sacramento (262)
Colorado (Colo.)	269,998	2,644	10	Denver (474)
Connecticut (Conn.)	12,973	3,142	242	Hartford (130)
Delaware (Del.)	5,328	588	110	Dover (23)
District of Columbia (D.C.)	174	689	3,960	Washington (689)
Florida (Fla.)	151,670	8,652	57	Tallahassee (86)
Georgia (Ga.)	152,489	5,032	33	Atlanta (406)
Hawaii	16,705	914	55	Honolulu (328)
Idaho	216,412	855	4	Boise (108)
Illinois (Ill.)	146,076	11,277	77	Springfield (86)
Indiana (Ind.)	93,994	5,306	56	Indianapolis (711)
Iowa	145,791	2,881	20	Des Moines (191)
Kansas (Kans.)	213,063	2,337	11	Topeka (121)
Kentucky (Ky.)	104,623	3,482	33	Frankfort (23)
Louisiana (La.)	125,674	3,906	31	Baton Rouge (198)
Maine	86,027	1,087	13	Augusta (21)
Maryland (Md.)	27,394	4,190	153	Annapolis (33)
Massachusetts (Mass.)	21,386	5,833	273	Boston (624)
Michigan (Mich.)	150,779	9,122	60	Lansing (124)
Minnesota (Minn.)	217,736	4,027	18	Saint Paul (271)
Mississippi (Miss.)	123,584	2,379	19	Jackson (193)
Missouri (Mo.)	180,486	4,801	27	Jefferson City (36)
Montana (Mont.)	381,087	764	2	Helena (28)
Nebraska (Nebr.)	200,018	1,565	8	Lincoln (165)
Nevada (Nev.)	286,298	637	2.2	Carson City (29)
New Hampshire (N.H.)	24,097	838	35	Concord (29)
New Jersey (N.J.)	20,295	7,359	363	Trenton (98)
New Mexico (N.Mex.)	315,115	1,202	3.8	Santa Fe (47)
New York (N.Y.)	128,402	18,102	141	Albany (107)
North Carolina (N.C.)	136,197	5,539	41	Raleigh (140)
North Dakota (N.D.)	183,022	651	3.6	Bismarck (41)
Ohio	106,765	10,669	100	Columbus (527)
Oklahoma (Okla.)	181,090	2,823	16	Oklahoma City (371)
Oregon (Oreg.)	251,181	2,373	9	Salem (83)
Pennsylvania (Pa.)	117,412	11,897	101	Harrisburg (55)
Rhode Island (R.I.)	3,144	922	293	Providence (159)
South Carolina (S.C.)	80,432	2,901	36	Columbia (111)
South Dakota (S.D.)	199,552	688	3.4	Pierre (12)
Tennessee (Tenn.)	109,412	4,276	39	Nashville-Davidson (459)
Texas (Tex.)	692,407	12,835	19	Austin (326)
Utah	219,932	1,264	6	Salt Lake City (165)
Vermont (Vt.)	24,887	483	19	Montpelier (8)
Virginia (Va.)	105,716	5,114	48	Richmond (223)
Washington (Wash.)	176,617	3,697	21	Olympia (26)
West Virginia (W.Va.)	62,629	1,846	29	Charleston (65)
Wisconsin (Wis.)	145,439	4,617	32	Madison (169)
Wyoming (Wyo.)	253,597	407	1.6	Cheyenne (47)
UNITED STATES OF AMERICA (U.S.A.)	9,363,125+	217,513	23	Washington (689)

+9,519,617 including Great Lakes area

Honolulu 328 (728), Austin 326 (394), Oakland 326, Newark 324, Louisville 320 (867), Tucson 302 (435), Albuquerque 289 (395), Charlotte 283 (454), Tampa 278 (541), St. Paul 271, Wichita 269 (359), Birmingham 267 (695), Norfolk 266 (764) and further 114 cities with a population of more than 100,000. The share of the urban population was 76.2%. **Census population** on 1 April 1970 was 203,211,926 inhabitants, of whom 177,748,975 were Whites, 22,580,289 Negroes and 2,882,662 other races (of whom 792,730 were Indians, 591,290 Japanese (on Hawaii 217,307), 435,062 Chinese, 343,060 Filipino (on Hawaii 93,915) ; in Alaska 54,704 were Indians, Aleutians and Eskimos; on Hawaii 298,160 were Hawaiians). The total population comprised 48.7% males and 51.3% females; 73.5% were urban and 26.5 were rural. The average annual rate of population increase was 0.86% (1970–75). The age composition was (1975): 0–14 years 25.3%, 15–64 years 64.3%, older than 65 years 10.4%. In 1976 the birth rate was 14.8 per 1,000, death rate 8.9 per 1,000, and infant mortality 15.1 per 1,000. The number of immigrants was 1.2 mn. in 1973–75, in 1976 it was 398,613 (18% from Europe, 37.6% from Asia, 14.5% from Mexico); in 1976 the number of emigrants was 793,092. There were 97.2 mn. economically active inhabitants in 1977, of which only 2.5% were in agriculture, 33.5% in industry and 64% in services. – **Currency:** US dollar = 100 cents.

1
2
3
4
5
6
7
B
D
125°
120°
115°
110°
105°
100°
50°
45°
40°
CANADA
British Columbia
Alberta
Saskatchewan
Manitoba
Washington
Oregon
Idaho
Montana
Wyoming
North Dakota
South Dakota
Nebraska
Kansas
Colorado
Utah
Nevada
California
PACIFIC OCEAN
Vancouver
Squamish
Vancouver I.
C. Flattery
Victoria
Burnaby
Hope
Bellingham
Revelstoke
Kelowna
Penticton
Procter
Trail
Republic
Calgary
Hanna
Bassano
Michel
Cranbrook
Lethbridge
Medicine Hat
Biggar
Kindersley
Leader
Saskatoon
Watrous
Swift Current
Regina
Moose Jaw
Watson
Sturgis
Yorkton
Melville
Big Quill Lake
Lake Winnipegosis
Minitonas
Winnipegosis
Dauphin
Birtle
L. Manitoba
Gypsumville
Lake Winnipeg
Hodgson
Victoria Beach
Portage-la-Prairie
Gladstone
Brandon
Winnipeg
Carman
Alida
Deloraine
Emerson
South Jct.
Beaver
Moclips
Aberdeen
Seattle
Everett
Bellevue
Olympia
Tacoma
Astoria
Tillamook
Vancouver
Yakima
Moses Lake
Grand Coulee
Spokane
Sandpoint
Coeur d'Alene
Lind
Pasco
Walla Walla
Moscow
Lewiston
Rexford
Kalispell
Flathead L.
Orion
Shelby
Havre
Arena
Mankota
Opheim
Weyburn
Flaxton
Portland
The Dalles
Salem
Toledo
Corvallis
Eugene
Coos Bay
Roseburg
Grants Pass
Crater Lake
Medford
Klamath Falls
Chemult
Bend
Pendleton
Baker
Seneca
Burns
Lakeview
Nampa
Boise
New Meadows
Salmon
Missoula
Augusta
Great Falls
Helena
Anaconda
Butte
Darby
Three Forks
Bozeman
Gardiner
Billings
Fort Peck
Fort Peck Lake
Brockway
Glendive
Terry
Miles City
Crow Agency
Minot
Surrey
Devils Lake
Warren
Grand Forks
Lake Sakakawea
Dickinson
Bismarck
Jamestown
Valley City
Mahnomen
Moorhead
Fargo
Fergus Falls
Wahpeton
West Yellowstone
Cody
Yellowstone L.
Sheridan
Gillette
Mackay
Hailey
Idaho Falls
Twin Falls
Pocatello
Mc Cammon
Lander
Casper
Edgemont
Crawford
Aberdeen
Faith
Belle Fourche
Rapid City
Pierre
Watertown
Huron
Pipestone
Mitchell
Wood
Sioux Falls
Yankton
Le Mars
Sioux City
C. Mendocino
Eureka
Weed
Alturas
Vya
Redding
Winnemucca
Elko
Wells
Great Salt Lake
Logan
Ogden
Lakeside
Granger
Rock Springs
Walcott
Laramie
Cheyenne
Scottsbluff
Alliance
Thedford
Norfolk
Fremont
Omaha
Ft. Bragg
Ukiah
Chico
Squaw Valley
Yuba City
Reno
Hazen
Lovelock
Pyramid L.
Carson City
L. Tahoe
Austin
Wendover
Salt Lake City
Provo
Utah L.
Sidney
North Platte
Grand Island
Lincoln
Santa Rosa
Vallejo
Sacramento
Stockton
Berkeley
Oakland
SAN FRANCISCO
SAN JOSE
Sunnyvale
Merced
Luning
Bishop
Ruth
Pioche
Eureka
Hiawatha
Craig
Fort Collins
Greeley
Denver
Brush
Lakewood
Aurora
Grand Junction
Limon
Holdrege
Hastings
Beatrice
Colby
Manhattan
Columbia
Snake
Missouri
Milk
Yellowstone
Bighorn
Powder
Little Missouri
Cheyenne
White
James
Niobrara
North Platte
South Platte
Platte
Republican
Humboldt
Sacramento
Klamath
Green
Sevier
South Saskatchewan
Red Deer
Bow
Qu'Appelle
Assiniboine
Souris
Red
Sheyenne
Kootenai
Clark Fork
Marias

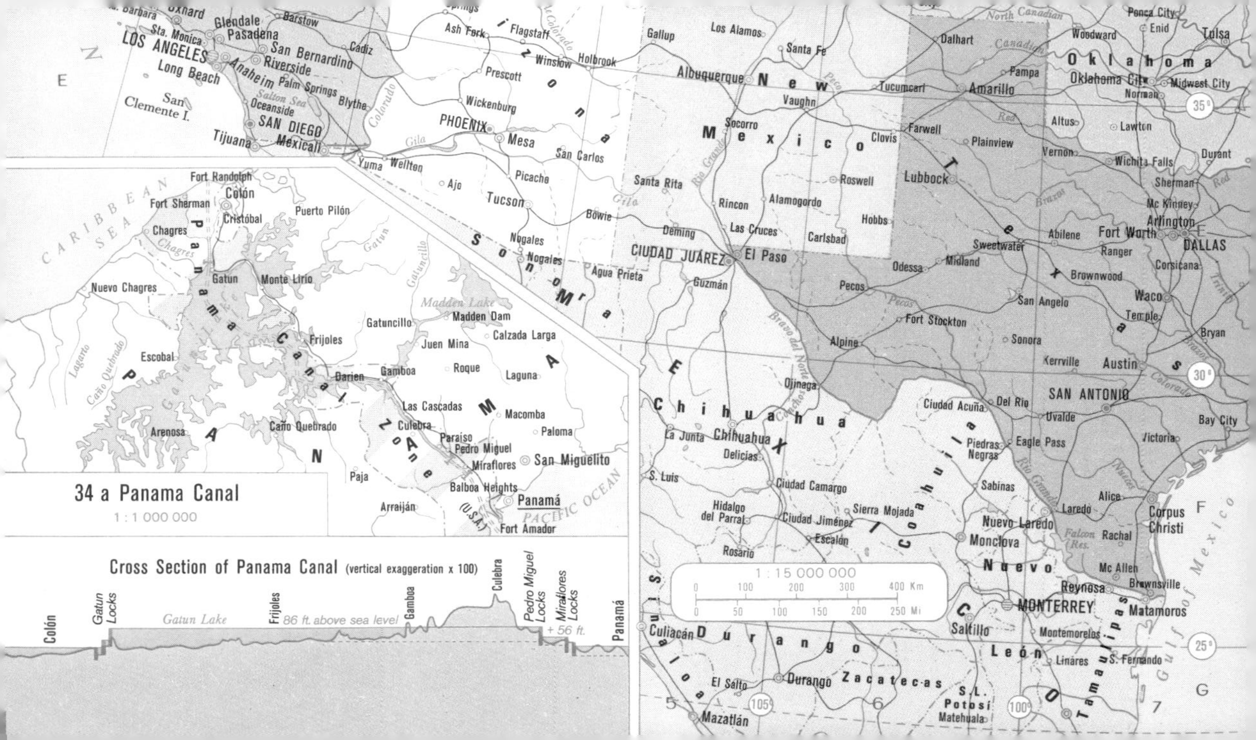
34 a Panama Canal
1 : 1 000 000
Cross Section of Panama Canal (vertical exaggeration x 100)
Colón
Gatun Locks
Gatun Lake
Frijoles
86 ft. above sea level
Gamboa
Culebra
Pedro Miguel Locks
Miraflores Locks
+ 56 ft.
Panamá
CARIBBEAN SEA
PACIFIC OCEAN
Panama Canal Zone
PANAMA
Fort Randolph
Colón
Fort Sherman
Cristóbal
Puerto Pilón
Chagres
Gatun
Monte Lirio
Nuevo Chagres
Madden Lake
Madden Dam
Calzada Larga
Gatuncillo
Frijoles
Juen Mina
Escobal
Roque
Darien
Gamboa
Laguna
Las Cascadas
Macomba
Arenosa
Caño Quebrado
Culebra
Paloma
Paraíso
Pedro Miguel
San Miguelito
Miraflores
Paja
Balboa Heights
Arraiján
Panamá
(U.S.A.)
Fort Amador
1 : 15 000 000
0 100 200 300 400 Km
0 50 100 150 200 250 Mi
LOS ANGELES
Long Beach
Sta. Monica
Glendale
Pasadena
San Bernardino
Riverside
Anaheim
Palm Springs
Blythe
Oceanside
SAN DIEGO
Tijuana
Mexicali
San Clemente I.
Barstow
Cadiz
Arizona
Ash Fork
Flagstaff
Winslow
Holbrook
Prescott
Wickenburg
PHOENIX
Mesa
San Carlos
Yuma
Wellton
Ajo
Picacho
Tucson
Bowie
Nogales
Sonora
Agua Prieta
Gallup
Los Alamos
Santa Fe
Albuquerque
New Mexico
Vaughn
Socorro
Tucumcari
Clovis
Roswell
Santa Rita
Rincon
Alamogordo
Hobbs
Deming
Las Cruces
Carlsbad
CIUDAD JUÁREZ
El Paso
Guzmán
Texas
Dalhart
Amarillo
Pampa
Farwell
Plainview
Lubbock
Odessa
Midland
Sweetwater
Abilene
Pecos
Fort Stockton
San Angelo
Brownwood
Alpine
Sonora
Kerrville
Austin
SAN ANTONIO
Oklahoma
Woodward
Enid
Ponca City
Tulsa
Oklahoma City
Midwest City
Norman
Altus
Lawton
Vernon
Wichita Falls
Durant
Sherman
Mc Kinney
Arlington
Fort Worth
DALLAS
Ranger
Corsicana
Waco
Temple
Bryan
Bay City
Victoria
Uvalde
Del Rio
Eagle Pass
Ciudad Acuña
Piedras Negras
Chihuahua
Ojinaga
La Junta
Delicias
S. Luis
Ciudad Camargo
Hidalgo del Parral
Ciudad Jiménez
Sierra Mojada
Escalón
Rosario
Coahuila
Sabinas
Nuevo Laredo
Laredo
Alice
Corpus Christi
Monclova
Falcon Res.
Rachal
Nuevo León
Mc Allen
Reynosa
Brownsville
MONTERREY
Matamoros
Saltillo
Montemorelos
Linares
S. Fernando
Tamaulipas
Sinaloa
Culiacán
Durango
El Salto
Zacatecas
S.L. Potosí
Matehuala
Mazatlán
Gulf of Mexico
Río Grande
Bravo del Norte
35°
30°
25°
100°
105°
E
F
G
5
6
7

Economy: developments in science, research and technology have created the most advanced industrial and agricultural country in the western world. The total volume of the gross domestic product reached 1,164,000 mn. dollars in 1976 (45.8% of the volume in all countries barring the socialist countries). The US share in the industrial production of all countries (excepting socialist) amounted to 37% (1977) and in agricultural production to 20%. The United States of America occupied the leading position (share in world production in 1977, +in 1976): – in mining: first place – molybdenum ore (58%), uranium (44%), natural gas (41%), phosphates (38%), salt (23%), copper (17%), sulphur; second place – vanadium (28%), coal (24%), crude petroleum (15%), lead (15%); – in industrial production: first place – motor spirit (55%), sulphuric acid (48%), chemical wood pulp (42%+), paper (40%), lorries (32%), aluminium (31%), aviation spirit (31%), electric energy (31%), passenger cars (29%+), lead (25%+), copper – refined (19%), -smelted (17%), oils (17%): second place – roundwood (24%+), crude steel (17%), tin; – in agricultural production: first place – grapefruits (65%), soybeans (60%), maize (corn) (46%), sorghum (36%), oranges (30%), beef meat (26%), cotton-lint (22%), lemons (18,5%), cheese, tomatoes, apples; – second place: hops (22%), oats (21%), cotton-seed, tobacco (16%), eggs (16%), wheat (14.5%), pork meat, milk, pineapples, cattle, wheat flour, etc.

Industry: these industries hold a leading position in the U.S.A.: energy, mining of fuels and ores, metallurgy, engineering (especially transport and electrotechnical), chemical industry (above all, petrochemistry and electrochemistry), production of plastics and artificial fibres, nuclear industry.

Mining (1977, in 1,000 tons): coal and anthracite 593,244 (1978, highest quality coking coal and anthracite – the Appalachians district, Pennsylvania, West Virginia, Kentucky, Illinois, Ohio), brown coal 26,232 (North and South Dakota, New Mexico), crude petroleum 462,830 (Texas 35%, Louisiana, California, Alaska, Oklahoma) – verified deposits 3,977 mn. tons (1978), natural gas 612,569 mn. cub.m (Texas 40%, Louisiana, Oklahoma, New Mexico) – verified deposits 5,943 billion cub.m (1978), iron ore 34,053 (region around Lake Superior 80%: Minnesota, Wisconsin and Michigan), copper ore 1,380 (Arizona 62%, Utah, New Mexico, Nevada), lead ore 534 (Missouri, Idaho, Colorado, Utah), zinc ore 415 (Missouri, Tennessee, New York, Colorado), molybdenum ore 51,362 tons (1976, Colorado, Arizona, New Mexico), bauxite 2,420 (1976, Arkansas, Alabama, Georgia), mercury 974 tons (Nevada, California, Idaho), gold 32,598 kg (1976, Nevada, South Dakota, Utah, Arizona), silver 1,187 tons (Idaho, Arizona, Colorado, Montana), vanadium 6,691 tons (1976, Arkansas, Colorado, Utah), tungsten 3,357 tons (1976, California, Colorado, Nevada), nickel ore 14,940 tons (1976), manganese ore 28,100 tons (1976), antimony ore 553 tons, asbestos 104 (California, Vermont, Arizona), uranium 9,800 tons (1976, New Mexico, Wyoming, Utah, Colorado), sulphur – native 5,954 (1976, Texas, Louisiana, California), phosphates 44,671 (1976, Florida, Idaho, North Carolina), mica 127 (1976, North Carolina, Alabama, New Mexico), potassium salt 2,195 (1976, New Mexico, Utah, California), salt 40,091 (1976, Louisiana, Texas, Ohio, New York), asphalt 1,825 (1976, Texas, Utah, Alabama), pyrites 862 (1976, Tennessee, Colorado, Arizona), ilmenite (titanium concentrates) 561 (Florida, New Jersey, New York).

Energy production (1977): capacity of all electric power stations 576 mn. kW, of which hydro-electric power plants 68 mn. kW (11.8%) and nuclear 50 mn. kW (8.7%). Electricity 2,211 billion kWh, of which nuclear 251 billion kWh and hydro-electric 220 billion kWh. Some of the largest thermal power plants: e.g. Monroe, Pittsburg (Cal.), Paradise, Johnsonville, Sammis, Stuart, Gavin, Houston, Oak Creek and others; largest hydro-electric power plants: e.g. the Grand Coulee, John Day, Chief Joseph, Niagara Falls System, Ludington, The Dalles, Hoover, St. Lawrence Power System; largest nuclear power plants: e.g. Browns Ferry, Oconee, Sequoyah, Peach Bottom, Salem, Zion, Diablo Canyon, Indian Point. Production of gas 2,290 mn. cub.m. **Metallurgy:** production (1977, in 1,000 tons): pig iron 74,970 and crude steel 125,300 (17% of the world production, Pennsylvania, Indiana, Ohio, the Appalachians district – Pittsburgh, Youngstown, Chicago, Cleveland, Bethlehem, Sparrows Point), steel products 82,670 (especially Pittsburgh, Chicago, Youngstown, Steubenville, Trenton), aluminium 4,402 (Rockdale, Bellingham, Evansville, Massena), copper - smelted 1,302, – refined 1,716 (Anaconda, Morenci, Garfield, Carteret, El Paso), lead 549 (El Paso, East Helena, Herculaneum, Omaha), zinc 412 (Josephtown, Anaconda, Corpus Christi, Torrance), magnesium 74 (1975, Freeport, Albany, Henderson), uranium processing (Bluewater, Moab, Uravan, Shirley Basin and others).

Engineering – production: motor vehicles – passenger 9.15 mn. and commercial 3.7 mn. units (1978, Detroit, Cleveland, Kenosha, San Francisco, Toledo, Kansas City, Flint), locomotives and railway carriages (Chicago, Erie, Greenville), aircraft (Los Angeles, Wichita, New York, Fort Worth, Columbus), vessels 1,068,370 GRT (1976, Baltimore, Boston, Norfolk, Pascagoula, San Diego, San Francisco, Chester), machinery equipment, machines and apparatus (Chicago, New York, Cleveland, Philadelphia, Boston, Los Angeles, Worcester, St. Louis, Detroit, Houston and others), radio 12.8 mn. and television receivers 7.8 mn. (1976, Los Angeles, Rochester, New York and others), refrigerators 5.8 mn. and washing machines 5 mn. units. The rocket and astronautical industries are concentrated on the Pacific coast (San Diego, Los Angeles, San Jose, Seattle), in Texas (Fort Worth, Dallas, Houston), Kansas (Wichita), Florida (Orlando), on the Atlantic coast (New York, Philadelphia, Hartford, Boston) and around the Great Lakes (Chicago, Detroit, Buffalo, Cleveland, Cincinnati and others).

Chemical industry (1977, in 1,000 tons): sulphuric acid 31,188, nitrogenous fertilizers 9,262, plastics 780 (1976), synthetic rubber 210 (1976). Petroleum refineries have an annual capacity of 838 mn. tons (1978, Houston, Beaumont, Port Arthur, Baton Rouge, Philadelphia, Toledo, Chicago, Tulsa, Los Angeles, San Francisco, New York and others). Leading world producer (in 1,000 tons): motor spirit 303,662, aviation spirit 53,770 and oils 315,849. Production of tyres 2,303 mn. units (1978, Akron). **The timber and paper industries** (1976) are concentrated in the states of Washington, Oregon, California, Montana and in the South- and North-east of the U.S.A. Roundwood 341.4 mn. cub.m, sawnwood 75.5 mn. cub.m. **Production** (1977, in 1,000 tons): chemical wood pulp 37,225, mechanical wood pulp 3,905, newsprint 3,192, other paper 50,886. Production of cement 67.3 mn. tons (1977, states of Pennsylvania, New York, California, Texas, Michigan).

The textile and clothing industries are concentrated in the East of the U.S.A. from the southern state of Alabama to the north-eastern state of Maine. Production (1976): cotton – fibres 1,368 mn. tons, – fabrics 3,996 mn.m (1977), woollen fibres 61,000 tons, linen – fibres 38,600 tons, – fabrics 71 mn. m; silk fabrics 5.5 mn. tons (1975); synthetic fibres 1.66 mn. tons (1977).

Food industry (1977, in 1,000 tons); meat 25,508 (of which poultry 7,181), milk 55,772, cheese 1,845, butter 496, wheat flour 11,770 (1976), sugar 5,429, canned fish 346 (1976), beer 185.3 mn. hl (1975), wine 1,422,000 hl. **Production** (1975): 627 billion cigarettes, 8.3 billion cigars, pipe tobacco 103,782 tons.

Agriculture: arable land covers 188,330,000 ha (i.e. 20.6% of the area of the country), meadows and pastures 26.5%, forests 31.8%; irrigated land 16.6 mn. ha (1976, i.e. 8.8% of the arable land), 7.8 mn. inhabitants engaged in agriculture (1977, i.e. 3.6% of the total population) and 2.4 mn. persons work in advanced agricultural production (i.e. 2.5% of economically active inhabitants). In 1977 4.4 mn. tractors and 805,000 combines and harvesters were used on highly mechanized and specialized farms. The number of farms declined from 6.3 mn. in 1940 to 2,830,490 in 1974. **Animal production** predominates in the region of the Great Lakes and in California (Wisconsin, Minnesota, New York, Pennsylvania, Ohio etc.), pasture cattle-farming on the prairies (Texas, Iowa, Nebraska, Montana). **Livestock** (1977, in 1,000 head): cattle 122,810 (dairy cows 10,984), pigs 54,934 (Iowa, Illinois, Indiana, Minnesota), sheep 12,766 (Texas, California, Colorado, Utah, Wyoming), horses 9,075, poultry 397,536; production (1977, in 1,000 tons): eggs 3,820, honey 91, fresh hides 1,251 (sheepskins 19.6). Fish catch 3 mn. tons (1976). **Vegetable production:** the world's largest corn-growing region, known as the "corn-belt", runs through the Mid-West (from the Great Lakes along the Canadian frontier to the Great Plains), the Mississippi Lowlands to the Appalachian Mts. Wheat, barley, oats, soybeans, fodder plants, sugar beet, potatoes and vegetables are all grown there. **Crops** (1977, in 1,000 tons): maize (corn) 161,485 (Illinois, Iowa, Indiana, Nebraska, Minnesota, Ohio), wheat 55,134 (Kansas, North Dakota, Oklahoma, Minnesota, Montana, Washington), oats 10,856 (Minnesota, North Dakota, Iowa, Wisconsin), barley 9,056 (North Dakota, California, Montana, Idaho), rice 4,501 (Arkansas, Texas, California, Louisiana), sorghum 20,083 (Kansas, Texas, Nebraska, Montana), potatoes 15,972 (Idaho, Washington, Maine, Oregon, California), sweet potatoes 568, legumens 881, soybeans 46,712 (Illinois, Iowa, Montana, Indiana, Ohio), groundnuts 1,670, flax seed 409, sunflower seed 1,248, cotton -seed 5,018, -lint 3,156 (Texas, California, Mississippi, Arizona), vegetables 25,014 (California, Texas, Ohio, Florida), tomatoes 7,943 (California, Ohio and others), fruit 25,176 (California, Washington, New York, Michigan), grapes 3,855 (California), sugarcane 25,089 (Hawaii, Louisiana, Florida), sugar beet 22,784 (California, Minnesota, North Dakota, Idaho); oranges (+ tangerines) 9,612 (+577) and lemons 923 and grapefruits 2,748 (California, Florida, Texas, Arizona); pineapples 626 (Hawaii), almonds 231 (36% of the world crops), walnuts 177, tobacco 877 (North Carolina, Kentucky, Virginia, South Carolina, Georgia), hops 25.
Communications: length of railways, decreasing, 320,394 km, carried 15.7 billion passenger-km and 1,146 billion ton-km freight (1976); surfaced roads 5,885,900 km (1977, of which 1,584,900 km highways and 661,300 km federal highways), passenger cars 113,667,000 and commercial vehicles (incl. buses) 30,077,000 (1977), 30% of all automobiles in the world in 1976. Petroleum pipelines 279,616 km, inter-state gas pipelines 305,550 km, other gas pipelines 1.5 mn. km. Civil aviation (1976); leading world position: 13,770 airports (of which 4,667 public), civil aircraft 205,900, length of federal air-routes 665,320 km, flown 3,732 mn. km, passengers carried 223.9 mn., recorded 228 billion passenger-km and 7.4 billion ton-km. Largest airports: Chicago, New York, Long Beach, San Francisco, Miami, Dallas, Washington, Atlanta. Sea-going merchant vessels (1,000 gross tons and over) 564 with tonnage of 15,542,000 GRT, of which 239 tankers of 10,1 mn. GRT (1977). All vessels (5 tons and over) 65,971 with tonnage 33.2 mn. GRT (1977). In 1975 246 mn. tons of goods were loaded and 416 mn. tons unloaded (199 mn. tons of crude petroleum and petroleum products). Largest ports: New York, New Orleans, Houston, Baton Rouge, Baltimore, Philadelphia, Norfolk, Beaumont, Tampa, Los Angeles, Boston; inland ports: Duluth, Detroit, Chicago, Cleveland, Buffalo. Tourism 15.7 mn. visitors (1975).
The foreign trade of the U.S.A. reached a total turnover of 275,790 mn. dollars in 1977 (i.e. 11.6% of the world turnover), of which exports amounted to 119,006 and imports to 156,784 mn. dollars. – **Exports:** machines, machinery equipment, transport vehicles (especially cars and aircraft), metal products and others 50.1%, agricultural products 15.7%, chemicals 12.7%, raw materials (ores, coal etc.) 9.8%, foodstuffs and tobacco products 6.2%. Exports mainly to Canada 21.4%, Japan 8.8%, Fed. Rep. of Germany 5.1%, United Kingdom 4.5%, Mexico 4%, Netherlands, Saudi Arabia, France, Venezuela. **Imports:** crude petroleum and fuels 29.1%, machines, apparatus, industrial products 28.3%, consumer goods 12.2%, foodstuffs and tropical fruit 9.5%, passenger cars 7.6% and others. Imports came mostly from Canada 20.1%, Japan 12.8%, Fed. Rep. of Germany 5%, Saudi Arabia 4.3%, Nigeria 4.1 %, United Kingdom 3.5%, Mexico, Venezuela, Libya, etc.

PUERTO RICO

Commonwealth of Puerto Rico, area 8,897 sq.km, population 3,319,000 (1977), **self-governing federal state of the United States of America** (since 25 July 1952).
Capital: San Juan 514,500 inhabitants (1976, with agglomeration 820,442 inhab.); **other towns** (1976, in 1,000 inhab.): Bayamón 204, Ponce 187, Carolina 161, Caguas 107, Mayagüez 99. **Density** 373 persons per sq.km (1977), average annual rate of population increase 2.9%, birth rate 22.6 per 1,000, death rate 6.8 per 1,000 (1970–75); urban population 65.2% (1975). – **Currency:** US dollar = 100 cents.
Economy (1976): arable land 18%, meadows and pastures 37.7%, forests 20%. **Agriculture:** crops (1977, in 1,000 tons): sugarcane 2,396, bananas 113, pineapples 40, citrus fruit 38, coffee 12, high quality tobacco 2.2; cattle 571,000, pigs 317,000, poultry 6.1 mn. head; production of eggs 23,386 tons; fish catch 80,890 tons (1976). **Production** (1976, in 1,000 tons): sugar 243 (1977), cement 1,390, capacity of petroleum refineries 16,580 (largest: Guayanilla, Bayamón and Ponce), motor spirit 4,115, oils 7,203, cigars, textiles, consumer goods, spirits. Electricity (1976) 17.2 billion kWh (of which 98% thermal). – **Communications:** roads 16,827 km and cars 732,336. Tourism 1.3 mn. visitors (1976). – **Exports:** sugar, tobacco, textiles.

BAHAMAS

The Commonwealth of the Bahamas, area 13,864 sq.km, population 226,000 (1978), **member of the Commonwealth** (Prime Minister Lynden Oscar Pindling since 1973).
Capital: Nassau 101,503 inhab. (1970). Average annual rate of population increase 2.8% (1970–75). **Currency:** Bahamian dollar = 100 cents.
Economy: chiefly tourism – 939,900 visitors (1976); sugarcane, pineapples, sisal, roundwood (400,000 cub.m); mining of salt 1.1 mn. tons (1975), fishing, light industry. – **Exports:** salt, crabs, wood, rum.

Manitoba
Lake Winnipeg
Winnipeg
Victoria Beach
Winnipeg
Kenora
Waugh
South Jct.
Nestor Falls
L. of the Woods
Rainy L.
Hudson
Sioux Lookout
Lac Seul
Berens
Otoskwin
Pipestone
Albany
Ogoki
Armstrong Station
L. Nipigon
Nipigon
Thunder Bay
I. Royale
Nakina
Kenogami
Longlac
Ameson
Maratbon
Michipicoten
Hearst
Oba
Franz
Missinaibi
Smoky-Falls
Kapuskasing
Moosonee
Fort Rupert
Abitibi
Cochrane
Ansonville
Timmins
L. Abitibi
Kirkland Lake
Sultan
Agawa
Kesagami L.
L. Evans
La Reine
Noranda
Barraute
Val-d'Or
Rés. Cabonga
Monet
Lac Mistassini
Chibougamau
Chutte-des Passes
L. Péribonca
Dolbeau
L. Saint-Jean
Alma
Jonquière
Chicoutimi
Saguenay
Rés. Gouin
St. Maurice
Sanmaur
Lac Édouard
La Tuque
Québec
Baie-Comeau
Betsiamites
Ste.-Anne-des-Monts
Matane
Mont-Joli
Gaspé
Campbellton
St. Lawrence
Tadoussac
Rivière-du-Loup
Edmundston
La Malbaie
New Brunswick
Bathurst
Tracadie
Presque Isle
Houlton
St. John
Fredericton
St. John
Lancaster
Mattawamkeag
Machias
Bangor
Greenville
Maine
Augusta
Lewiston
Portland
Biddeford
Ste.-Foy
Beauceville-Est
Thetford Mines
Sherbrooke
Lac-aux-Sables
Grand Mère
Trois-Rivières
Mont Laurier
Maniwaki
MONTRÉAL
Laval
Magog
Newport
Hull
Ottawa
Cornwall
Coniston
North Bay
Ottawa
Mattawa
L. Nipissing
Sudbury
Sault Sainte Marie
Blind River
Sault Ste. Marie
Ludgate
Pembroke
Whitney
Parry Sound
Manitoulin I.
Georgian Bay
Orillia
Kaladar
Ogdensburg
Kingston
Peterborough
Belleville
Lake Placid
L. Champlain
Plattsburgh
Burlington
Montpelier
N.H.
Vermont
Rutland
Concord
Manchester
Keene
Lowell
Cambridge
BOSTON
Mass.
Fall River
New Bedford
R.I.
Providence
Worcester
Pittsfield
Springfield
Conn.
Hartford
New Haven
Waterbury
Bridgeport
Long I.
Yonkers
NEW YORK
New York
Watertown
Glens Falls
Schenectady
Troy
Albany
Hudson
Utica
Oswego
Syracuse
Cortland
Binghamton
Rochester
Lake Ontario
Oshawa
NORTH YORK
TORONTO
Hamilton
Owen Sound
Kitchener
Brantford
London
Niagara Falls
Buffalo
Dunkirk
Olean
Elmira
Lake Huron
Alpena
Port Austin
Saginaw B.
Port Huron
Sarnia
Flint
Saginaw
Standish
Traverse City
Mackinaw City
Lake Michigan
Green B.
Escanaba
Iron Mountain
Menominee
Green Bay
Marquette
Ironwood
Ashland
Lake Superior
Copper Harbor
Duluth
Superior
Grand Marais
Virginia
Bemidji
Red L.
Warren
Grand Forks
Red
Moorhead
Fargo
Fergus Falls
Wahpeton
Minnesota
Mississippi
Brainerd
Mille Lacs L.
St. Cloud
Turtle Lake
St. Paul
Minneapolis
Willmar
Eau Claire
Bloomington
Watertown
New Ulm
Mankato
Pipestone
Rochester
Austin
Des Moines
S. D.
Sioux Falls
Le Mars
Sioux City
Missouri
Nebr.
Fremont
Omaha
Council Bluffs
Lincoln
Beatrice
Harlan
Des Moines
Chariton
Iowa
Fort Dodge
Mason City
Waterloo
Cedar Rapids
Iowa City
Maryville
Kirksville
St. Joseph
Atchison
Manhattan
Hannibal
Moberly
Mexico
Wisconsin
Wausau
Winona
La Crosse
Appleton
Sheboygan
Fond du Lac
Wauwatosa
MILWAUKEE
Madison
Racine
Kenosha
Dubuque
Rockford
Waukegan
Davenport
Rock Island
Aurora
Joliet
Galesburg
Macomb
Peoria
Illinois
Champaign
Danville
Springfield
Decatur
Effingham
Mississippi
CHICAGO
Gary
Hammond
South Bend
Fort Wayne
Lafayette
Muncie
Anderson
INDIANAPOLIS
Terre Haute
Seymour
Indiana
Michigan
Reed City
Muskegon
Grand Rapids
Lansing
Kalamazoo
Jackson
DETROIT
Warren
Windsor
Ann Arbor
Port Rowan
Lake Erie
Toledo
Lorain
CLEVELAND
Parma
Akron
Canton
Lima
Ohio
COLUMBUS
Springfield
Dayton
Cincinnati
Covington
Maysville
Portsmouth
Steubenville
Erie
Ashtabula
Warren
Youngstown
Oil City
Bradford
Pittsburgh
Pennsylvania
Williamsport
Wilkes-Barre
Scranton
Allentown
Reading
Harrisburg
Altoona
Johnstown
PHILADELPHIA
Jersey City
Newark
Elizabeth
Trenton
Camden
N.J.
Atlantic City
Wilmington
BALTIMORE
Dover
Annapolis
Del.
Delaware Bay
Maryland
Salisbury
Chesapeake Bay
Hagerstown
Wheeling
Winchester
Potomac
Arlington
WASHINGTON
Fredericksburg
West Point
Waynesboro
Clarksburg
Parkersburg
West Virginia
Charleston
Ohio
OCEAN
1 2 3 4 5 6 7
A B C D
95° 90° 85° 80° 75° 70° 65°
50° 45° 40°

35 a Hawaiian Is.
1 : 15 000 000
Kauai
Mana
Kapaa
Lihue
Niihau
Kaula I.
Kauai Channel
Oahu
Waialua
Kaneohe
Ewa
Honolulu
Kaiwi Channel
Kepuhi
Halawa
Lahaina
Wailuku
Hana
Molokai
Lanai
Kahoolawe
Maui
Alenuihaha Channel
Hawi
Puako
Honokaa
Hilo
Kalapana
Hawaii
Pahala
PACIFIC OCEAN
Oklahoma
Tulsa
Muskogee
Mc Alester
Durant
Hugo
Texas
Sherman
Mc Kinney
DALLAS
Arlington
Tyler
Corsicana
Palestine
Waco
Lufkin
Bryan
HOUSTON
Pasadena
Beaumont
Port Arthur
Galveston
Galveston Bay
Bay City
Victoria
Longview
Mt. Pleasant
Texarkana
Arkansas
Fayetteville
Marshall
Hoxie
Ft. Smith
Heavener
Hot Springs
North Little Rock
Little Rock
Pine Bluff
Camden
Louisiana
Shreveport
Ruston
Monroe
Alexandria
De Ridder
Lake Charles
Lafayette
Baton Rouge
NEW ORLEANS
Houma
Buras
Marsh I.
L. Pontchartrain
Hammond
Bogalusa
Mississippi
MEMPHIS
Holly Springs
Clarksdale
Greenville
Greenwood
Jackson
Vicksburg
Natchez
Meridian
Laurel
Hattiesburg
Gulfport
Biloxi
Tupelo
Corinth
Tennessee
Dyersburg
Jackson
Nashville
Tullahoma
Sparta
Chattanooga
Knoxville
Alabama
Huntsville
Sheffield
Decatur
Gadsden
Birmingham
Anniston
Columbus
Tuscaloosa
Selma
Opelika
Montgomery
Troy
Dothan
Mobile
Mobile Bay
Georgia
Dalton
Rome
Atlanta
La Grange
Macon
Columbus
Athens
Greenwood
Augusta
Dublin
Albany
Tifton
Thomasville
Waycross
Brunswick
Savannah
Florida
Pensacola
Cottondale
Panama City
Mexico Beach
St. George I.
Tallahassee
Live Oak
JACKSONVILLE
Gainesville
St. Augustine
Daytona Beach
Ocala
Orlando
Port Canaveral
Cape Canaveral
Tampa
Lakeland
St. Petersburg
Tampa Bay
Bradenton
Sarasota
Fort Pierce
W. Palm Beach
L. Okeechobee
Fort Myers
Pompano Beach
Ft. Lauderdale
Hollywood
Hialeah
Miami
Everglades
Homestead
Cape Sable
Florida Keys
Key West
Dry Tortugas
Straits of Florida
Gulf of Mexico
South Carolina
Carolina
Hickory
Charlotte
Asheville
Spartanburg
Rock Hill
Greenville
Columbia
Florence
Georgetown
Yemassee
Charleston
Sanford
Kinston
New Bern
Fayetteville
Morehead City
Wilmington
C. Fear
ATLANTIC
BAHAMAS
Grand Bahama
Freeport
Little Abaco I.
Great Abaco
Cornwall
Dunmore Town
Eleuthera
Nassau
New Providence
Nicolls Town
Andros Town
Andros I.
Kemp's Bay
Cat I.
San Salvador (Watling I.)
Long I.
Clarence Town
Tropic of Cancer
CUBA
LA HABANA
Matanzas
Colón
Santa Clara
1 : 15 000 000
0 100 200 300 400 Km
0 50 100 150 200 250 Mi
35°
30°
25°
85°
80°
75°
22°
20°
160°
158°
156°

BARBADOS

Area 431 sq.km, population 254,000 (1977), sovereign state, member of the Commonwealth (Prime Minister J. M. Adams since 1976).

Capital: Bridgetown 8,789 inhab. (1970, with agglomeration 88,097). **Population:** Negroes 89%. – **Currency:** Barbadosian dollar = 100 cents.
Economy (1977, in 1,000 tons): 54% of island area covered by sugarcane plantations, crops: 1,147; most important products: sugar 124, world-famous rum, electrical goods; mining: natural gas 10.3 mn. cub.m (1978); fish catch 4,000 tons. Tourism 224,300 visitors (1976). – **Exports** (1975): sugar (45%), molasses, rum, fish.

BELIZE

Area 22,965 sq.km, population 149,000 inhab. (1977), **sovereign state, member of the Commonwealth** (Prime Minister George Cadle Price since 1981).

Capital: Belmopan 4,000 inhab. (1974), **largest town** and harbour Belize 39,257 inhab. (1970). – **Currency:** Belize dollar = 100 cents. **Economy** (1977, in 1,000 tons): sugarcane 840, sugar 94, citrus fruit 37, bananas, coconuts, fishing, hardwoods 116,000 cub.m (1976). – **Exports:** sugar, canned fruit, wood, chicle, fish.

COSTA RICA

República de Costa Rica, area 50,900 sq.km, population 2,111,000 inhab. (1978), **republic** (President Rodrigo Carazo Odio since 1978).

Administrative units: 7 provinces. **Capital:** San José 233,691 inhab. (1977, with agglomeration 428,041); **other towns** (in 1,000 inhab.): Alajuela 85, Puntarenas 75, Cartago 62. Average annual rate of population increase 2.8% (1970–75). – **Currency:** colón = 100 centimos.
Economy (1977, in 1,000 tons): high quality coffee 79, bananas 1,230, sugarcane 2,160, cocoa beans 8, rice 130, citrus fruit 73, coconuts 25, palm oil 24, sugar 194; cattle 1.9 mn. head; rare hardwoods 3.5 mn. cub.m; fish catch 12,700 tons (1976). Electricity 1,764 mn. kWh (1977, 85% hydro energy). – **Communications** (1976): railways 1,042 km, roads 5,197 km.– **Exports** (1976): coffee 29%, bananas 24%, sugar, cocoa, wood.

CUBA

República de Cuba, area 114,524 sq.km, population 9,889,000 inhab. (1977), **republic** (President Dr Fidel Castro Ruz since 1976).

Administrative units: 14 provinces. **Capital:** La Habana 1,861,442 inhab. (1975); **other towns** (1975, in 1,000 inhab.) Santiago de Cuba 352, Camagüey 222, Holguín 152, Guantánamo 150, Santa Clara 147. **Population:** Creoles 70%, mulattoes 17%, Negroes 12%. Urban population 62% (1976). 25.3% inhabitants employed in agriculture (1977). – **Currency:** Cuban peso = 100 centavos.
Economy: monocultural agriculture, mineral mining and industry developing. **Agriculture:** crops (1977, in 1,000 tons): sugarcane 57,000 (7.8% of world crops), tobacco 49, tropical fruit 588 (of which citrus fruit 207, bananas 92, pineapples 22), coffee 25, rice 420, vegetables 477, potatoes 147, sweet potatoes 257; **livestock** (1977, in 1,000 head): cattle 5,644, pigs 1,506, horses 840, sheep 346, poultry 18.9 mn., eggs 87,000 tons; fish catch 203,996 tons (1976).
Mining (1976, in 1,000 tons, metal content): nickel 37.3 (4.6% of world output), manganese about 30, chromium 12, copper, crude petroleum 144, salt 150. **Industrial production** (1976, in 1,000 tons): sugar 6,485 (1977, second world producer, 7% of world production); 14.4 billion cigarettes, 259 mn. cigars; woven cotton fabrics 156 mn. sq.m, sulphuric acid 383. Electricity 7,198 mn. kWh (1976). – **Communications** (1974): railways 14,872 km, roads 27,074 km. Merchant shipping 668,000 GRT (1977). – **Exports:** sugar (84% of turnover), tobacco and tobacco products, ores, fish, fruit. Chief trading partners: U.S.S.R., Japan, German Dem. Rep., United Kingdom.

DOMINICA

Area 751 sq.km, population 80,000 (1977), **sovereign state, member of the Commonwealth** (President Aurelius Marie since 1980).

Capital: Roseau 10,157 inhab. (1970). – **Currency:** Eastern Caribbean dollar = 100 cents.
Economy (1977, in 1,000 tons): bananas 37, cocoa beans 3.5, citrus fruit, coconuts 18. – **Exports:** coconut oil, bananas, laurel oil, citrus fruit and fruit juices.

DOMINICAN REPUBLIC

República Dominicana, area 48,442 sq.km, population 5,124,000 (1978), **republic** (President Silvestre Antonio Guzmán Fernández since 1978).

Administrative units: 26 provinces and 1 national district. **Capital:** Santo Domingo 671,402 inhab. (1970, with agglomeration 922,528 inhab. 1975) – oldest town in America established by Europeans in 1496; **other towns** (1975, in

,000 inhab.); Santiago de los Caballeros 209, San Pedro de Macorís 62, San Francisco de Macorís 58, Barrahona 51. **'opulation:** mulattoes 68%, Creoles 20% and Negroes 12%. Average annual rate of population increase 3.3% (1970–75). ‧8% of inhabitants engaged in agriculture (1977). – **Currency:** peso = 100 centavos.
:conomy: Agriculture (1977, in 1,000 tons): sugarcane 11,048, coffee 60, cocoa beans 37, tropical fruit 1,293 of which citrus fruit 85, avocados and mangoes 298, bananas 310); cattle 2 mn. head and poultry 7.9 mn. head; oundwood 1.8 mn. cub.m (1976). **Mining and industry** (1976, in 1,000 tons): bauxite 621, nickel (metal content) 4.4, gold 12,870 kg; production of sugar 1,361; 3.2 billion cigarettes. Electricity: 2,690 mn. kWh (1976). – **Communica-ions:** railways 1,444 km, roads 10,467 km. – **Exports:** sugar (50%), cocoa and coffee (13%), bauxite, tropical fruit.

:L SALVADOR

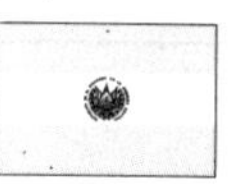

'epública de El Salvador, area 21,041 sq.km, population 4,255,000 (1977), **republic** (Military unta headed by Col. A. A. Majano and Col. J. A. Gutierrez since 1979).

dministrative units: 14 departments. **Capital:** San Salvador 336,008 inhab. (1974); **other towns** (1974, in 1,000 hab.): Santa Ana 105, San Miguel 67, Mejicanos 63. **Population:** mestizos 74%, Indians 12%. 52% of inhabitants ngaged in agriculture (1977). – **Currency:** colón = 100 centavos.
conomy (1977, in 1,000 tons): good quality coffee 180, maize (corn) 377, cotton-seed 119 and-lint 71, sugarcane ,300, citrus fruit 57, bananas 53; raising of cattle 1.2 mn. head and poultry 3.4 mn. head; roundwood 3.9 mn. cub.m, alsam gum (the world's principal source). Production of sugar 286,000 tons (1976). Electricity 1,332 mn. kWh (1977). **'ommunications** (1976): railways 696 km, roads 10,973 km. – **Exports** (1975): coffee (33%), sugar (16%), cotton.

;RENADA

rea 344 sq.km (including part of Grenadine Is. with area 33 sq.km, **population 108,000** (1977), **ember of the Commonwealth** (Prime Minister Maurice Bishop since 1979).
apital: Saint George's 29,860 inhab. (1970). **Population:** Negroes 53%, mestizos. **Density** 314 persons per sq.km. – **urrency:** Eastern Caribbean dollar = 100 cents. **Economy:** spices, mainly nutmeg and mace, cloves and vanilla; ananas 15,000 tons, cocoa beans, coconuts, good quality cotton. – **Exports** (1976): spices (36%), cocoa (30%), bananas.

;UATEMALA

epública de Guatemala, area 108,889 sq.km, population 6,621,000 (1978), **republic** (President en. Romeo Lucas García since 1978).

dministrative units: 22 departments. **Capital:** Guatemala 717,322 inhab. (1973); **other towns** (1973, in 1,000 hab.): Totonicapán 54, Quezaltenango 53. **Population:** Indians 54%, mestizos 38%. 57% of inhabitants engaged in griculture (1977). – **Currency:** quetzal = 100 centavos.
:onomy (1977, in 1,000 tons): coffee 147, bananas 550, sugarcane 6,800, cotton-seed 228 and-lint 135, maize orn) 756, chicle 900 tons; cattle 2.2 mn. head, pigs 667,000, sheep 612,000, poultry 11.2 mn.; rare hardwoods 7 mn. cub.m (1977); mining of antimony 1,120 tons (metal content), production of sugar 540,000 tons (1977). ectricity 1,250 mn. kWh (1976). – **Communications** (1975): railways 824 km, roads 13,450 km. – **Exports:** coffee 1%), cotton, sugar, chicle, essential oils, bananas.

AITI

épublique d'Haiti, area 27,750 sq.km, population 4,833,000 (1978), **republic** (President Jean-'laude Duvalier since 1971).
Iministrative units: 9 departments. **Capital:** Port-au-Prince 458,675 inhab. (1974). **Population:** Negroes 60%, ulattoes 30% and Creoles 10%. 79% of inhabitants live in the countryside. – **Currency:** gourde = 100 centimos.
:onomy (1977, in 1,000 tons): high quality coffee 33, sugarcane 2,801, bananas 51, mangoes 332, rice 100, maize orn) 250; breeding of pigs 1.8 mn. head, cattle 755,000, horses 392,000; mining of bauxite 684,000 tons; round-ood 4 mn. cub.m. – **Exports** (1975): coffee (42%), bauxite, sugar.

ONDURAS

pública de Honduras, area 112,088 sq.km, population 2,918,000 (1978), **republic** (Provisional esident Gen. Policarpo Juan Paz García since 1978).

Iministrative units: 18 departments. **Capital:** Tegucigalpa 273,894 inhab. (1974); **other towns** (1974, in 1,000 inhab.): n Pedro Sula 134, La Ceiba 44, Puerto Cortés 30. Average annual rate of population increase 3.5% (1970–75). – **rrency:** lempira = 100 centavos.
onomy: Agriculture (1977, in 1,000 tons): bananas 1,320, sugarcane 1,660, coffee 57, maize (corn) 377, palm 'nels, citrus fruit; cattle 1.8 mn. head, pigs 525,000, poultry 8 mn. head. **Mining** (1976, in 1,000 tons): lead 23, c 21 (metal content), salt 30; production of sugar 107 (1977); roundwood 3.8 mn. cub.m. – **Communications** 75): railways 991 km, roads 6,595 km. – **Exports** (1976): bananas (27%), coffee (26%), wood, ores, hides.

SAN DIEGO
PHOENIX
Yuma
Mexicali
Arizona
Ensenada
Puerto Peñasco
S. Felipe
Tucson
Nogales
Santa Ana
Hermosillo
Guaymas
New Mexico
Roswell
Lubbock
Santa Rita
Alamogordo
Deming
Carlsbad
Sweetwater
El Paso
Agua Prieta
CIUDAD JUÁREZ
San Angelo
Alpine
Little Rock
Arkansas
Texarkana
Wichita Falls
Red
Fort Worth
DALLAS
Shreveport
Louisiana
Texas
Waco
Palestine
Alexandria
Brazos
Colorado
Mississippi
UNITED S
Bryan
Beaumont
Austin
HOUSTON
Port Arthur
Galveston
SAN ANTONIO
Del Rio
Ojinaga
Rio Grande
Bravo del Norte
Chihuahua
Piedras Negras
Sabinas
Sierra Mojada
Nuevo Laredo
Laredo
Corpus Christi
I. Cedros
I. Tiburón
Golfo de California
Santa Rosalía
La Junta
Sánchez
Ciudad Obregón
Alamos
Loreto
Hidalgo del Parral
Ciudad Jiménez
Monclova
Reynosa
Brownsville
Matamoros
Topolobampo
Gómez Palacio
Saltillo
MONTERREY
La Paz
Culiacán
Torreón
Tropic of Cancer
Durango
El Salto
S. José d. C.
C. San Lucas
Mazatlán
Matehuala
Ciudad Victoria
Zacatecas
Aguascalientes
Ciudad de Valles
Ciudad Madero
Tampico
Tepic
Rio Gr. de Santiago
San Luis Potosí
GUADALAJARA
LEÓN
Pánuco
Poza Rica de Hildalgo
Is. de Revillagigedo (Mex.)
Celaya
Querétaro
Pachuca
Jalapa Enríquez
Colima
Morelia
MÉXICO
Manzanillo
Uruapan
Toluca
Veracruz
Bahía de Campeche
Cuernavaca
PUEBLA
Orizaba
Coatzacoalcos
Villahermo
Balsas
Río Balsas
Chilpancingo
Minatitlán
Teapa
Acapulco
Oaxaca
S. Cristóba las Casas
Juchitán
Salina Cruz
Tuxtla Gutiérrez
Puerto Angel
Quezaltenango
Tapachula
Champerico
Escuintla
Pto. de S. José
Campec
MEXICO
PACIFIC OCEAN
Gulf of Mexico
Clipperton (Fr.)
1 : 25 000 000
0 200 400 600 Km
0 100 200 300 400 Mi
MEXICO:
States:
1 Aguascalientes
2 Baja California Norte
3 Baja California Sur
4 Campeche
5 Chiapas
6 Chihuahua
7 Coahuila
8 Colima
9 Durango
10 Guanajuato
11 Guerrero
12 Hidalgo
13 Jalisco
14 México
15 Michoacán
16 Morelos
17 Nayarit
18 Nuevo León
19 Oaxaca
20 Puebla
21 Querétaro
22 Quintana Roo
23 San Luis Potosí
24 Sinaloa
25 Sonora
26 Tabasco
27 Tamaulipas
28 Tlaxcala
29 Veracruz
30 Yucatán
31 Zacatecas
Distrito Federal (México)
C.S. Antonio
Pinar
Caleta Larga
36

Economy (1977, in 1,000 tons): cotton -seed 221, -lint 138, coffee 62, sugarcane 2,578, maize (corn) 222, rice 45, citrus fruit 53, pineapples 35, bananas 15; cattle 2.8 mn. head, poultry 4.3 mn. head. **Mining** (1976): gold 1,951 kg, silver 6,000 kg. Sugar (production) 224,000 tons (1977). – **Communications:** railways 373 km, roads 12,500 km. – **Exports** (1975): cotton (26%), coffee, sugar.

PANAMA

República de Panama, area 75,650 sq.km, population 1,826,000 (1978), **republic** (President Aristides Royo since 1978).

Administrative units: 9 provinces. **Capital:** Panamá 415,790 inhab. (1976); **other towns** (in 1,000 inhab.): San Miguelito 124, Colón 95, David 41. **Population:** mestizos 52%, Whites 18%, Indians 14%, Negroes, mulattoes. Average annual rate of population increase 3.1% (1970–76). – **Currency:** balboa = 100 centesimos.
Economy (1977, in 1,000 tons): bananas 1,002, maize (corn) 77, rice 190, oranges 62, coconuts 25, sugarcane 2,396; sugar 182; cattle 1.4 mn. head and poultry 4.3 mn. head (1977); fish catch 171,641 tons (1976). Roundwood 1.5 mn. cub. m (1976) – forests cover 56% of the land area. Electricity: 1,508 mn. kWh (1976). – **Communications:** railways 185 km, roads 7,324 km. Merchant shipping 19,458,000 GRT (third place in world; tankers 6,524,000 and ore vessels 4,289,000 GRT). – **Exports:** petroleum products, bananas, sugar.

PANAMA CANAL ZONE

Area 1,676 sq.km, population 38,000 (1978), **territory under the administration of the U.S.A. till 2000.**
Seat of administration: Balboa Heights 2,801 inhab. (1970). **The Panama Canal** was built in 1903–14, length 81.6 km, depth 12.5–13.7 m, 6 locks lifting ships to 26 m above sea-level. In 1976 12,157 ships passed through the canal with a cargo of 117.2 mn. tons (of which 6,169 ships with 66 mn. tons from the Atlantic to the Pacific Ocean and 5,988 ships with 51.2 mn. tons from the Pacific to the Atlantic Ocean).

SAINT LUCIA

Area 616 sq.km, population 117,000 (1978), **member of the Commonwealth** (Prime Minister Allan Louisy since 1979).

Capital: Castries 45,000 inhab. (1975). **Population:** Negroes 68%, mestizos , Indos. **Density** 190 persons per sq.km. – **Currency:** Eastern Caribbean dollar = 100 cents.
Economy (1977, in 1,000 tons): bananas 54, coconuts 40, copra, cocoa beans, citrus fruit and citrus juices, ginger. Tourism is important. – **Exports:** bananas, copra, juices.

SAINT VINCENT

Area 389 sq.km, population 106,000 (1978), **member of the Commonwealth** (Prime Minister R. Milton Cato since 1979).

Capital: Kingstown 23,800 inhab. (1976). **Population:** Negroes 72%, mestizos. – **Currency:** Eastern Caribbean dollar = 100 cents.
Economy (1977, in 1,000 tons): world producer of arrowroot (used in pharmacy), bananas 28, coconuts 22, copra, nutmeg, ginger and other spices. – **Exports:** arrowroot, spices, copra.

ANTIGUA

Area 442 sq.km, population 72,000 (1978), **sovereign state, member of the Commonwealth** (Prime Minister Vere C. Bird since 1981).

Islands: Antigua (280 sq.km), Barbuda (160 sq.km) and Redonda (2 sq.km). **Capital:** St. John's 13,000 inhab. (1975). **Economy:** sugarcane, cotton. **Production:** sugar, rum, spirits. Tourism 62,971 visitors (1976).

BRITISH TERRITORIES:

BERMUDA – The Bermuda Colony, area 53 sq.km, population 62,000 inhab. (1978), **British colony** (Governor Sir Peter Ramsbotham since 1976).
Capital: Hamilton 3,000 inhab. (1976). **Density** 1,170 persons per sq.km. – **Currency:** Bermuda dollar = 100 cents. **Economy:** tourism 558,874 visitors (1976); early vegetables, bananas, citrus fruit, tobacco, flowers, fish. Merchant shipping 1,450,387 GRT (1975). Strategic naval and air base of the United Kingdom and the U.S.A.

add page 158

JAMAICA

Area 10,991 sq.km, population 2,140,000 (1978), **member of the Commonwealth** (Prime Minister Edward Seaga since 1980).

Capital: Kingston 111,879 inhab. (1970, with agglomeration 475,548); **other towns** (1970, in 1,000 inhab.) Montego Bay 43, Spanish Town 42. **Population:** Negroes 66%, mestizos 15%, mulattoes 13%. Only 23% of inhabitants engaged in agriculture (1977). – **Currency:** Jamaica dollar = 100 cents.
Agriculture (1977, in 1,000 tons): sugarcane 3,228, bananas 146, citrus fruit 107, coconuts 87, Jamaica pepper 3,873 tons (1975); cattle 282,000 head, poultry 3.9 mn. **Mining** of bauxite 11,424,000 tons (1977 15% of world production). **Production:** aluminium, cement 365,000 tons, sugar 297,000 tons, world-famous rum, light industry. Electricity: 2,378 mn. kWh (1976). – **Communications:** railways 401 km, roads 4,326 km Tourism 327,700 visitors (1976). – **Exports** (1974): aluminium (53%), bauxite (21%), sugar, bananas, rum

MEXICO

Estados Unidos Mexicanos, area 1,972,54? sq.km, population 69,381,000 (1979) **federal republic** (President José Lopez Portillo since 1976).

Administrative units: 31 federal states and 1 federal district. **Capital:** México 8,988,230 inhab. (1978, with agglomeration 13,993,866); **other towns** (1978, in 1,000 inhab.): Netzahualcoyotl 2,068, Guadalajara 1,81? (with agglom. 2,343), Monterrey 1,054 (with agglom. 1,923), Puebla 678, Ciudad Juárez 597, León 59? Tijuana 535, Acapulco 421, Chihuahua 370, Mexicali 338, San Luis Potosí 315, Cuernavaca 307, Culiacán 30? Hermosillo 300, Veracruz 295, Torreón 269, Mérida 263, Aguascalientes 248, Saltillo 246, Tampico 24? Morelia 239, Toluca 223, Reynosa 219, Durango 219, Nuevo Laredo 214, Jalapa Enríquez 191, Poza Rica de Hidalgo 189, Matamoros 186, Mazatlán 178, Querétaro 176, Ciudad Obregón 173, Villahermosa 165 and further 18 cities with more than 100,000 inhabitants. **Population:** mestizos 52%, Indians 29%, Whites and Creoles 18%. **Density** 34 persons per sq.km (1978), average annual rate of population increase 3.3%, birth rate 42 per 1,000, death rate 8.6 per 1,000 (1970–75); urban population 63.6% (1976). Economically active 18.3 mn., 38.7% engaged in agriculture (1977). – **Currency:** Mexican peso = 100 centavos.
Economy: agricultural and industrial country with important mineral mining, among the most developed countries in Latin America. **Agriculture:** arable land covers 14% (of this: 4.8 mn. ha irrigated), meadows and pastures 33.8% and forests 36% of the land area (1976). **Crops** (1977, in 1,000 tons): cereals 15,74? (of this: wheat 2,451, rice 481, barley 409, maize (corn) 8,991, sorghum 3,390), potatoes 653, beans 74? (16% of world production), soybeans 490, groundnuts 56, sesame 127 (6.4% of world production – states Michoacán, Guerrero, Oaxaca), cotton -seed 590, -lint 325; coconuts 980, copra 135, palm kernels 3? vegetables 2,480, sugarcane 31,500, fruit 5,713 (oranges and tangerines 1,283, lemons 611, strawberries 8? avocados 293, mangoes 443, pineapples 437, bananas 1,210, grapes 260), coffee 246 (states Veracruz Chiapas, Oaxaca, Guerrero). **Livestock** (1977, in million head): cattle 28.9 (dairy cows 4.2), pigs 12 sheep 7.9, horses 6.6 (11% of world production and second greatest producer), mules and asses 6.? goats 8.3, poultry 147.7 (turkeys 13); eggs 474,120 tons, honey 56,993 tons, cowhides 65,432 tons; fish catch 572,285 tons (1976). Roundwood 14.8 mn. cub.m (1976).
Mining (1977, in 1,000 tons, metal content): crude petroleum 60,840 (1978, along the Gulf of Mexico: states Veracruz, Chiapas, Tabasco), natural gas liquids 3,409, natural gas 16,932 mn. cub.m (1978), silver 1,462 ton (second place, 14% of world output, states Hidalgo, Chihuahua), lead 161 and zinc 265 (Chihuahua, Nuevo Léon, Hidalgo); coal 5,650 (1976), iron ore 4,212 (states Durango, Colima), copper 89, manganese 16? (1976), gold 5,064 kg (1976), antimony 2,698 tons, mercury 333 tons, graphite 61 (1975), bismuth 718 ton (1974), tungsten 235 tons (1976), barytes 300 (1975), fluorite 1,089 (1975), sulphur 2,152, natural phosphates 224 (1976), salt 4,391 (1976). **Production** (1977, in 1,000 tons): pig iron 4,428, crude steel 5,412, lead 16? zinc 171, aluminium 42, copper 73 (1978); cars 277,200 units, tyres 6.1 mn., television 729,000 and radio receivers 1,135,000 (1976); sulphuric acid 2,028, synthetic rubber 80, motor spirit 7,978 and oils 19,50? (1976 – capacity of petroleum refineries 47.9 mn. tons), nitrogenous fertilizers 650 (1976), chemical woo pulp 317, paper 1,275 (1976), woven cotton fabrics 123 and fibres 158 (1975), cement 13,092; 46.6 billio cigarettes (1976), beer 19.4 mn. hl and wine 315,000 hl (1976), meat 1,431, flour 1,680, sugar 2,72? milk 5,731, cheese 85, canned fish 66 (1976). Electricity: capacity 12.8 mn. kWh (1976), productio 50,052 mn. kWh (1977). – **Communications:** railways 19,665 km (1977), roads 175,540 km (of which 10,283 km Panamerican Highway, 1975), motor vehicles – passenger 2.7 and commercial 1.2 mn (1977). Merchant shipping 674,000 GRT (1976). Civil aviation – 102 mn. km flown, 1.4 mn. passenger carried (1976). – **Exports** (1977): crude petroleum (21%), coffee (10.7%), textile fibres (8.8%), fruit and vegetables, fish and canned fish, sugar, ores. Chief trading partners: U.S.A. (65%), Japan, Fed. Rep. of Germany, Canada, Brazil, United Kingdom.

NICARAGUA

República de Nicaragua, area 148,000 sq.km, population 2,395,000 (1978), **republic** (Provisional President F. Urcuyo since 1979).
Administrative units: 16 departments. **Capital:** Managua 409,810 inhab. (1973); **other towns** (197? in 1,000 inhab.): León 91, Matagalpa 69, Granada 57, Masaya 56. **Population:** mestizos 71%, Whites 15% Indians. Average annual rate of population increase 3.3% (1970–75). – **Currency:** córdoba = 100 centavos

SOUTH AMERICA

The larger part of the South American continent lies in the Southern Hemisphere where it is isolated from the other :ontinents, save North America to which it is linked by the narrow Isthmus of Panama (width 48 km). Christopher Columbus s generally recognized as one of the first explorers to have reached South America, and on his third expedition (1498–1500), he discovered Trinidad and the mouth of the Orinoco; in 1499 a Spanish expedition led by Alonso de Ojeda 'eached the mouth of the Amazonas (Amazon), and in 1500 the Portuguese P.A. Cabral reached the shores of Brazil.

South America is the fourth largest continent; it measures **17,834,000 sq.km,** has **236 mn. inhabitants** (1978))ut with a density of only 13 persons per sq.km. **Geographical position:** northernmost point: Punta Gallinas 12°27′ N.Lat.; southernmost point-on the mainland: Cape Froward 53°54′ S.Lat., on the Peninsula de Brunswick, of the entire continent-s. Diego Ramírez 56°32′ S.Lat. in Drake Passage; easternmost point: C.Branco 34°45′ W.Long.; westernmost point: Cape Punta Pariñas 81°22′ W.Long., incl. Galapagos Is. 92°01′ W.Long. The maximum length of the South American :ontinent is 7,350 km, as far as Cape Horn 7,550 km, and its maximum width is 5,170 km.

The topography of South America has few distinguishing horizontal features. The coastline is 28,700 km long and slands make up but 1% of the area of the continent. The largest islands include Tierra del Fuego (71,500 sq.km) and :he Falkland Is. (11,961 sq.km) in the South Atlantic Ocean, the Chilean islands of Western Patagonia (18,000 sq.km) and the Galapagos Is. (7,844 sq.km) to the West of Ecuador in the Pacific Ocean, Ilha de Marajó (42,000 sq.km) and the Ilha Caviana (5,000 sq.km) in the mouth of the Amazonas and Trinidad in the Atlantic Ocean (4,827 sq.km).

The vertical features of the continent are most striking in the West where the Andes rise to nearly 7,000 m,)ut otherwise low-lying land prevails. Lowlands not higher than 300 m make up more than 50% of the land area, and higher plateaux only 15%. The simple shape of the continent is due to its geological structure and geomorpho-ogical evolution. Above the geologically old depressions in the Pacific Ocean, from west of the Isthmus of Panama as far as the southern point of the continent, Tertiary and Quaternary folding raised the narrow strip of high nountain ranges of the South American Cordilleras, called the Andes, to a length of some 9,000 km. They are divided into 30 main ranges, of which the most important are: the north and north-western Andes reaching their ighest point in P. Cristóbal Colón (5,775 m), the Ecuadorian Andes with the extinct crater of Chimborazo (6,297 m), he Peruvian Andes with the highest peaks Nevado de Huascarán (6,807 m) and Nudo Coropuna (6,613 m). In the Boli-vian Cordillera Real, Nev. Ancohuma reaches the height of 6,550 m. The highest point of the continent, Aconcagua 6,959 m) rises in the Argentinian-Chilean Cordilleras, and the strongly glaciated Patagonian Cordilleras rise to 4,058 m n San Valentín. The core of the continent is formed by the vast, ancient Brazilian-Guyanian shield, which s composed of the oldest rocks and effusive plutonic rock, deeply eroded and denuded, which reach their highest point n the Brazilian peak of Pico de Neblina (3,014 m). Between these two main units, the Andes and the shield, stretch xtensive lowlands: those of the Orinoco, the Amazonas and the Río de La Plata where the basins of the great rivers orm terraces and plains with slopes and plateaux – chapadas – at the watersheds. The lowest point is Salinas Grandes (−40 m) on the Valdés peninsula (Argentina). Volcanic activity in the Andes dates back to the end of he Tertiary period and it continues even now. There are some 40 extinct volcanoes and roughly 50 active ones :he highest: V. Guallatiri, 6,060 m). The Chile-Peru Trench (8,066 m) off the west coast of the continent is the epicentre f powerful earthquakes.

The river network is one of the densest in the world. The annual mean discharge amounts to 8,050 cub.km. Almost 88% of the land drains into the Atlantic Ocean, where the Amazonas (Amazon), the longest river in the world, as its mouth (with the Ucayali-Apurímac: length 7,025 km, drainage area 7,050,000 sq.km, annual discharge 3,800 cub.km, naximum flow 220,000 cub.m per sec., minimum 115,000 cub.m per sec.). The Amazonas has 20 tributaries longer han 1,500 km. Other major rivers include the Paraná, Orinoco, São Francisco and Magdalena. There are few **lakes,** he largest being L. de Maracaibo (14,343 sq.km), Lagoa dos Patos (10,145 sq.km), and L.Titicaca (6,850 sq.km). he Andes are characterized by valleys that have no outlet to the sea, extensive salt swamps – salars – and, in he South, lakes of glacial origin.

The main part of South America belongs to zones with tropical and subtropical **climates,** while the narrower outh belongs to the temperate zone. Everywhere the climate changes with altitude, as do the natural and cultivated orms of vegetation. The highest annual mean temperature was measured at Maracaibo (Venezuela) 28.9°C, the lowest t Cristo Redentor in the Paso de Bermejo (Argentina) −1.8°C; absolute maximum temperature – Rivadavia 48.9°C Argentina), minimum – Sarmiento (−33°C). Highest annual mean precipitation – Buenaventura 7,155 mm, lowest – rica 0.8 mm. Mean January and July temperatures in °C (and annual rainfall in mm): Maracaibo 26.7 and 29.5 (549), ogotá 14.2 and 13.9 (1,059), Quito 13.1 and 12.9 (1,246), Manaus 25.9 and 26.9 (2,001), Recife 27.1 and 24.2 1,498), Lima 21.7 and 15.6 (29), La Paz 11.6 and 8.7 (574), Arica 27.1 and 19.3 (0.8), Goiás 23.6 and 22.5 1,689), Rio de Janeiro 25.4 and 20.2 (1,076), São Paulo 27.7 and 21.2 (1,361), Asunción 28.8 and 18.2 (1,344), órdoba 32.1 and 18.6 (707), Buenos Aires 29.5 and 14.5 (1,008), Santiago 29.4 and 14.5 (351), Sarmiento 17.7 and 6 (142), Punta Arenas 14.4 and 4.4 (425).

The natural vegetation of the neotropical region: the largest evergreen rain forests in the world (Hylea), scrub teppe, savanna (llanos), Xerophilous Chaco woodlands, pampas, deserts with cacti, Andean desert (punas and aramos), Antarctic flora. The neotropical zone contains some **animal life:** pumas, jaguars, tapirs, mountain llamas, uanaco, vicuñas, alpaca, ant-eaters, sloths, armadillos, howler, chatter monkeys, etc. It has the richest bird life in he world, 3,500 species, including Harpie eagles, condors, toucans with brilliant plumage, humming-birds, parrots, s well as the largest snakes in the world – anacondas, rattle snakes, caymans-large numbers of insects and g fishes – piraibas, pirañas, multi-coloured aquarium fishes, etc.

The oldest **inhabitants** of South America are Indians. The inflow of White Europeans began in the 16th century nd later Black Africans (and fewer Asians) arrived; which led to a marked intermingling of races: mestizos – dian and White, mulattoes – Negro and White, and zamboes – Indian and Negro. In 1978 there were 36 mn. inhabitants, i.e. 5.6% of the world's population, of which 48.9% lived in Brazil, 11.2% in Argentina, 10.8% in olombia and 7% in Peru. At almost 3% there is a high natural increase of population. 68.6% of the inhabitants ve in towns; in Chile the figure is as high as 81% and there has been a rapid expansion of the metropolises, sually the capital cities. There are 15 towns with over a million inhabitants (largest São Paulo, Rio de Janeiro, Buenos res) and 14 towns with more than 500,000 inhabitants.

CARIBBEAN SEA
Lesser Antilles
Martinique
Grenada
Tobago
Trinidad
ATLANTIC OCEAN
Panama Canal
Pta. Gallinas
P.Cristóbal Colón
5775
L. de Maracaibo
Caracas
Cord.de Mérida
P. Bolívar
5002
Llanos
Orinoco
Isthmus of Panama
Cordillera Occidental
Central
Oriental
Magdalena
5493
Meta
Bogotá
Nev. del Huila
5750
Guaviare
Vaupés
Salto Angel
Mt. Roraima
2810
Guiana Highlands
Georgetown
Paramaribo
C. Orange
Pico de Neblina
3014
Branco
Caviana Is.
Macapá
I. de Marajó
Equator
Belém
Cotopaxi
5897
6297
Chimborazo
Sangay 5230
Putumayo
Caquetá
Japurá
Negro
Amazonas
Manaus
Iquitos
Tefé
Marañón
Pta. Pariñas
Ucayali
Juruá
Purus
Selvas
Madeira
Tapajós
Xingu
Iriri
Tocantins
Parnaíba
Teles Pires
Juruena
Araguaia
Nev. de Huascarán
6807
6632
Co. Yerupaja
Lima
Apurímac
Andes
Nev. Auzangate
6384
L. Titicaca
N. Coropuna
6613
5672
Ubinas
6550
Nev. Ancohuma
La Paz
Beni
Guaporé
Mamoré
1995
Brazilian Highlands
Planalto do Mato Grosso
Brasília
1678
Sa. Geral de Goiás
São Francisco
1850
São Fr.
Cach. Paulo Afonso
6866
Arica
6060
L. de Poopó
V. Guallatiri
Salar de Uyuni
Gran Chaco
Corumbá
Paraguay
Paranaíba
Grande
Paraná
P. d. Bandeira
2890
Agulhas Negras
2787
Tietê
Rio de Janeiro
São Paulo
Salto das Sete Quêdas
Asunción
Pilcomayo
Bermejo
Rivadavia
Lascar
5990
V. Llullaillaco
6723
6870
Ojos del Salado
6872
Co. Bonete
Des. de Atacama
8066
Tropic of Capricorn
PACIFIC OCEAN
I. San Félix
I. San Ambrosio
550
Saltos do Iguaçu
Sa. do Mar
2000
Salado
Entre Rios
Uruguay
Lagoa dos Patos
Pôrto Alegre
6380
Aconcagua
6959
Sa. de Córdoba
Córdoba
2884
Santiago
Islas Juan Fernández
I. A. Selkirk
I. Rób. Crusoe
V. Maipo
5323
Buenos Aires
Río de la Plata
Montevideo
Lagoa Mirim
4709
Pampas
V. Lanín
3776
Negro
Colorado
Bahía Blanca
2661
V. Osorno
G. S. Matías
Pen. Valdés
-40
Chubut
5021
I. de Chiloé
Arch. de los Chonos
Sarmiento
G. S. Jorge
San Valentín
4058
C. Tres Puntas
Patagonia
ATLANTIC OCEAN
6212
55
65
I. Wellington
Str. of Magellan
Pta. Arenas
Falkland Is.
West-Falkland
East-Falkland
Tierra del Fuego
C. Froward
2469
Drake Pas.
C. Horn
Is. D. Ramírez
5840
South Georgia
1 : 50 000 000
0 250 500 750 1 000 1 250 Km
0 200 400 600 800 Mi
80° 70° 60° 50° 40° 30° 90°
10° 0° 20° 30° 40° 50°
1 2 3 4 5
A B C D E F G H

CARIBBEAN SEA
ATLANTIC OCEAN
PACIFIC OCEAN
COLOMBIA
VENEZUELA
GUYANA
SURINAM
French Guiana
ECUADOR
PERU
BRAZIL
BOLIVIA
PARAGUAY
CHILE
ARGENTINA
URUGUAY
ST. LUCIA
Martinique (Fr.)
Neth. Antilles
ST. VINCENT
GRENADA
Port of Spain
TRINIDAD AND TOBAGO
Sta. Marta
Cartagena
MARACAIBO
CARACAS
Valencia
Barquisimeto
MEDELLÍN
S. Cristóbal
Bucaramanga
Pto. Carreño
Ciudad Bolívar
Georgetown
Paramaribo
Cayenne
BOGOTÁ
Pto. Ayacucho
Boa Vista
CALI
Neiva
Pasto
Mitú
Macapá
Equator
I. de Marajó
Óbidos
Manaus
BELÉM
Santarém
São Luís
Parnaíba
Arica
Fonte Boa
Amazonas
Iquitos
Leticia
Tefé
FORTALEZA
Codó
Marabá
Manicoré
Eirunepé
Barra do São Manuel
Carolina
Teresina
Macau
Natal
João Pessoa
Crato
Paulistana
RECIFE
Juázeiro
Pôrto Velho
Rio Branco
Huánuco
Guajará-Mirim
Rondônia
Maceió
Pôrto Nacional
Cerro de Pasco
Cobija
Aracaju
Callao
LIMA
Cuzco
Trinidad
SALVADOR
Mato Grosso
BRASÍLIA
Ica
LA PAZ
Cochabamba
Cuiabá
GOIÂNIA
Ilhéus
Arequipa
Puno
Mollendo
Oruro
Santa Cruz
Sucre
Pirapora
Montes Claros
Arica
Pisagua
Iquique
Potosí
Corumbá
Uberaba
BELO HORIZONTE
Campo Grande
Vitória
Campinas
RIO DE JANEIRO
Tropic of Capricorn
Antofagasta
SÃO PAULO
Niterói
Salta
Asunción
Santos
CURITIBA
Formosa
Copiapó
S. M. de Tucumán
Resistencia
Florianópolis
I. San Ambrosio (Chile)
La Serena
PÔRTO ALEGRE
CÓRDOBA
Santa Fe
Lag. dos Patos
Mendoza
Rio Grande
Islas Juan Fernández (Chile)
Valparaíso
ROSARIO
SANTIAGO
BUENOS AIRES
MONTEVIDEO
Talca
La Plata
Río de la Plata
Talcahuano
Concepción
Zapala
Mar del Plata
Bahía Blanca
Valdivia
Puerto Montt
San Carlos-d. Bar.
G. S. Matias
I. de Chiloé
Rawson
Pto. Aisén
Comodoro Rivadavia
Colonia-Las Heras
G. S. Jorge
Puerto Deseado
I. Wellington
Falkland Is. (U. K.)
Río Gallegos
Stanley
Strait of Magellan
Punta Arenas
Tierra del Fuego
Ushuaia
Drake Passage
South Georgia (U. K.)
1 : 50 000 000
0 250 500 750 1 000 1 250 Km
0 200 400 600 800 Mi

LONGEST RIVERS

Name	Length in km	River basin in sq.km
Amazonas (-Ucayali-Apurímac)	7,025	7,050,000
Paraná (-Grande)	4,380	4,250,000
Madeira (-Mamoré)	4,100	1,360,000
Purus	3,380	1,100,000
Juruá	3,285	
Tocantins (-Araguaia)	3,100	1,180,000
São Francisco	2,900	631,670
Tocantins	2,700	840,000
Araguaia	2,600	340,000
Paraguay	2,550	1,150,000
Japurá (-Caquetá)	2,520	.
Negro (-Vaupés)	2,380	.
Uruguay (-Canoas)	2,200	420,000
Xingu	2,100	450,000
Orinoco	2,060	1,085,000
Ucayali (-Apurimac)	1,980	375,000
Tapajós	1,950	460,000

HIGHEST MOUNTAINS

Name (Country)	Height in m
Aconcagua (Arg.)	6,959
Co.Bonete (Arg.)	6,872
Ojos del Salado (Arg.-Chile)	6,870
Nev.de Huascarán (Peru)	6,807
Co.Tupungato (Chile)	6,800
Mte.Pissis (Arg.)	6,779
Co.Mercedario (Arg.)	6,770
V.Llullaillaco+ (Chile-Arg.)	6,723
Nev.Cachi (Arg.)	6,720
Co.Yerupaja (Peru)	6,632
Nudo Coropuna+ (Peru)	6,613
Nev.Ancohuma (Bol.)	6,550
Nev.Sajama (Bol.)	6,520
Nev.Illimani (Bol.)	6,457
Nev.Auzangate (Peru)	6,384
Chimborazo (Ecuador)	6,297

+ inactive volcano

LARGEST LAKES

Name	Area sq.km	Altitude in m
Lago de Maracaibo	14,343	Sea level
Lagoa dos Patos	10,145	Sea level
Salar de Uyuni	10,000	3,660
Lago Titicaca	6,850	3,812
Lagoa Mirim	2,965	1
Lago de Poopó	2,530	3,690
Lago Buenos Aires	2,400	217
Lago Argentino	1,500	187

LARGEST ISLANDS

Name	Area in sq.km
Tierra del Fuego	71,500
Ilha de Marajó	42,000
Isla de Chiloé	8,394
Wellington	6,750
East Falkland	6,682
West Falkland	5,258
Trinidad	4,827
Isla Isabela	4,278

ACTIVE VOLCANOES

Name (Country)	Altitude in m	Latest eruption
V. Guallatiri (Chile)	6,060	1960
Lascar (Chile)	5,990	1968
Cotopaxi (Ec.)	5,897	1975
Ubinas (Peru)	5,672	1969
Sangay (Ec.)	5,230	1976
Cotacachi (Ec.)	4,939	1955
Puracé (Col.)	4,756	1977
Reventador (Ec.)	3,485	1976

FAMOUS NATIONAL PARKS

Name (Country)	Area in sq.km
Nahuel Huapi (Arg.)	7,850
Los Glaciares (Arg.)	6,430
Lanín (Arg.)	4,200
Los Alerces (Arg.)	2,850
Pilcomayo (Arg.)	2,820
Iguaçu (Braz.-Arg.)	2,530
L. Mar Chiquita (Arg.)	1,800
Sierra Nevada (Ven.)	1,600

SOUTH AMERICA

Country	Area in sq.km	Population	Year	Density per sq.km
Argentina	2,780,092	26,393,000	1978	9
Bolivia	1,098,581	6,113,000	1978	6
Brazil	8,511,965	115,397,000	1978	14
Chile	756,945	10,656,000	1977	14
Colombia	1,138,914	25,582,000	1978	22
Ecuador	283,561	7,814,000	1978	28
Falkland Islands and Dependencies (U.K.)	16,264	2,000	1977	0.1
French Guiana (Fr.)	91,000	59,000	1977	0.6
Guyana	214,969	819,000	1978	4
Netherlands Antilles	988	245,000	1978	248
Paraguay	406,752	2,889,000	1978	7
Peru	1,285,216	16,819,000	1978	13
Surinam	163,265	448,000	1977	2.7
Trinidad and Tobago	5,128	1,133,000	1978	221
Uruguay	177,508	2,864,000	1978	16
Venezuela	912,050	13,304,000	1978	15

South America has two distinct types of **economy:** one typical of the developing countries and the other typical of advanced, more industrialized countries, like Brazil, Argentina, Venezuela. The economies of countries with highly specialized industries fluctuate in accordance with world economics. Agriculture takes the form of monoculture on plantations: the most important crops are corn, manioc, sugarcane, soya beans, sunflower seed, flax, coffee, bananas, oranges, while cattle breeding and sheep rearing are also important. The mining industry is highly developed and extremely important, but most of the products are exported as raw or semi-finished materials. Manufacturing is developing slowly as yet and plays only an insignificant part. The continent is densely covered in forests (52% of the total area) and it has the richest sources of water power in the world (especially Brazil), but so far it has been put to limited use, for only in recent decades have major dams and hydro power-stations been built, mainly in Brazil. **Transport and foreign trade** are of exceptional importance to the economic development of South America.

ARGENTINA

República Argentina, area 2,780,092 sq.km, population 26,393,000 (1978), **federal republic** (President Gen. Leopoldo Fortunato Galtieri since 1981).

Administrative units: 22 provinces, 1 federal district and 1 national territory. **Capital:** Buenos Aires 2,976,000 inhab. (1974, +metropolitan area Gran Buenos Aires 8,925,000 inhab.); **other towns** (1970 census, in 1,000 inhab.): Córdoba 799, Rosario 750, La Matanza+ 659, Morón+ 486, Lanús+450, Lomas de Zamora+ 411, La Plata 408, Gen. San Martín+ 361, Quilmes+ 355, Avellaneda+ 338, S.M.d. Tucumán 322, Gen. Sarmiento+ 315, Caseros+ 313, Mar del Plata 302, Vicente López+ 285, San Isidro+ 250, Santa Fe 245, Almirante Brown+ 242, Merlo+ 189, Bahía Blanca 182, Salta 176, Tigre+ 152, Resistencia 143, Corrientes 137, Berazategui+ 128, Mendoza 119, San Juan 113. **Population:** Argentinians (only 30,000 Indians). **Density** 9 persons per sq.km, (1978); average annual rate of population increase 1.4%; urban population 80.7% (1975). 14% of inhabitants employed in agriculture (1977). – **Currency:** Argentine peso = 100 centavos. **Economy:** agricultural and industrial country. **Agriculture:** 12.8% of the land area covered by arable land, 53% meadows and pastures, 22% forests; irrigated land 1.8 mn. ha. **Crops** (1977, in 1,000 tons): wheat 5,300 (in 1976 – 11,000), maize (corn) 8,300, barley 494, oats 570, sorghum 6,730 (third world producer), rice 320, potatoes 1,777, soybeans 1,400, groundnuts 600, sunflower 900, flax (seed) 640 (22% of world crop — leading world producer), cotton – seed 308, – lint 166, vegetables 2,084, fruit 6,614 (citrus fruit 1,484, grapes 3,400), tea 34, tobacco 90. **Livestock** (1977, in million head): cattle 59.6 (dairy cows 2.7), pigs 4.2, sheep 34, horses 3.5, poultry 38.4; fish catch 281,700 tons (1976); production (1977, in 1,000 tons): meat 3,568, milk 5,200, eggs 205, grease wool 167, cowhides 430, sheepskins 43.3. Roundwood: 11.5 mn. cub.m (1976).
Mining (1976, in 1,000 tons, metal content): crude petroleum 22,488, natural gas 7,710 mn. cub.m and coal 528 (all 1977), iron ore 163.4 (1977), zinc 40.5, lead 33, silver 54 tons, gold 369 kg, uranium 60 tons (1975), sulphur 14, salt 1,200, beryllium. Electricity: 30,328 mn. kWh (of this 2,572 nuclear). **Industry:** food processing predominates, light industry and metallurgy are important. **Production** (1977, in 1,000 tons): pig iron 1,380, crude steel 2,676, lead 40 (1976), zinc 31.3 (1976), motor spirit 4,223, cement 6,036, sulphuric acid 234; fibres – cotton 94.9, synthetic 43–flour 2,529 (1976), sugar 1,667, cheese 225, wine 24.9 mn. hl (1976); 33.8 billion cigarettes (1976).
Communications (1976): railways 39,782 km, roads 311,900 km (4,835 Panamerican Highway), passenger cars 2.2 mn. Merchant shipping 1,677,000 GRT (1977). Civil aviation (1976): 68.4 mn. km flown and 3.3 mn. passengers carried.
Exports (1976): vegetable (wheat, oil seeds etc.) and animal (meat, hides, wool) products predominate, flour and foodstuffs (67.5% of turnover), machines and raw materials. Chief trading partners: U.S.A., Brazil, Fed. Rep. of Germany.

BOLIVIA

República de Bolivia, area 1,098,581 sq.km, population 6,113,000 (1978), **republic** (President Celso Torrelio Villa since 1981).

Administrative units: 9 departments. **Capital:** La Paz 654,713 (seat of the government) and Sucre 106,590 inhab. (1976); **other towns** (1976, in 1,000 inhab.): Santa Cruz 237, Cochabamba 194, Oruro 124, Potosí 105. **Population:** Indians 52%, mestizos 27% and Creoles 21%. 52% of inhabitants engaged in agriculture (1977). – **Currency:** Bolivian peso = 100 centavos.
Economy: principal branch – mineral **mining** (1976, in 1,000 tons, metal content): antimony 17 (prime world producer, 24% of world output – Potosí), tin 31 (second world producer 15% of world output – Oruro, Potosí), tungsten 3,832 tons (Atocha), crude petroleum 1,890, natural gas 1,735 mn. cub.m, zinc 49, lead 18, silver 183 tons (1977), gold 896 kg. **Agriculture:** crops (1977, in 1,000 tons): cereals 523 (maize /corn/, rice), potatoes 679, sugarcane 679, tropical fruit 654; **livestock** (1977, in million head): cattle 3.3, pigs 1.2, sheep 7.9, llamas and alpacas 1.8, poultry 6.1. Electricity 1,130 kWh (1976). – **Communications** (1975): railways 3,579 km, roads 37,075 km. – **Exports** (1976): tin (38%), crude petroleum, other metals (30%). Chief trading partners: U.S.A., Argentina, Brazil, United Kingdom.

BRAZIL

República Federativa do Brazil, area 8,511,965 sq.km, population 115,397,000 (1978), **federal republic** (President Gen. João Baptista de Oliveira Figueiredo since 1979).

Administrative units: 23 federal states, 4 federal territories, 1 federal district. **Capital:** Brasília 763,254 inhab. (1975); **other towns** (1975, in 1,000 inhab.): São Paulo 7,199 (with agglomeration 10 mn.), Rio de Janeiro 4,858 (with agglom. 8.3 mn.), Belo Horizonte 1,557, Recife 1,250, Salvador 1,237, Fortaleza 1,110, Pôrto Alegre 1,044, Nova Iguaçu 932, Belém 772, Curitiba 766, Duque de Caxias 537, São Gonçalo 534, Goiânia 518, Santo André 515, Campinas 473,

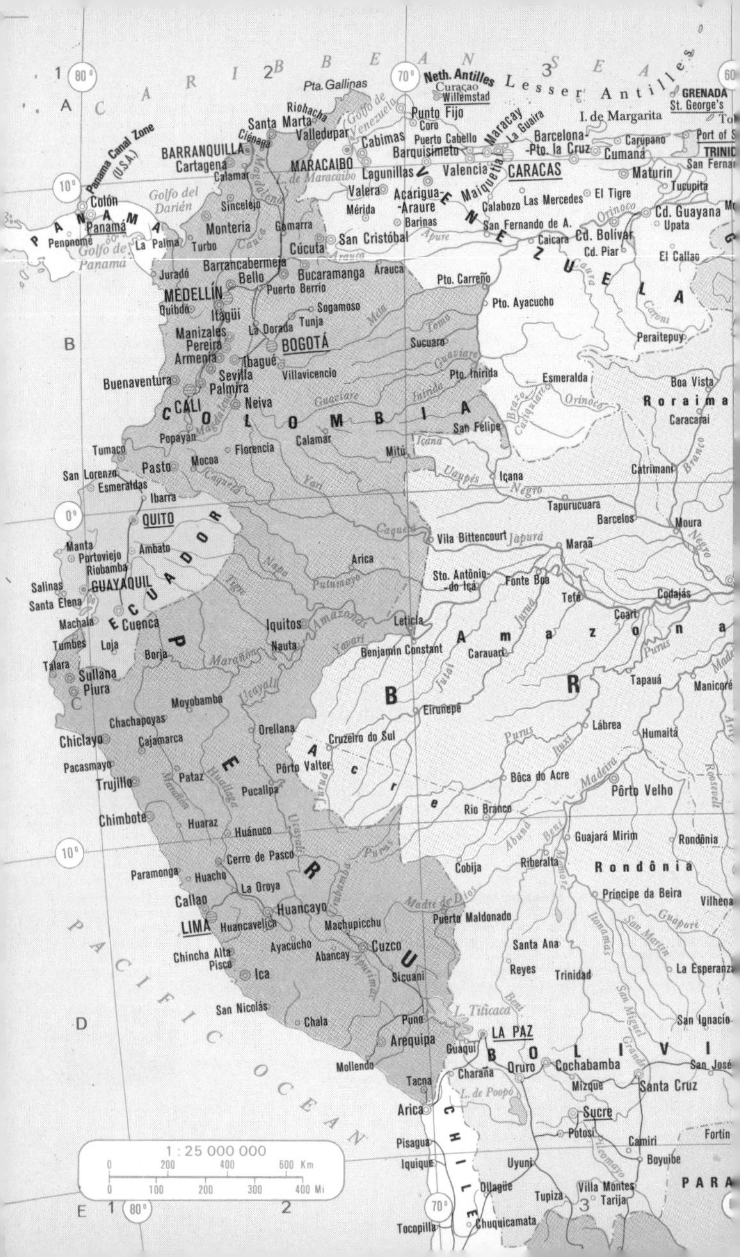C A R I B B E A N S E A
Neth. Antilles
Curaçao
Willemstad
Lesser Antilles
GRENADA
St. George's
Pta. Gallinas
Riohacha
Golfo de Venezuela
Punto Fijo
Coro
I. de Margarita
Santa Marta
Ciénaga
Valledupar
Cabimas
Puerto Cabello
Maracay
La Guaira
Barcelona-
-Pto. la Cruz
Carúpano
Cumaná
Port of Spain
TRINIDAD
San Fernando
Panama Canal Zone (U.S.A.)
BARRANQUILLA
Cartagena
Calamar
MARACAIBO
L. de Maracaibo
Barquisimeto
Lagunillas
Valencia
Maiquetía
CARACAS
Maturín
Tucupita
Golfo del Darién
Colón
Sincelejo
Magdalena
Valera
Mérida
Acarigua-
-Araure
Calabozo
Las Mercedes
El Tigre
Orinoco
Cd. Guayana
Upata
P A N A M Á
Panamá
Penonomé
Golfo de Panamá
La Palma
Turbo
Montería
Cauca
Gamarra
Cúcuta
San Cristóbal
Barinas
Apure
San Fernando de A.
Caicara
Cd. Bolívar
Cd. Piar
El Callao
Arauca
V E N E Z U E L A
Barrancabermeja
Jurado
Bello
Bucaramanga
Puerto Berrío
MEDELLÍN
Quibdó
Itagüí
Sogamoso
Tunja
Pto. Carreño
Pto. Ayacucho
Caura
Caroní
Meta
Tomo
Manizales
Pereira
Armenia
La Dorada
BOGOTÁ
Sucuaro
Peraitepuy
Ibagué
Guaviare
Villavicencio
Pto. Inírida
Esmeralda
Boa Vista
Buenaventura
Sevilla
Palmira
CALI
Neiva
Inírida
Brazo Casiquiare
Orinoco
Roraima
Caracaraí
C O L O M B I A
Popayán
San Felipe
Calamar
Tumaco
Mocoa
Florencia
Mitú
Içana
Branco
San Lorenzo
Esmeraldas
Pasto
Caquetá
Yari
Uaupés
Içana
Negro
Catrimani
Ibarra
Tapurucuara
Barcelos
QUITO
Caquetá
Vila Bittencourt
Japurá
Maraã
Moura
Manta
Portoviejo
Ambato
Riobamba
E C U A D O R
Napo
Arica
Salinas
Santa Elena
GUAYAQUIL
Tigre
Putumayo
Sto. Antônio-
-do Içá
Fonte Boa
Juruá
Tefé
Codajás
Machala
Cuenca
Iquitos
Amazonas
Leticia
Coari
Tumbes
Loja
Nauta
Yavari
Benjamin Constant
Carauari
A m a z o n a s
Purus
Talara
Borja
Marañón
Jutaí
Sullana
Piura
Moyobamba
Ucayali
B R
Tapauá
Manicoré
Eirunepé
Chachapoyas
Orellana
Lábrea
Humaitá
Chiclayo
Cajamarca
Cruzeiro do Sul
Purus
Ituxi
Pacasmayo
Pôrto Valter
A c r e
Bôca do Acre
Madeira
Pôrto Velho
Trujillo
Pataz
Huallaga
Pucallpa
Juruá
Roosevelt
Chimbote
Huaraz
Ucayali
Rio Branco
Abunã
Beni
Huánuco
Guajará Mirim
Rondônia
Purus
Mamoré
Cerro de Pasco
Cobija
Riberalta
Rondônia
Paramonga
Huacho
Urubamba
La Oroya
Príncipe da Beira
Vilhena
P E R U
Callao
Huancayo
Madre de Dios
Puerto Maldonado
Guaporé
LIMA
Huancavelica
Machupicchu
Itonamas
San Martín
Chincha Alta
Pisco
Ayacucho
Abancay
Cuzco
Santa Ana
Ica
Apurímac
Sicuani
Reyes
Trinidad
La Esperanza
P A C I F I C O C E A N
Beni
San Miguel
Grande
San Nicolás
Chala
Puno
L. Titicaca
San Ignacio
Arequipa
LA PAZ
Guaqui
B O L I V I A
Mollendo
Charaña
Oruro
Cochabamba
San José
Tacna
L. de Poopó
Mizque
Santa Cruz
Arica
C H I L E
Sucre
Potosí
Fortín
Pisagua
Camiri
1 : 25 000 000
0 200 400 600 Km
0 100 200 300 400 Mi
Iquique
Uyuni
Pilcomayo
Boyuibe
Ollagüe
Villa Montes
PARAGUAY
Tupiza
Tarija
Tocopilla
Chuquicamata
1 2 3
A B C D E
80°
70°
60°
10°
0°
10°

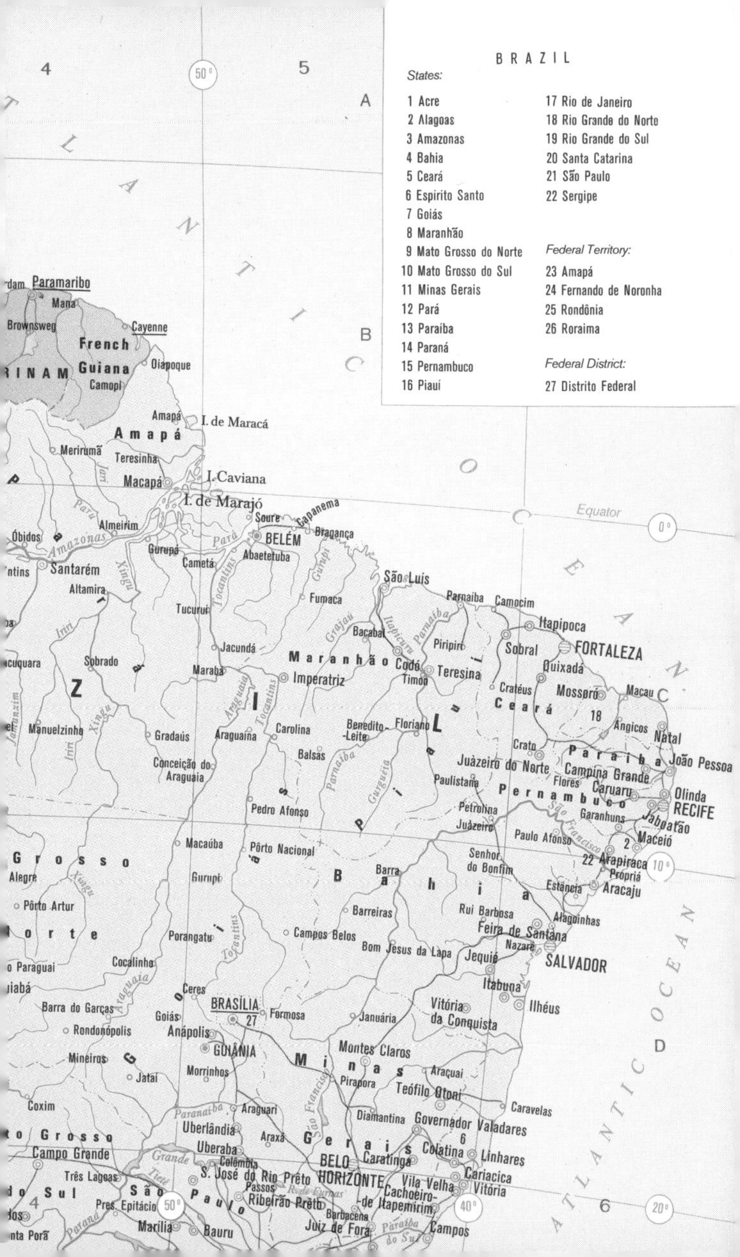

BRAZIL
States:
1 Acre
2 Alagoas
3 Amazonas
4 Bahia
5 Ceará
6 Espírito Santo
7 Goiás
8 Maranhão
9 Mato Grosso do Norte
10 Mato Grosso do Sul
11 Minas Gerais
12 Pará
13 Paraíba
14 Paraná
15 Pernambuco
16 Piauí
17 Rio de Janeiro
18 Rio Grande do Norte
19 Rio Grande do Sul
20 Santa Catarina
21 São Paulo
22 Sergipe
Federal Territory:
23 Amapá
24 Fernando de Noronha
25 Rondônia
26 Roraima
Federal District:
27 Distrito Federal
ATLANTIC OCEAN
Equator
Paramaribo
Cayenne
French Guiana
Amapá
I. de Maracá
Macapá
I. Caviana
I. de Marajó
BELÉM
Santarém
São Luís
Teresina
FORTALEZA
Natal
João Pessoa
Olinda
RECIFE
Maceió
Aracaju
SALVADOR
BRASÍLIA
GOIÂNIA
BELO HORIZONTE
Vitória
Campo Grande
Maranhão
Ceará
Paraíba
Pernambuco
Bahia
Minas Gerais
São Paulo
Mato Grosso do Sul

Santos 396, Manaus 389, Osasco 377, Niterói 376, São João do Meriti 366, Natal 344, Campos 337, São Luís 330, Maceió 324, Guarulhos 311, Teresina 290, João Pessoa 288, Londrina 283, São Bernardo do Campo 267, Ribeirão Prêto 259, Jaboatão 259, Olinda 251. **Population:** Brazilians – Whites 62%, mestizos 28%, Negroes 8% and Indians 2%. **Density** 14 persons per sq.km (1978), average annual rate of population increase 2.8%, birth rate 37 per 1,000, death rate 8.8 per 1,000 (1970–75); urban population 61% (1976). Economically active inhabitants 36.6 mn., 40.5% engaged in agriculture (1977). – **Currency:** cruzeiro = 100 centavos.

Economy: agricultural and industrial country moving towards the group of countries with a highly developed economy. It has enormous mineral resources – the world's richest deposits of iron ore, tin, manganese, bauxite, tungsten and precious stones. **Agricultural production** is among the greatest in the world; highly-productive monocultural plantations predominate. Arable land takes up 4.4% of the land area, meadows and pastures 19.5%, forests 60% (1976). **Crops** (1977, in 1,000 tons): – leading world producer (% of world crops): coffee 950 (22%), bananas 6,188 (17%), manioc 26,511 (24%), castor beans 223 (32%), sisal 225 (48%); – second largest producer in the world: oranges and tangerines 7,510 (23%), sugarcane 120,095 (16.3%); – third largest producer in the world: maize (corn) 19,122, beans 2,327 (18%), soybeans 12,100 (16%), cocoa beans 228 (16%); other products: wheat 2,066, rice 8,941, potatoes 1,900, sweet potatoes 1,815, cotton – seed 1,180, – lint 570, coconuts 237, palm kernels 230, vegetables 3,868, grapes 663, pineapples 550, tobacco 357, jute 96, natural rubber 25. **Livestock** (1977, in million head): cattle 97 (milk cows 14.8), pigs 36.8, sheep 17.3, horses 5.1, poultry 307; fish catch 950,000 tons (1976). **Production** (1977, in 1,000 tons): meat 3,893, milk 11,426, eggs 500, cowhides 354. Roundwood; 164 mn. cub.m (1976).

Mining (1976, in 1,000 tons, metal content): iron ore 45,130 (1978, Minas Gerais), manganese 900 (Bahia, Amapá), bauxite 1,036 (1977, Minas Gerais), magnesite 419 (Ceará, Bahia), nickel 3,100 tons (Goiás), chromite 161 (Bahia, Minas Gerais, Goiás), tungsten 1,500 tons (Rio Grande do Norte), tin 6,395 tons (1977, Rondônia), coal 3,500 (1977), crude petroleum 8,136 (Bahia, Alagoas, Sergipe), natural gas 1,640 mn. cub.m, gold 4,922 kg (Minas Gerais), diamonds 270,000 carats, precious and semi-precious stones (Minas Gerais, Goiás, Bahia), uranium, zinc 61, mica, asbestos, graphite 27,347 tons (Minas Gerais), phosphate rock 463, salt 2,200. **Industrial production** (1977, in 1,000 tons): pig iron 9,324, crude steel 12,180, phosphorus fertilizers 822, synthetic rubber 188 (1978), 18.7 m. tyres, cement 21,132, paper 1,518 (1975), aluminium 122 (1976); motor vehicles: passenger 463,200 and commercial 454,800 units, television receivers 1.8 mn. units; woven fabrics: cotton 1,170 mn. m, rayon 308 m. sq.m; flour 2,292 (1976), sugar 8,900, wine 280; 112 billion cigarettes (1976). Electricity (1976): capacity 21.8 mn kW (of which 84% hydro-electric power stations – e.g. Urubupungá, Jupiá, Furnas, Paulo Afonso etc.), production 99,864 mn. kWh (1977).

Communications: railways 30,546 km (1973), roads 1,347,700 km (72,700 km hard-surfaced, 6,225 km Panamerican Highway); motor vehicles: – passenger 4.6 mn., – commercial 1.1 mn. (1975). Merchant shipping 3.3 mn. GRT (1977). Civil aviation – 173.3 mn. km flown and 8.8 mn. passengers carried (1976). – **Exports** (1976): agricultural products 51% of turnover (coffee 27.4%), industrial products 23.2%, iron ore 20%. Chief trading partners: U.S.A., Fed. Rep. of Germany, Japan, Saudi Arabia, Iraq, Italy.

CHILE

República de Chile, area 756,945 sq.km, population 10,656,000 (1977), **republic** (President Gen. Augusto Pinochet since 1973).

Administrative units: 25 provinces. **Capital:** Santiago 3,186,000 inhab.; **other towns** (1976, in 1,000 inhab.): Valparaíso 249, Viña del Mar 229, Talcahuano 184, Concepción 181, Antofagasta 150, Temuco 138, Talca 115, Arica 112, Rancagua 108, Chillán 102. **Population:** Chileans (only 120,000 Indians). **Density** 14 persons per sq.km (1977), average annual rate of population increase 1.8%, urban population 81%. 20% of inhabitants engaged in agriculture. – **Currency:** Chilean peso = 100 centesimos.

Economy: developing agricultural and industrial country with important mineral mining. **Agriculture** – crops (1977, in 1,000 tons): wheat 1,215, maize (corn) 355, potatoes 928, sugar beet 2,208, vegetables 1,068, fruit 1,395 (grapes 1,012); **livestock** (1977, in million head): cattle 3.4, sheep 5.7, pigs 0.7, poultry 19.5; fish catch 1,264,200 tons (1976); fresh hides 35,840 tons, milk 1.1 mn. tons. Roundwood 8.9 mn. cub.m (1976). **Mining** (1977, in 1,000 tons, metal content): copper 1,056 (third world producer, 13.2% of world output – El Teniente, El Salvador and Chuquicamata), coal 1,256, crude petroleum 960, natural gas 3,400 mn. cub.m, iron ore 6,186 (1976), saltpeter 713 (1974), iodine 1,963 tons (1975), molybdenum 10,898 tons (1976), gold 4,062 kg and silver 227 tons (1976), vanadium 1,088 tons (1976), salt 435. **Production** (1977, in 1,000 tons): copper – smelted 594, – refined 639 (1978), pig iron 420, crude steel 504, cement 1,140, motor spirit 980, chemical wood pulp 515 (1976), sugar 315, wine 600. Electricity 10,164 mn. kWh (1978).

Communications (1976): railways 9,757 km (1973), roads 75,197 km. Merchant shipping 678,556 GRT. Civil aviation: 20.6 mn. km flown and 490,000 passengers carried (1976). – **Exports** (1976): copper (72%), raw materials, chemical wood pulp, fish. Chief trading partners: U.S.A., Argentina, Fed. Rep. of Germany, Japan, United Kingdom.

COLOMBIA

República de Colombia, area 1,138,914 sq.km, population 25,582,000 (1978), **republic** (President Júlio César Turbay Ayala since 1978).

Administrative units: 22 departments, 4 intendencies, 4 commissaries and 1 Capital District. **Capital:** Bogotá 3,618,750 inhab. (1977); **other towns** (1973 census, in 1,000 inhab., +with agglomeration): Medellín 1,071 (+1,417), Cali 898 (+1,077), Barranquilla 662 (+727), Cartagena 293 (+313), Bucamaranga 292 (+341), Cúcuta 220 (+270), Manizales 200 (+231), Ibagué 176 (+205), Pereira 174 (+212), Montería 149, Palmira 140, Armenia 137, Santa Marta 129, Pasto 119, Buenaventura 116, Bello 113, Neiva 105, Sevilla 102. **Population:** mestizos 48%, mulattoes 24%, Whites and Creoles 20%, Negroes 5%, Indians 3%. **Density** 22 persons per sq.km (1978), average annual rate of population increase

3.2%, birth rate 40.6 per 1,000, death rate 8.8 per 1,000 (1970–75); urban population 67% (1976). 30% of inhabitants engaged in agriculture. – **Currency:** Columbian Peso = 100 centavos.
Economy: developing agricultural and industrial country with mineral mining. **Agriculture** – crops (1977, in 1,000 tons): chief products – coffee 558 (second world producer, 13% of world crops), bananas 1,300, sugarcane 20,800, rice 1,329, maize (corn) 753, sorghum 406, potatoes 1,609, manioc 2,113, soybeans 109, sesame (seed) 21, cotton –seed 260, –lint 158, sisal 45, palm kernels 17.4, vegetables 1,072, citrus fruit 375, pineapples 150, cocoa beans 28, tobacco 64; **livestock** (1977, in million head): cattle 24.4, sheep 2.2, pigs 1.9, horses 1.5, poultry 54.2; fresh hides 82,750 tons; fish catch 75,000 tons (1976). Forests cover 74% of the land area, roundwood 23 mn. cub.m.
Mining (1977, in 1,000 tons, metal content): gold 9,454 kg (1976, dep. Antioquia), coal 3,840, crude petroleum 7,114, natural gas 1,850 m. cub.m, iron ore 460, platinum 688 kg (1975), precious stones, salt 1,112, sulphur 31 (1975).
Production (1977, in 1,000 tons): motor spirit 2,320, cement 3,300, passenger car assembly 26,504 units (1976), sugar 823, meat 768, milk 2,300, eggs 140; 18.3 billion cigarettes (1976). Electricity 15,292 mn. kWh (1976, 65% hydro-electric). – **Communications** (1976): railways 3,403 km, roads 59,171 km (1973), passenger cars 465,238. Merchant shipping 247,000 GRT. Civil aviation (1976): 49.9 mn. km flown and 3.7 mn. passengers carried. – **Exports** (1976): coffee (56%), cotton (10%), petroleum, sugar, bananas, tobacco. Chief trading partners: U.S.A., Fed. Rep. of Germany, Japan, Venezuela, Spain.

ECUADOR

República del Ecuador, area 283,561 sq.km, population 7,814,000 (1978), **republic** (President Osvaldo Hurtado since 1981).
Administrative units: 20 provinces (inc. Arch. de Colón = Galapagos Is.). **Capital:** Quito 599,828 inhab. (1974 census); **other towns** (in 1,000 inhab.): Guayaquil 823, Cuenca 105, Ambato 77, Machala 68, Manta 63, Esmeraldas 60.
Population: mestizos 41%, Indians 39%, Creoles 11%. **Density** 28 persons per sq.km; average annual rate of population ncrease 3.2%. 46.5% of inhabitants engaged in agriculture (1977). – **Currency:** sucre = 100 centavos.
Economy: agriculture predominates. **Agriculture:** crops (1977, in 1,000 tons): bananas 2,384, coffee 77, cocoa beans 72, citrus fruit 375, pineapples 116, sugarcane 6,600, rice 316, potatoes 504, palm kernels 4; **livestock** (1977, in million head): cattle 2.9, pigs 2.9, sheep 2.2, poultry 13; fish catch 223,400 tons (1976). Roundwood 4 mn. cub.m (1976). **Mining:** crude petroleum 10.2 mn. tons (1978), gold 343 kg. **Production** (1977, in 1,000 tons): milk 850, sugar 375, palm oil 22, cement 616 (1976). – **Communications:** railways 1,517 km, roads 21,300 km. – **Exports** (1975): petroleum (20%), bananas (16%), cocoa, coffee, sugar, balsawood. Chief trading partners: U.S.A., Japan, Panama, Fed. Rep. of Germany, Colombia, Peru.

GUYANA

Co-operative Republic of Guyana, area 214,969 sq.km, population 819,000 (1978), **co-operative republic, member of the Commonwealth** (President Linden Forbes Sampson Burnham since 1980).

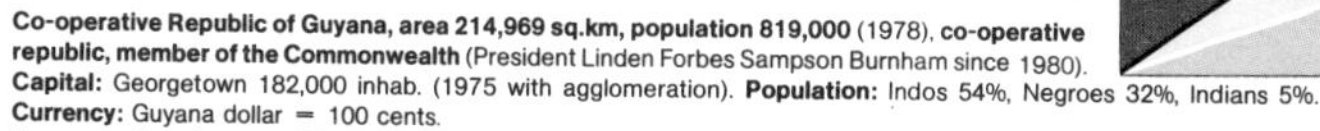

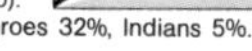

Capital: Georgetown 182,000 inhab. (1975 with agglomeration). **Population:** Indos 54%, Negroes 32%, Indians 5%. – **Currency:** Guyana dollar = 100 cents.
Economy: forests cover 92% of the land area. **Mining** (1976, in 1,000 tons): bauxite 3,204, gold 487 kg, diamonds 15,000 carats. **Agriculture** (1977, in 1,000 tons): rice 300, sugarcane 3,300, coconuts 33; cattle 290,000 head; poultry 10.8 mn. head. **Production** (1976): sugar 343,000 tons, aluminium, rum. – **Exports** (1976): sugar 35%, bauxite, rice and aluminium; chief trade with United Kingdom, U.S.A., Trinidad – Tobago and Jamaica.

NETHERLANDS ANTILLES

De Nederlandse Antillen, area 988 sq.km, population 245,000 (1978) – 3 main islands: Curaçao, Aruba, Bonaire; 3 Leeward Islands: St. Maarten, St. Eustatius, Saba. **Autonomous state of the Netherlands** (Prime Minister Dominico F. Martina since 1979).
Capital: Willemstad 94,133 inhab. (with agglomeration). – **Currency:** guilder. **Economy:** processing of petroleum imported from Venezuela; production (1976, in million tons): motor spirit 1.8, oils and naphtha 27.7 (capacity of refineries 40 mn. tons per year); electricity 1,500 mn. kWh. – **Exports** (1976): petroleum products 92% and chemicals; chief trade with U.S.A., Venezuela and Saudi Arabia.

PARAGUAY

República del Paraguay, area 406,752 sq.km, population 2,889,000 (1978), **republic** (President Gen. Alfredo Stroessner since 1954).
Administrative units: 20 departments. **Capital:** Asunción 388,959 inhab. (1972, with agglomeration 473,013 inhab.); **other towns** (in 1,000 inhab.): Coronel Oviedo 59, Concepción 53. **Population:** Indians (Guaraní) 58%, mestizos 38%; 54% of inhabitants engaged in agriculture. – **Currency:** guaraní = 100 centimos.
Economy: backward agricultural country. **Agriculture:** (1977, in 1,000 tons): manioc 1,700, maize (corn) 372, soybeans 375, cotton – seed 143, – lint 75, castor beans 15, sugarcane 1,450, bananas 268, citrus fruit 209, tobacco 46; extensive cattle breeding (1977, in million head): cattle 5.7, pigs 1.1, poultry 9.8. **Production** (1977, in 1,000 tons): tung oil 17, orange oil (70% of world production), palm oil 4.8, sugar 70, meat 205; tannin 18, rare hardwoods 4.3 mn. cub.m (1976). Electricity 700 mn. kWh (1976). – **Communications** (1974): railways 441 km, roads 6,680 km.
Exports (1975): meat and meat products, wood, vegetable oils, cotton, tannin and essential oils. Chief trading partners: Argentina, Brazil, U.S.A. and Fed. Rep. of Germany.

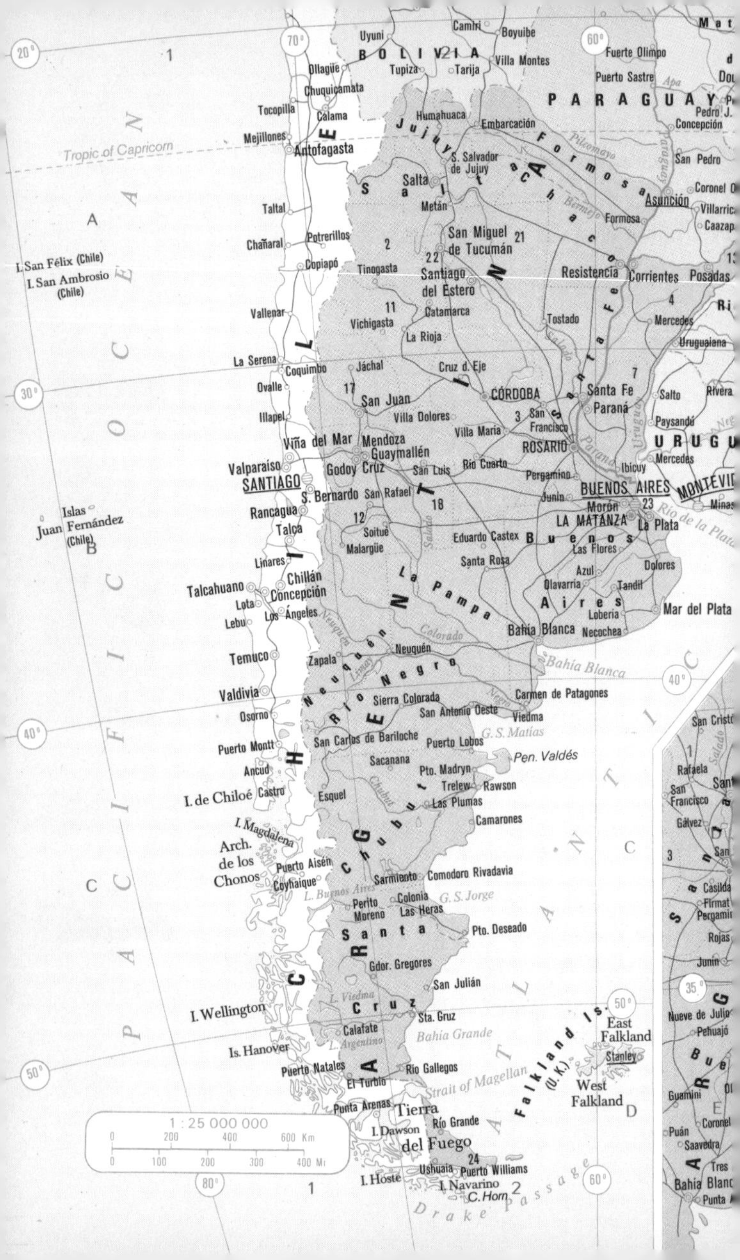
BOLIVIA
Uyuni
Camiri
Boyuibe
Tupiza
Tarija
Villa Montes
Fuerte Olimpo
Puerto Sastre
PARAGUAY
Pedro J.
Concepción
Ollagüe
Chuquicamata
Tocopilla
Calama
Mejillones
Antofagasta
Tropic of Capricorn
Humahuaca
Embarcación
Jujuy
Formosa
Pilcomayo
S. Salvador de Jujuy
San Pedro
Salta
Chaco
Coronel O
Asunción
Villarrica
Caazapá
Taltal
Metán
Bermejo
Formosa
Chañaral
Potrerillos
San Miguel de Tucumán
I. San Félix (Chile)
I. San Ambrosio (Chile)
Copiapó
Tinogasta
Santiago del Estero
Resistencia
Corrientes
Posadas
Vallenar
Vichigasta
Catamarca
Tostado
Mercedes
La Rioja
Uruguaiana
La Serena
Coquimbo
Jáchal
Cruz d. Eje
Ovalle
San Juan
CÓRDOBA
Santa Fe
Paraná
Salto
Rivera
Illapel
Villa Dolores
San Francisco
Paysandú
Villa María
Viña del Mar
Mendoza
ROSARIO
URUGUAY
Guaymallén
Valparaíso
Godoy Cruz
San Luis
Río Cuarto
Pergamino
Ibicuy
Mercedes
SANTIAGO
S. Bernardo
San Rafael
Junín
BUENOS AIRES
MONTEVIDEO
Islas Juan Fernández (Chile)
Rancagua
Morón
Minas
LA MATANZA
La Plata
Río de la Plata
Talca
Soitue
Malargüe
Salado
Eduardo Castex
Buenos Aires
Las Flores
Linares
Santa Rosa
Dolores
Chillán
Azul
Talcahuano
Concepción
La Pampa
Olavarría
Tandil
Lota
Los Ángeles
Lobería
Mar del Plata
Lebu
Neuquén
Colorado
Bahía Blanca
Necochea
Temuco
Zapala
Neuquén
Bahía Blanca
Limay
Río Negro
Valdivia
Carmen de Patagones
Sierra Colorada
Osorno
San Antonio Oeste
Viedma
Puerto Montt
San Carlos de Bariloche
G. S. Matías
Puerto Lobos
Pen. Valdés
Ancud
Sacanana
Pto. Madryn
Trelew
Rawson
I. de Chiloé
Castro
Esquel
Chubut
Las Plumas
Camarones
I. Magdalena
Arch. de los Chonos
Puerto Aisén
Coyhaique
Sarmiento
Comodoro Rivadavia
L. Buenos Aires
Perito Moreno
Colonia Las Heras
G. S. Jorge
Santa Cruz
Pto. Deseado
Gdor. Gregores
San Julián
L. Viedma
I. Wellington
Sta. Cruz
Calafate
Bahía Grande
L. Argentino
Is. Hanover
Falkland Is. (U.K.)
East Falkland
Stanley
Puerto Natales
Río Gallegos
El Turbio
Strait of Magellan
West Falkland
Punta Arenas
Tierra del Fuego
I. Dawson
Río Grande
Ushuaia
Puerto Williams
I. Hoste
I. Navarino
C. Horn
Drake Passage
PACIFIC OCEAN
ATLANTIC OCEAN
CHILE
ARGENTINA
1 : 25 000 000
0 200 400 600 Km
0 100 200 300 400 Mi
San Cristóbal
Salado
Rafaela
San Francisco
Gálvez
Casilda
Firmat
Pergamino
Rojas
Junín
Nueve de Julio
Pehuajó
Guaminí
Coronel
Puán
Saavedra
Tres
Bahía Blanca
Punta

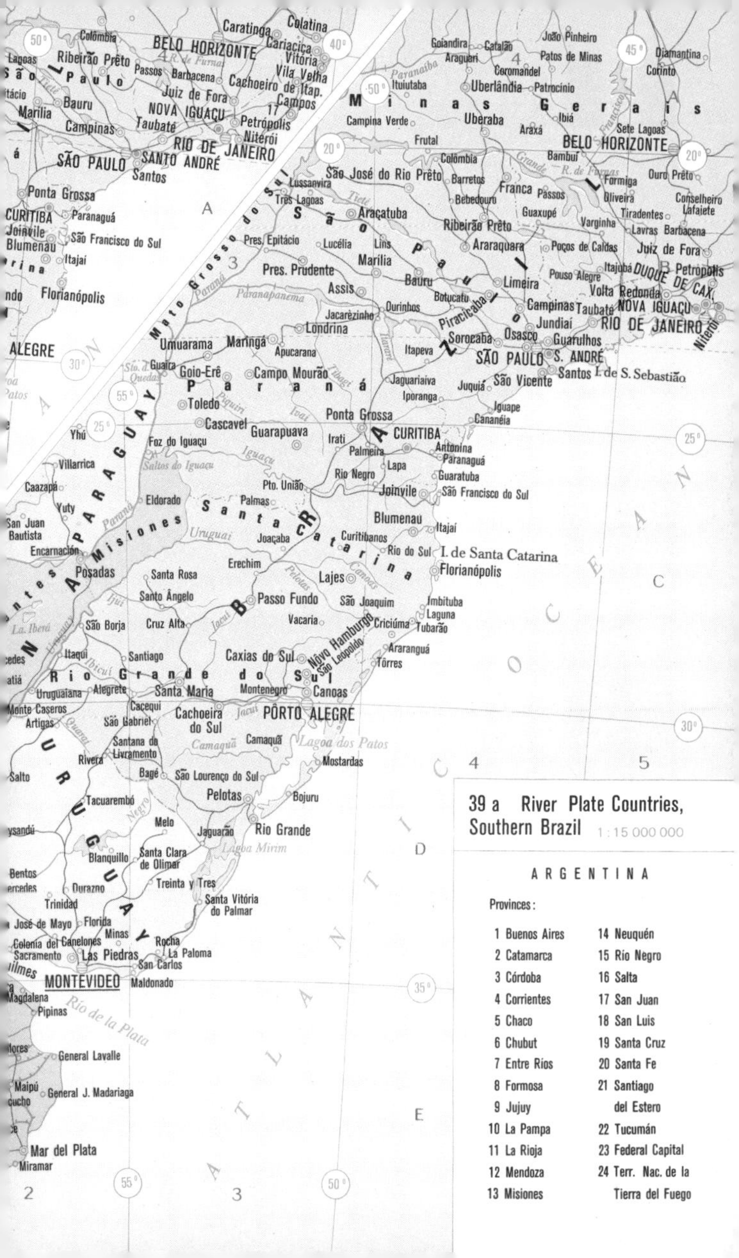

39 a River Plate Countries, Southern Brazil 1 : 15 000 000
ARGENTINA
Provinces:
1 Buenos Aires
2 Catamarca
3 Córdoba
4 Corrientes
5 Chaco
6 Chubut
7 Entre Ríos
8 Formosa
9 Jujuy
10 La Pampa
11 La Rioja
12 Mendoza
13 Misiones
14 Neuquén
15 Río Negro
16 Salta
17 San Juan
18 San Luis
19 Santa Cruz
20 Santa Fe
21 Santiago del Estero
22 Tucumán
23 Federal Capital
24 Terr. Nac. de la Tierra del Fuego
Minas Gerais
São Paulo
Paraná
Santa Catarina
Rio Grande do Sul
BRAZIL
PARAGUAY
URUGUAY
Misiones
Mato Grosso do Sul
ATLANTIC OCEAN
BELO HORIZONTE
RIO DE JANEIRO
SÃO PAULO
SANTO ANDRÉ
NOVA IGUAÇU
DUQUE DE CAXIAS
CURITIBA
PÔRTO ALEGRE
MONTEVIDEO
Uberlândia
Uberaba
Ribeirão Prêto
Campinas
Santos
Londrina
Maringá
Ponta Grossa
Joinvile
Blumenau
Florianópolis
Lajes
Passo Fundo
Caxias do Sul
Santa Maria
Pelotas
Rio Grande
Posadas
Encarnación
Lagoa dos Patos
Lagoa Mirim
Río de la Plata
Mar del Plata
I. de Santa Catarina
I. de S. Sebastião

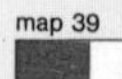

PERU

República del Perú, area 1,285,216 sq.km, population 16,819,000 (1978), **republic** (President Fernando Belaúnde Terry since 1980).

Administrative units: 23 departments and 1 constitutional province (Callao). **Capital:** Lima 2,973,845 inhab. (1972, with agglomeration 3.3 mn.); **other towns** (in 1,000 inhab.): Arequipa 321, Callao 316, Trujillo 242, Chiclayo 190, Chimbote 159, Cuzco 131, Piura 127, Huancayo 116, Iquitos 111, Ica 109, Sullana 104. **Population:** Indians 49% (Quechuas, Aymarás, Panos etc.), mestizos 33%, Whites 10%, Negroes and mulattoes. **Density** 13 persons per sq.km (1978), average annual rate of population increase 3.1%, birth rate 41.6 per 1,000, death rate 10.5 per 1,000 (1970–75); urban population 55.8% (1975). 40% of inhabitants engaged in agriculture (1977). – **Currency:** sol = 100 centavos.
Economy: developing agricultural and industrial country with important mining of minerals. **Agriculture:** only 2.6% of the land is cultivated and of this 1/3 is irrigated. **Crops** (1977, in 1,000 tons): cereals 1,662 (maize 700, rice 580), potatoes 1,600, manioc 450, legumes 108, cotton – seed 125 and – lint 60, sugarcane 8,900, coffee 60, fruit 1,650 (citrus fruit 314, avocados and mangoes 185); **livestock** (1977, in million head): sheep 14.5, cattle 4.1, pigs 2, llamas, alpacas and vicuñas 2.6, horses 0.6, poultry 30; fresh hides 22,602 tons, grease wool 12,000 tons; fish catch 4.3 mn. tons (1976). Forests cover 58% of the land, roundwood 7.3 mn. cub.m.
Mining (1977, in 1,000 tons, metal content): iron ore 3,090, crude petroleum 7,716 (1978), natural gas 700 mn. cub.m, copper 341 (Cerro de Pasco, La Oroya), zinc 478 (Junín, 7.6% of world output), lead 182, silver 1,183 tons (11.2% of world output), gold 2,511 kg (1976), vanadium, bismuth 800 tons, antimony 820 tons, molybdenum 850 tons (1976), tungsten 1,057 tons (dep. Ancash, 1976). **Production** (1977, in 1,000 tons): copper – smelted 316, – refined 182, lead 79, zinc 67, pig iron 295 and crude steel 370, motor spirit 1,451 (1976), cement 1,995, sugar 928, meat 343, milk 840. Electricity 8,650 mn. kWh (1976, 75% hydro-electric).
Communications: railways 3,218 km, the Lima-La Oroya line reaches the highest world altitude 4,829 m at Galero; roads 56,940 km (1976, Panamerican Highway 3,340 km), passenger cars 290,000. Merchant shipping 555,000 GRT (1977). Civil aviation (1976): 24.1 mn. km flown, and 1.4 mn. passengers carried. – **Exports** (1975): copper and processed copper (37%), sugar (16%), cotton (15%), petroleum, silver, iron ore, coffee. Chief trading partners (1977): U.S.A., Japan, Fed. Rep. of Germany, Ecuador, Venezuela, United Kingdom, Brazil.

SURINAM

Republiek van Suriname, area 163,265 sq.km, population 448,000 (1977), **republic** (President Dr Henck Chin a Sen since 1980).

Capital: Paramaribo 102,300 inhab. (with agglomeration 151,500 inhab., 1971). **Population:** Indos 37%, Creoles 31%, Indonesians, Negroes, Indians. – **Currency:** Surinama guilder = 100 cents.
Economy: forests cover 89% of the land area. **Mining** of bauxite 4,860,000 tons (1977); production of aluminium (export over 1 mn. tons yearly), electricity 1,335 mn. kWh (1976, 88% hydro). **Agriculture** (1977, in 1,000 tons): rice 182, bananas 41, sugarcane 157; poultry 946,000 head. – **Exports:** aluminium (48%), bauxite, rice. Chief trading partners: U.S.A., Netherlands, United Kingdom, Japan.

TRINIDAD AND TOBAGO

Republic of Trinidad and Tobago, area 5,128 sq.km, population 1,133,000 inhab. (1978), **republic, member of the Commonwealth** (President Ellis Clarke since 1976).

Capital: Port of Spain 60,450 inhab. (with agglomeration 350,000 inhab., 1973). **Population:** Negroes 43%, Indos 40%, mestizos. **Density** 221 persons per sq.km (1978). – **Currency:** Trinidad and Tobago dollar = 100 cents.
Economy: agricultural country with mining of petroleum 11,856,000 tons (1978), natural gas 2,025 mn. cub.m and asphalt 112,000 tons (1977). **Agriculture:** crops (1977, in 1,000 tons): sugarcane 1,891, rice 21, cocoa beans, coffee, coconuts 79, copra 9, citrus fruit 27; cattle 75,000 head; poultry 7.1 mn. head. **Production** (1976): electricity 1,367 mn. kWh, sugar 207,000 tons, rum and spirits. – **Communications:** roads 6,115 km. – **Exports:** petroleum and petroleum products, agricultural products. Chief trading partners: U.S.A., United Kingdom, Canada, Japan.

URUGUAY

República Oriental del Uruguay, area 177,508 sq.km, population 2,864,000 (1978), **republic** (President Gregorius Alvarez since 1981).

Administrative units: 19 departments. **Capital:** Montevideo 1,229,748 inhab. (1975); **other towns** (in 1,000 inhab.) Paysandú 80, Salto 80, Mercedes 53, Rivera 46, Minas 40, Melo 38. – **Population:** Whites (descendants of immigrant Spaniards) 90%, mestizos, Negroes, Indians. **Density** 16 persons per sq.km (1978); average annual rate of population increase 1.2%, urban population 81% (1975). – **Currency:** nuevo peso = 100 centésimos.
Economy: agricultural country with extensive animal production. **Agriculture:** arable land 11%, meadows and pastures 77% of the land area. **Livestock** (1977, in million head): cattle 10.2, sheep 17.8, horses 0.5, poultry 7.5 grease wool 62,600 tons, hides 59,440 tons. **Crops** (1977, in 1,000 tons): wheat 160 (505 in 1976), rice 228 maize (corn) 121, sorghum 120, potatoes 130, sunflower 34, flax (seed) 46, sugar beet 627, sugarcane 675, citrus fruit 89, grapes 80. **Production** (1977, in 1,000 tons): sugar 133, wine 44, meat 383, milk 780, flax processing production of shoes, cement 676 (1976). Electricity 2,800 mn. kWh (1976). – **Communications** (1976): railways 2,975 km, roads 51,745 km, passenger cars 151,600 (1974). Merchant shipping 193,000 GRT. – **Exports** (1976): meat and meat products (25%), wool (20%), hides, vegetable oils, woven fabrics. Chief trading partners: Brazil, U.S.A. Fed. Rep. of Germany, Argentina, Kuwait, Iraq.

VENEZUELA

República de Venezuela, area 912,050 sq.km, population 13,304,000 (1978), **federal republic** (President Luis Herrera Campins since 1979).

Administrative units: 20 states, 2 federal territories, 1 federal district (the capital) and 1 federal dependency. **Capital:** Caracas 1,754,527 inhab. (1973, with agglomeration 2,479,743 inhab. in 1975); **other towns** (census 1971, in 1,000 inhab.): Maracaibo 652, Valencia 367, Barqusimeto 331, Maracay 255, San Cristobal 152, Ciudad Guyana 144, Cumaná 120, Cabimas 118. **Population:** mestizos 62%, Whites 33%, Indians 5%. **Density** 15 persons per sq.km (1978); average annual rate of population increase 2.9%, birth rate 36 per 1,000, death rate 7 per 1,000 (1970–75); urban population 64.7%. 20% of inhabitants engaged in agriculture (1977). – **Currency:** bolivar = 100 céntimos.
Economy: agricultural and industrial country with important mining of minerals, especially petroleum. **Agriculture:** arable land 6%, meadows and pastures 19%, forests 54% of the land area. **Crops** (1977, in 1,000 tons): rice 508, maize (corn) 800, sorghum 325, manioc 370, sesame (seed) 80, cotton – seed 55, – lint 30, coconuts 165, sugarcane 6,000, bananas 1,050, oranges 258, coffee 40, cocoa beans 17. **Livestock** (1977, in million head): cattle 9.7, pigs 2.0, horses 0.5, poultry 29.8; fish catch 145,700 tons (1976); fresh hides 36,480 tons. Roundwood 8 mn. cub.m (1976).
Mining (1977, in 1,000 tons): crude petroleum 113,040 (1978) (in the Maracaibo Basin and lower reaches of the R. Orinoco) natural gas 11,332 mn. cub.m, good quality iron ore 8,763 (Cerro Bolívar, El Pao, San Isidro), gold 514 kg (1976), diamonds 833,000 carats (of which 77% industrial in 1976), salt 301. **Production** (1976, in 1,000 tons): motor spirit 5,664, naphtha 3,231, oils and other petroleum products 8,254 (capacity of refineries 77,750), asphalt 391, pig iron 427, crude steel 754, aluminium 54, car assembly 163,000 units; sugar 460, meat 566 and milk 1,134 (1977), cement 3,838; 18.8 billion cigarettes. Electricity 23,276 mn. kWh (1976). – **Communications:** railways 175 km, roads 44,279 km (1972), motor vehicles – passenger 955,200,-commercial 370,000 (1975). Merchant shipping 639,000 GRT (1977). Civil aviation (1976): 40.4 mn. km flown and 2.7 mn. passengers carried. – **Exports** (1975): petroleum (66%), petroleum products (27%), iron ore, natural gas, diamonds, agricultural products. Chief trading partners: U.S.A., Netherlands Antilles, Canada, Fed. Rep. of Germany, Japan, United Kingdom.

FALKLAND ISLANDS

Falkland Islands and Dependencies, area 16,264 sq.km, population 2,000 (1977), **British crown colony** to which Argentina lays claim as Islas Malvinas.
Administrative units: East and West Falkland (12,172 sq.km), **dependencies:** South Georgia 3,755 sq.km, South Sandwich Is. (337 sq.km). **Capital:** Stanley 1,081 inhab. (1972). – **Currency:** Falkland pound. **Economy** (1977): sheep raising 638,116 head; grease wool, skins, fishing. – **Exports:** wool, skins.

FRENCH GUIANA

Guyane Française, area 91,000 sq.km, population 59,000 (1977), **French overseas department.**
Capital: Cayenne 27,000 inhab. (1972). – **Currency:** franc = 100 centimos. **Economy:** forests cover almost 90% of the land area. **Production:** roundwood 51,000 cub.m; gold; crops: sugarcane, bananas, manioc, rice, rum, electricity 65 mn. kWh. – **Exports:** wood, rum, essential oils, predominantly to France

AUSTRALIA AND OCEANIA

The smallest and least densely inhabited continent on the Earth is Australia with the island world of Oceania. It is very remote from the other continents: 13,000 km from South America and 350 km from Asia. It lies in the southern hemisphere surrounded by the waters of the Pacific and the Indian Oceans. The name derives from the Latin "australis", i.e. "southern" (Terra Australis Incognita). Australia was first discovered by the Dutch explorer W. Jansz, who in 1606 landed on the west coast of the Cape York Peninsula. In the forties of the 17th century Abel Tasman discovered Tasmania, New Zealand (1642), Tonga, Fiji and Bismarck Archipelago. The British seafarer Captain James Cook proved in 1769 that New Zealand consisted of islands, and in 1770 he discovered the eastern shores of Australia, which he named New South Wales. The British were not slow to colonize it. In 1788 the first convict colony was established and in the 19th century settlers came in the wake of the gold rush. The exploration of the inland areas was completed by the end of the century, and by 1884 the British had occupied Papua. On 1 January 1901 the individual British colonies on the Australian continent federated under the name of the Commonwealth of Australia. The colonization of New Zealand began in the first half of the 19th century despite resistance from the native Maoris.

The area of Australia is **7,686,848 sq.km** (8,511,000 sq.km including Oceania); it has **14,248,000 inhabitants** (with Oceania 23 mn.). **Geographical position:** the northernmost point of the mainland is Cape York 10°41′ S.Lat. (Mata Kawa I. 9°11′ S.Lat.); southernmost point: South East Point 39°07′ S.Lat. (South East Cape 43°39′ S.Lat. – Tasmania); easternmost: Cape Byron 153°39′ E.Long.; westernmost: Steep Point 113°09′ E.Long. Width of the continent 3,200 km, length 4,100 km, length of the coastline 19,700 km. Oceania – the largest island area on Earth – extends in the central and south-western part of the Pacific Ocean: in the North, Kure I. (Midway Is.) 28°25′ N.Lat., in the South, Campbell I. 52°30′S.Lat., in the East; Isla Sala y Gómez 105°28′ W.Long. and in the West, Pulau Misool I. (Indonesia) 129°43′ E.Long. Largest island: New Guinea (785,000 sq.km). Largest peninsula: Arnhem Land (243,000 sq.km).

The geological structure of Australia is very simple; **the surface** has three basic units. First, the Western Australian Hills (the oldest part) which cover one half of the continent, at an average height of 200–500 m rising to 1,524 m on Mt. Liebig in the Macdonnell Ranges; its deserts extend over an area of roughly 1.7 mn. sq.km. Second, the Central Australian Lowlands with the Great Artesian Basin and the Lake Eyre Basin, which has no outlet to the sea; the lowest point on the continent is Lake Eyre (–16 m) in the middle. Third, the Great Dividing Range, known as the Australian Cordilleras, which is the result of Hercynian folding in the East. It reaches its highest point at Mt. Kosciusko, 2,230 m, in the Australian Alps. The mountainous islands of New Zealand are varied in character. In the Southern Alps there are glaciers (Tasman Gl. 156 sq.km); the highest point is Mt. Cook (3,764 m). On North Island there is volcanic activity (Ruapehu, 2,796 m), hot springs and geysers. The islands of Oceania can be divided

Bali
Lombok
3726
Raba
Ende
Dili
Baucau
Mataram
Sumbawa
Flores
Timor
Denpasar
Waingapu
Sumba
Kupang
P. Roti
INDONESIA
120°
130°
10°
TIMOR SEA
ARAFURA
Melville I.
Bathurst I.
Darwin
Rum Jungle
Arnhem Land
Pine Creek
Katherine
Roper
Daly
C. Londonderry
J. Bonaparte Gulf
INDIAN OCEAN
Ord
Victoria
Birdum
Borroloola
776
Wyndham
Kununurra
Yampi Sound
Kimberley Plat.
983
Victoria River Downs
Daly Waters
Mt. Ord
Newcastle Waters
936
Mt. Wells
Broome
Derby
Fitzroy Crossing
Halls Creek
Fitzroy
Northern Territory
Tanami
Tennant Creek
Wallal Downs
20°
Port Hedland
Goldsworthy
Dampier
Marble Bar
Great Sandy Desert
Roebourne
L. Mackay
1524
North-West C.
Fortescue
Western Australia
Mt. Liebig
Macdonnell Ranges
Onslow
Deepdale
Mt. Meharry
Alice Springs
Tom Price
Paraburdoo
1251
Learmonth
Ashburton
Newman
Gibson Desert
L. Amadeus
L. Mc Leod
1219
Mt. Augustus
Musgrave Ranges
Finke
1106
AUSTRALIA
Carnarvon
Mt. Woodroffe 1440
Kulgera
Gascoyne
L. Carnegie
Meekatharra
Wiluna
Oodnadatta
Murchison
Steep Pt.
Cue
South Australia
Sandstone
Laverton
Great Victoria Desert
Coober Pedy
Mullewa
Yalgoo
Magnet
Leonora
Geraldton
Menzies
Tarcoola
Dongara
L. Barlee
Kalgoorlie-Boulder
Rawlinna
Forrest
Nullarbor Plain
30°
L. Moore
Coolgardie
Kambalda
Eucla
Penong
L. Gairdner
Southern Cross
L. Cowan
Ceduna
Northam
Merredin
Norseman
Great Australian Bight
Buckleboo
5395
PERTH
Kwinana
582
Rockingham
Mt. Cooke
Narrogin
Ravensthorpe
Esperance
Port Lincoln
Bunbury
Busselton
Bridgetown
Augusta
Bluff Knoll
1109
Albany
INDIAN OCEAN
40°
1 : 25 000 000
0 200 400 600 Km
0 100 200 300 400 Mi

PAPUA NEW GUINEA
Gulf of Papua
Kerema
Mt. Victoria 4073
Popondetta
Port Moresby
Daru
Torres Strait
Thursday I.
C. York
Cape York Pen.
Weipa
Aurukun
Coen
Cooktown
Laura
Mitchell
Cairns
Normanton
Bartle Frere 1611
Innisfail
Croydon
Forsayth
Greenvale
Townsville
Charters Towers
Bowen
Flinders
Hughenden
Mackay
1277
Winton
Blair Athol
Longreach
Emerald
Rockhampton
Gladstone
Thomson
Barcoo
Yaraka
Tambo
Theodore
Bundaberg
Charleville
Maryborough
Fraser I.
Quilpie
Roma
Chinchilla
Gympie
Warrego
Balonne
Ipswich
Toowoomba
BRISBANE
Cunnamulla
Dirranbandi
Warwick
Southport
Goondiwindi
C. Byron
Culgoa
Glen Innes
Lismore
Bourke
Barwon
Walgett
Grafton
New South Wales
Macquarie
Armidale
Coffs Harbour
Darling
1615
Cobar
Tamworth
Port Macquarie
1585
Barrington Tops
Dubbo
Maitland
Roto
Orange
Newcastle
Wentworth
Lachlan
West Wyalong
SYDNEY
Griffith
Bathurst
Blue Mts.
Wollongong
Murrumbidgee
Goulburn
Wagga Wagga
Canberra
Swan Hill
Albury
Echuca
Murray
Australian Alps
Mt. Kosciusko 2230
Bendigo
Bombala
Ballarat
Victoria
MELBOURNE
Orbost
Geelong
South East Pt.
Bass Strait
King I.
Furneaux Group
Devonport
Banks Str.
Burnie
1617
Zeehan
Launceston
Queenstown
Tasmania
Hobart
South East C.
150°
Great Dividing Range
Great Barrier Reef
CORAL SEA
Coral Sea Islands Territory
D'Entrecasteaux Is.
Alotau
Louisiade Arch.
Tagula I.
Bougainville (Pap.-N. G.)
Choiseul
SOLOMON IS.
S. Isabel
New Georgia
Auki
Malaita
Honiara
Guadalcanal
S. Cristóbal
Solomon Is.
Rennell I.
160°
10°
PACIFIC OCEAN
Îs. Chesterfield (N.-Cal.)
20°
New Caledonia (Fr.)
Tropic of Capricorn
TASMAN SEA
North C.
Opua
Whangarei
170°
Takapuna
Auckland
Manukau
Hamilton
Waikato
East C.
Whakatane
North Island
Rotorua
Taupo
Gisborne
New Plymouth
Ruapehu
2796
Napier
Wanganui
Hastings
40°
C. Farewell
Cook Strait
Palmerston North
Nelson
Lower Hutt
Picton
Wellington
Westport
Greymouth
Hokitika
South Island
Christchurch
Mt. Cook
Southern Alps
Mt. Aspiring
3764
3035
Timaru
L. Te Anau
Kingston
Oamaru
L. Manapouri
Dunedin
Invercargill
Stewart I.
Southwest C.
180°
40 a New Zealand
1 : 25 000 000

AUSTRALIA AND OCEANIA

LARGEST ISLANDS

Name	Area in sq.km
New Guinea	785,000
South I. (New Zealand)	150,461
North I. (New Zealand)	114,688
Tasmania	64,408
Birara (New Britain)	34,750
New Caledonia	16,058
Viti Levu	10,497
Hawaii	10,414
Tombara (New Ireland)	9,842
Bougainville	9,792
Guadalcanal	6,470
Vanua Levu	5,816

LARGEST RIVERS

Name	Length in km	River basin in sq.km
Murray-Darling	3,490	1,072,000
Darling (-Barwon)	2,720	710,000
Murrumbidgee	2,160	84,020
Fly	1,150	.
Lachlan	1,126	67,500
Cooper Creek (-Thomson)	960	

LARGEST LAKES

Name	Area in sq.km	Altitude in m
L. Eyre	9,500	−16
L. Torrens	5,900	30
L. Gairdner	5,500	110
L. Frome	2,410	80
L. Barlee	1,450	370
L. Mc Leod	1,300	5
L. Cowan	1,035	380
L. Taupo	606	369

HIGHEST MOUNTAINS

Name (Country)	Altitude in m
Puntjak Djaja (Irian-Indon.)	5,030
P. Mandala (Irian)	4,760
P. Trikora (Irian)	4,750
Mt. Wilhelm (Pap.N.Guinea)	4,509
Mauna Kea (Hawaii-U.S.A)	4,205
Mt. Cook (N.Z.)	3,764
Mt. Tasman (N.Z.)	3,498
Mt. Sefton (N.Z.)	3,157
Mt. Orohena (Tahiti)	2,235
Mt. Kosciusko (Austr.)	2,230

ACTIVE VOLCANOES

Name (Country, Island)	Altitude in m	Latest eruption
Mauna Loa (Hawaii)	4,168	1975
Ruapehu (North I. N.Z.)	2,796	1975
Ulawun (Birara)	2,300	1973
Ngauruhoe (North I. N.Z.)	2,291	1975
Manam (Pap. N.Guinea)	1,830	1977
Kilauea (Hawaii)	1,222	1975

Country	Area in sq.km	Population	Year	Density per sq.km
American Samoa (U.S.A.)	197	34,000	1977	173
Australia	7,686,848	14,248,000	1978	2
Canton and Enderbury (U.K.-U.S.A.)	70	(uninhabited)	.	.
Christmas Island (Austral.)	135	3,255	1977	24
Cocos Islands (Austral.)	14	444	1977	32
Cook Islands (N.Z.)	241	18,112	1976	75
Fiji	18,272	608,000	1978	33
French Polynesia (Fr.)	4,000	137,382	1977	34
Guam (U.S.A.)	549	104,000	1977	189
Hawaiian Islands (U.S.A.)	16,705	914,200	1978	55
Kiribati	860	64,500	1977	75
Midway Islands (U.S.A.)	5	2,300	1978	460
Nauru	21	7,254	1977	345
New Caledonia and Depend. (Fr.)	19,058	136,000	1977	7
New Zealand	268,675	3,130,083	1977	12
Niue Island (N.Z.)	259	3,954	1976	15
Norfolk Island (Austral.)	35	2,000	1977	57
Northern Mariana Islands (U.S.A.)	404	18,000	1977	45
Pacific Islands (U.S.A.)	1,410	110,000	1977	78
Papua New Guinea	461,691	2,905,000	1977	6
Pitcairn Island (U.K.)	5	82	1974	16
Samoa	2,842	153,000	1978	54
Solomon Islands	28,446	213,000	1978	7
Tokelau Islands (N.Z.)	10	1,575	1977	158
Tonga	699	92,000	1978	132
Tuvalu	26	6,500	1977	250
Vanuatu	14,763	99,000	1977	7
Wake Island (U.S.A.)	8	2,000	1977	250
Wallis and Futuna Is. (Fr.)	255	9,800	1977	38

› two groups: one of low, flat coral islands which arose on the atolls and submarine coral reefs and the other hilly, volcanic or continental islands. The highest active volcano is Mauna Loa, 4,168 m. Inland New Guinea is untainous – it contains the highest peak of the continent, Puntjak Djaja (5,030 m) – and has swampy lowlands.

Australia lacks a developed **network of rivers.** 54% of the land has no outlet to the sea, 38% drains into the Indian ean and 8% into the Pacific Ocean. There is great fluctuation in water level, chiefly in the periodical and seasonal ers. The mean annual discharge amounts to 610 cub.km. The biggest river is the Murray (-Darling) 3,490 km with ver basin of 1,072,000 sq.km, and a mean annual discharge of 1,900 cub.m per sec. New Zealand and New Guinea 'e numerous rivers with ample water throughout the year. The most remarkable of the numerous **lakes** is L. Manaıri in New Zealand with an area of 142 sq.km, depth 445 m, and whose bottom is 263 m below sea level.

Climate: Oceania (which, except for the most southerly islands, has a mean annual temperature above 20°C) l two fifths of the Australian continent lie within the tropical belt, and three fifths of Australia lie in the subɔical belt – as does the North of New Zealand, which has a generally mild oceanic climate. The warmest place ustralia is Marble Bar with a mean temperature of 34°C, in Oceania Canton I. 28.6°C; Cloncurry (Australia) has absolute ximum temperature at 52.8°C and Lake Tekapo minimum at −15.6°C (New Zealand), in Oceania Canton I. 36.7°C and nolulu (Hawaiian Is.) 13.9°C. Maximum annual rainfall Ninati 6,350 mm (New Guinea), Tully 7,773 mm (Queensland) l world record is held by Waialeale 11,684 mm (Kauai in the Hawaiian Is.). Mean January and July temperatures in °C d annual rainfall in mm) in selected places: Port Moresby 27.6 and 25.8 (1,038), Darwin 28.6 and 25.0 (1,491), rns 27.7 and 20.8 (2,253), Broome 29.7 and 21.1 (582), Cloncurry 31.1 and 17.8 (457), Brisbane 25.1 and 14.7 35), Alice Springs 28.6 and 11.7 (252), Perth 23.3 and 13.0 (881), Adelaide 23.0 and 11.1 (536), Sydney 21.9 l 11.7 (1,181), Auckland 19.2 and 10.6 (1,247), Dunedin 14.4 and 5.8 (937), Yap 27.1 and 27.6 (3,108), Canton I. 28.4 l 28.7 (938), Honolulu 22.1 and 25.4 (697), Suva 26.3 and 22.9 (3,240), Papeete 25.7 and 24.1 (1,872).

The fauna and flora have a unique primordial character. Almost 75% of all plants are endemic. This includes alyptus forests, scrub-land, acacias with brightly coloured blossom, salt-scrub flora, and in the regions of heavy rain- there are tropical rain forests or subtropical forests with typical tree-sized ferns. New Zealand has evergreen ad-leaved forests and characteristic coniferous Kauri. In the drier regions to the East of South Island steppes are predominant form. Oceania has large numbers of palm trees and forests rich in rare coloured timber varieties. fauna supports a wide range of marsupials and monotremes, mainly kangaroos, koala bears, phalangers, gliders, nbats and the primitive platypus and the echidna. The largest Australian bird is the brown emu; there are varieties of parrot, black swans, the kookaburra, the New Zealand kiwi, poisonous snakes, the rare Hatteris lizard. New Guinea we find crocodiles, rare birds of paradise. Nature conservation has a long tradition: Australia's largest ional Park is the Kosciusko N.P. (5,800 sq.km) and in New Zealand the Fiordland N.P. (11,200 sq.km).

The original **inhabitants** of Australia – the black Australians and the New Zealand Maoris – are today greatly uced in number. On the other hand, in Oceania the original inhabitants form the main part of the population Papuas, pygmy tribes, Melanesians, Micronesians, Polynesians and others). The number of Australian Whites rowing (in mn.): 1900 – 3.8, 1950 – 8.3, 1970 – 12.6, 1978 – 14.3. The number of New Zealanders has more ı trebled between 1900 and 1978, reaching 3.1 mn.. 86.2% of the population of Australia and 83.8% of New Zealand de in towns. The largest towns are Sydney with 3 mn. and Melbourne with 2.6 mn. inhabitants (1978).

'STRALIA

Commonwealth of Australia, area 7,686,848 sq.km, population 14,248,000 (1978), **independent member of the British Commonwealth** (Prime Minister Malcolm Fraser since 1977, ernor-General Sir Zelman Cowen appointed by the Queen in 1977).

ıinistrative units: 6 states (New South Wales, Victoria, Queensland, South Australia, Western Australia, Tasmania), rritories (Australian Capital Territory, Northern Territory, Coral Sea Islands Territory). **Capital:** Canberra 196,538 inhab. sus 1976, with agglomeration 215,414 inhab.); **other towns** (1976, in 1,000 inhab. with agglom.): Sydney 3,021, ɔourne 2,604, Brisbane 958, Adelaide 900, Perth 805, Newcastle 363, Wollongong 211, Hobart 162. **Population:** Whites of European descent of whom 92% British; Aborigines 106,000 (1971). **Density** 2 persons per sq.km (1978); age annual rate of population increase 1.9%, birth rate 21 per 1,000, death rate 8.1 per 1,000 (1970–75), urban ulation 86.2% (1975). – **Currency:** Australian dollar = 100 cents.

nomy: advanced industrial and agricultural country with rich deposits and important mining of minerals, neering, food, metallurgical and chemical industries. **Mining** (1978, +1976 in 1,000 tons, metal content): coal 84 (N.S.Wales, Queensland), brown coal 32,868 (Victoria), iron ore 61,187 (1977, second world producer, tern Australia – Newman, Tom Price, Goldsworthy and others), bauxite 24,828 (leading world producer, Weipa), 402 and zinc 470 (third world producer, Mt. Isa, Broken Hill), copper 220, manganese 765+, gold 15,462 kg+, r 722 tons+, tungsten 2,124 tons, nickel 85.7 (1977), antimony 1,584 tons (1977), asbestos 57+, crude petroleum 74, natural gas 7,242 mn. cub.m, salt 5,350 (1977), uranium 360 tons+. **Production** (1978, in 1,000 tons): pig iron 2, crude steel 7,536, copper smelted 166 and refined 151, aluminium 277, lead 217, zinc 294; motor spirit 10,258 7) and oils 13,733 (1976) with refineries capacity 35.5 mn. tons (1976); cement 5,016, tyres 6.3 mn. units, plastics sulphuric acid 2,024, cars 385,200 units, cotton fibres 22.4 and woven fabrics 49 mn. sq.m, 31.6 billion cigarettes; lstuffs (1977): flour 2,529, meat 2,884, milk 5,897, butter 118, sugar 3,342, beer 19.2 mn. hl. Electricity: installed acity of electric power stations 20 mn. kW; production 85,968 mn. kWh (1978).

culture is of major importance: sheep rearing, wool production and wheat cultivation. Arable land only 5.9% ıe land area, meadows and pastures 60%, forests 18%. **Crops** (1977, in 1,000 tons): wheat 9,371, barley 2,560, 530, oats 962, sorghum 956, potatoes 729, sunflower 74, flax 34, cotton – seed 48,-lint 28, vegetables 931, rcane 23,493, grapes 746, citrus fruit 418, pineapples 102, bananas 92, tobacco 16. **Livestock** (1977, in million): cattle 31.5 (dairy cows 2.2), pigs 2.2, sheep 135.4 (second producer and 13.2% of world sheep population); try 43.4; grease wool 7,027,000 tons (leading world producer) and scoured wool 420,000 tons, hides 203,500 tons, p skins 94,719 tons; eggs 174,920 tons, honey 21,400 tons.

munications (1976): railways 40,753 km, roads 884,700 km; motor vehicles – passenger 5.12 mn. and commercial

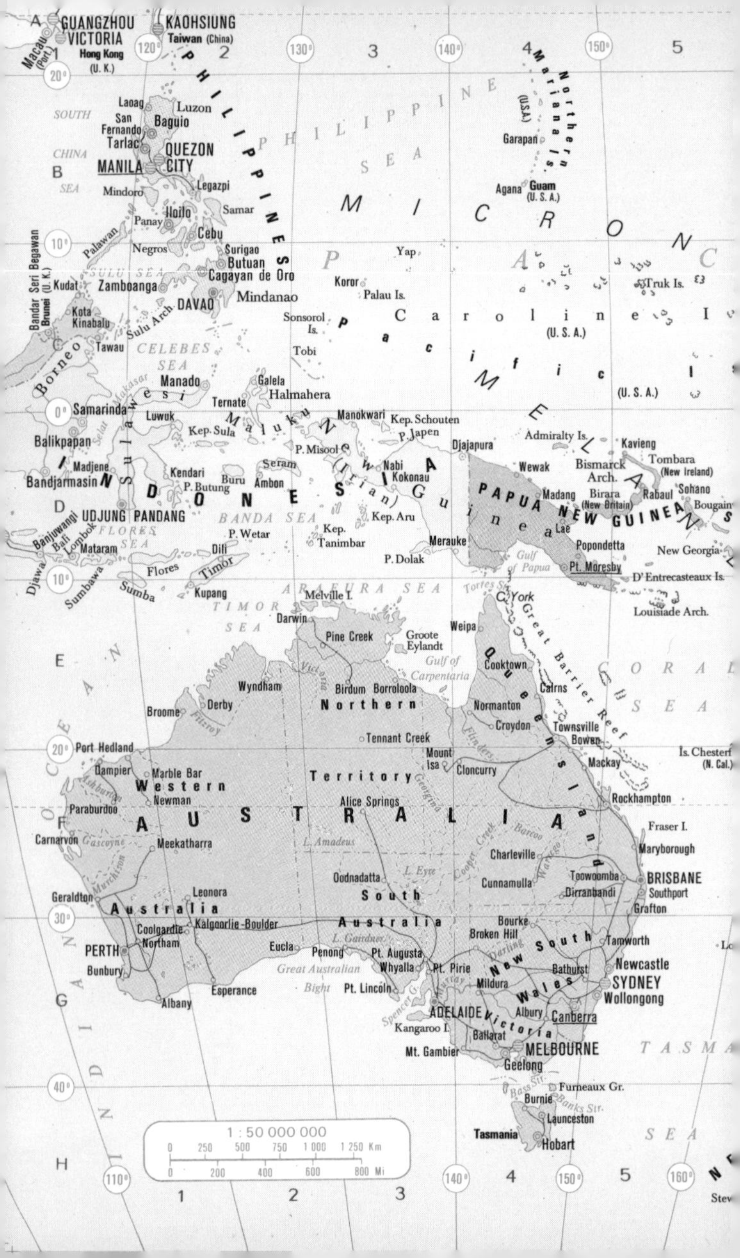
GUANGZHOU
VICTORIA
Hong Kong (U.K.)
Macau (Port.)
KAOHSIUNG
Taiwan (China)
PHILIPPINES
PHILIPPINE SEA
Northern Mariana Is. (U.S.A.)
Garapan
Guam (U.S.A.)
Agana
Laoag
Luzon
San Fernando
Baguio
Tarlac
QUEZON CITY
MANILA
SOUTH CHINA SEA
Mindoro
Legazpi
Samar
Panay
Iloilo
Cebu
Negros
Palawan
Surigao
Butuan
Cagayan de Oro
SULU SEA
Zamboanga
DAVAO
Mindanao
Sulu Arch.
Bandar Seri Begawan
Brunei (U.K.)
Kudat
Kota Kinabalu
Tawau
Borneo
CELEBES SEA
MICRONESIA
Yap
Koror
Palau Is.
Caroline Is. (U.S.A.)
Truk Is.
Sonsorol Is.
Tobi
Pacific Is. (U.S.A.)
Galela
Halmahera
Manado
Ternate
Sulawesi
Samarinda
Luwuk
Maluku
Manokwari
Kep. Schouten
P. Japen
Djajapura
MELANESIA
Admiralty Is.
Kavieng
Tombara (New Ireland)
Bismarck Arch.
Wewak
Balikpapan
Kep. Sula
P. Misool
New Guinea (Irian)
Nabi
Kokonau
Seram
Kendari
P. Butung
Buru
Ambon
Madjene
Bandjarmasin
INDONESIA
PAPUA NEW GUINEA
Madang
Birara (New Britain)
Rabaul
Sohano
Bougainville
UDJUNG PANDANG
BANDA SEA
Kep. Aru
Banjuwangi
Bali
Lombok
FLORES SEA
Mataram
P. Wetar
Kep. Tanimbar
Merauke
Lae
Popondetta
New Georgia
Djawa
Sumbawa
Flores
Dili
Timor
P. Dolak
Gulf of Papua
Pt. Moresby
D'Entrecasteaux Is.
Sumba
Kupang
ARAFURA SEA
Melville I.
Torres Str.
C. York
TIMOR SEA
Darwin
Louisiade Arch.
Great Barrier Reef
Weipa
Pine Creek
Groote Eylandt
Gulf of Carpentaria
Cooktown
CORAL SEA
Victoria
Queensland
Wyndham
Birdum
Borroloola
Cairns
Broome
Derby
Fitzroy
Northern Territory
Normanton
Croydon
Flinders
Townsville
Bowen
Tennant Creek
INDIAN OCEAN
Port Hedland
Mount Isa
Cloncurry
Is. Chesterfield (N. Cal.)
Dampier
Marble Bar
Mackay
Ashburton
Western Australia
Newman
Georgina
Alice Springs
Rockhampton
Paraburdoo
AUSTRALIA
Fraser I.
Carnarvon
Gascoyne
Meekatharra
L. Amadeus
Cooper Creek
Barcoo
Charleville
Maryborough
Murchison
Warrego
Oodnadatta
L. Eyre
Cunnamulla
Toowoomba
BRISBANE
Southport
Dirranbandi
Grafton
South Australia
Geraldton
Leonora
Kalgoorlie-Boulder
Coolgardie
Northam
Bourke
Broken Hill
L. Gairdner
New South Wales
Tamworth
PERTH
Eucla
Penong
Pt. Augusta
Darling
Bunbury
Whyalla
Pt. Pirie
Bathurst
Newcastle
SYDNEY
Wollongong
Great Australian Bight
Pt. Lincoln
Murray
Mildura
Esperance
Albany
Spencer G.
ADELAIDE
Albury
Canberra
Kangaroo I.
Ballarat
TASMAN SEA
Mt. Gambier
MELBOURNE
Geelong
Bass Str.
Furneaux Gr.
Burnie
Banks Str.
Launceston
Tasmania
Hobart
1 : 50 000 000
0 250 500 750 1 000 1 250 Km
0 200 400 600 800 Mi

roundwood 10 mn. cub.m (1976). **Mining** (1976, in 1,000 tons): coal 2,315, crude petroleum 477, natural gas 1,399 mn. cub.m (1977), gold 76 kg. **Production** (1976, +1977, in 1,000 tons): aluminium 139.8, motor spirit 1,414, oils 1,948, phosphorus fertilizers 375, paper 634, chemical wood pulp 582, meat 1,101+, milk 6,635+, butter 277+, cement 798 (1978). Electricity (1978): 21,348 mn. kWh (of which 76% is hydro-electric).
Communications (1977): railways 4,658 km, roads 95,026 km, 1,213,460 passenger cars and 236,720 lorries. – **Exports** (1977): meat (23%), wool (21%), butter and cheese (11%), hides and skins, milk and milk products, non-ferrous metals, newsprint, fish and others. Chief trading partners (1977): United Kingdom 18.6%, Australia 16.4%, Japan 13.9%, U.S.A., Fed. Rep. of Germany, Iran, U.S.S.R., Netherlands.

NEW ZEALAND TERRITORIES:

COOK ISLANDS, area 241 sq.km, population 18,112, annexed state with internal self-government; capital: Avarua (on Rarotonga). **Economy:** citrus fruit, bananas, coconuts.
NIUE ISLAND, area 259 sq.km, population 3,954 (1977), **annexed state with internal self-government; capital:** Alofi. **Economy:** coconuts, copra, tropical fruit.
TOKELAU ISLANDS, area 10 sq.km, population 1,575 (1976), **overseas territory,** 3 atolls. **Economy:** coconuts, copra.

PAPUA NEW GUINEA

area 461,691 sq.km, population 2,905,000 (1977), **independent state, member of the Commonwealth** (Prime Minister Sir Julius Chan since 1980).
The territory of the state is formed by the eastern part of the island of New Guinea and 2,890 islands and islets; the largest are Birara (New Britain), Tombara, the northern part of Solomon Is. (the largest Bougainville), Admiralty Is. and Louisiade Arch. **Administrative units:** 18 provinces. **Capital:** Port Moresby 76,507 inhab. (1971); **other towns** (in 1,000 inhab.): Lae 39, Rabaul 27, Madang 17, Wewak 15.
Population: Papuans and Melanesians. – **Currency:** kina = 100 toca.
Economy: mining (1976): copper 176,500 tons, gold 20,770 kg, silver 75 tons, platinum; **Agriculture:** crops (1977, in 1,000 tons): sweet potatoes 416, coconuts 757, copra 132, manioc 88, palm oil 32, vegetables 235, bananas 876, sugarcane 372, coffee 45; **livestock** (1977, in 1,000 head): pigs 1,184, poultry 1,110; fish. Forests cover 80% of the land area; tropical woods 5.9 mn. cub.m. Electricity: production 1,039 mn. kWh. – **Communications:** roads 18,188 km, chief airport Port Moresby. – **Exports** (1975): copper (56%), copra and coconut oil (11%), cocoa (9.5%), coffee.

SAMOA

Samoa i Sisifo, area 2,842 sq.km, population 153,000 (1978), **kingdom** (King Malietoa Tanumafili II since 1963), **member of the Commonwealth.**
Capital: Apia (on Upolu) 32,201 inhab. (1976). – **Currency:** tala. **Economy** (1977): crops (in 1,000 tons): coconuts 215, copra 19, bananas 36, cattle, pigs, fish.

add page 178

1.2 mn. Merchant shipping (1977): 1,347,800 GRT (of which 591,000 GRT are ore vessels); civil aviation (451 airports) – 187.4 mn. km flown (1976). – **Exports** (1977): iron ore, iron, and steel 12%, coal 11.7%, cereals 11.5%, textile fibres and products 10%, meat 8.6%, sugar. Chief trading partners (1977): Japan 27.7%, U.S.A. 14.4%, United Kingdom 7.6%, Fed. Rep. of Germany, New Zealand, Italy, Canada and others.

AUSTRALIAN OVERSEAS TERRITORIES:

CHRISTMAS ISLAND, area 135 sq.km, population 3,255 (1977), in the Indian Ocean on 10°25′ S.Lat. and 105°40′ E.Long.; **population:** Chinese 56% and Malays 29%. **Seat of administration:** Flying Fish Cove; **economy:**mining and export of phosphates 1,118,910 tons (1976).
COCOS ISLANDS, The Cocos (Keeling) Islands, area 14.2 sq.km, population 444 inhab. (1977); 27 coral islands in the Indian Ocean on 12°05′ S.Lat. and 96°53′ E.Long. **Seat of administration:** Home Island; **economy:** coconuts 3,000 tons, oil and copra.
NORFOLK ISLAND, area 35 sq.km, population 2,000 (1977), in the Pacific Ocean on 29°04′ S.Lat. and 167°57′ E.Long.; **seat of administration:** Kingston; **economy:** vegetables, citrus fruit and especially tourism.

FIJI

Area 18,272 sq.km, population 608,000 (1978), **independent state, member of the Commonwealth** (Prime Minister Ratu Kamisese K. T. Mara since 1970, Governor-General Sir George Cakobau).
Administrative units: 4 divisions. 106 of the 844 islands and islets are inhabited; the largest, Viti Levu 10,497 sq.km and Vanua Levu 5,534 sq.km. **Capital:** Suva 63,622 inhab. – **Currency:** Fiji dollar. **Economy** (1977, in 1,000 tons): sugarcane 2,387, sugar 376, coconuts 270, copra 29, rice 18, manioc 92, ginger; raising of cattle and poultry; fish catch; mining of gold 1,897 kg (1975). – **Exports:** sugar, coconut oil, gold.

HAWAIIAN ISLANDS – U.S.A.

State of Hawaiian Islands, area 16,705 sq.km, population 914,200 (1978). Since 18 March 1959 50th **state of the U.S.A.** in the North Pacific Ocean.
The archipelago is formed by more than 20 islands, 7 of which are inhabited. The largest are Hawaii 10,414 sq.km, Maui 1,886 sq.km, Oahu 1,549 sq.km, Kauai 1,427 sq.km. **Capital:** Honolulu 328,000 inhab. (1978, with agglomeration 728,000). **Economy:** farming – sugarcane, sugar, pineapples, vegetables, flowers, cattle raising, poultry, fishing; petroleum refinery at Honolulu. – **Communications:** 5,900 km roads (426,219 cars), important air and naval crossroads. Tourism is of great importance : 2.8 mn. visitors.

KIRIBATI

Republic of Kiribati, area 860 sq.km, population 64,500 (1977), **republic, member of the Commonwealth** (President Jeremiah Tabai since 12 July 1979).
Includes 3 large groups of the Pacific coral islands: Kiribati Is. (Gilbert Is.), Phoenix Is. 28 sq.km, Line Is. 567 sq.km (includes the world's largest atoll Christmas I. 359 sq.km) and Banaba (Ocean I.) 5.2 sq.km. **Capital:** Bairiki (on Tarawa I.) 17,188 inhab. **Population:** Micronesians. – **Currency:** Kiribati dollar. **Economy:** mining of phosphates (Banaba) 529,000 tons (1975), coconuts 83,000 tons, copra 11,000 tons, fishing. – **Exports:** phosphates 95%.

NAURU

Republic of Nauru, area 21 sq.km, population 7,254 (1977), **republic, member of the Commonwealth** (President Bernard Dowiyogo since 1978).
Coral island in the Pacific Ocean on 0°32′ S.Lat. and 166°55′ E.Long. **Capital:** Yaren 430 inhab. – **Currency:** Australian dollar = 100 cents. **Economy:** important mining of phosphates 1,535,000 tons for export (1975)

NEW ZEALAND

Area 268,675 sq.km, population 3,130,083 (1976), **independent state, member of the Commonwealth** (Prime Minister Robert David Muldoon since 1975, Governor-General Sir David Stuart Beattie since 1980).
Administrative units: 13 statistical areas, overseas territories. **Capital:** Wellington 139,566 inhab. (1976, with agglomeration 349,628); **other towns** (1976, in 1,000 inhab., + with agglom.): Christchurch 172 (+326), Auckland 151 (+797), Manukau 139, Hamilton 88 (+155), Dunedin 83 (+120). **Population:** New Zealanders and Maoris (8.2%). **Density** 11 persons per sq.km (1976); urban population 83.8%; 10% of inhabitants employed in agriculture. – **Currency:** New Zealand dollar = 100 cents.
Economy: advanced agricultural country with important animal production. Arable land covers only 3.2%, meadows and pastures 51% and forests 25% of the land area. **Agriculture:** crops (1977, in 1,000 tons): wheat 370, barley 316, maize (corn) 256, oats 71, potatoes 257, vegetables 379, fruit 227 (apples 137); **livestock** (1977, in 1,000 head): sheep 58,800 (fourth world population), cattle 9,472 (dairy cows 2,074), pigs 536, poultry 6,562, grease wool 302,500 tons, hides and skins 176,000 tons; fish catch 70,400 tons;

OLOMON ISLANDS

ea 28,446 sq.km, population 213,000 (1978), **independent state, member of the Common-**
ealth (Prime Minister Peter Kenilorea since 7 July 1978, Governor-General Baddeley Devesi).
dministrative units: 4 divisions. **Capital:** Honiara (on Guadalcanal) 14,942 inhab. (1976). **Population:** Melanesians
%, Polynesians. – **Currency:** Solomon dollar. **Economy:** forests cover 93% of the land area; roundwood 421,000 cub.m
976); crops (1977, in 1,000 tons): coconuts 183, copra 24, sweet potatoes 49, rice, cocoa, palm oil; raising of cattle,
gs, poultry; fish catch 18,600 tons. – **Exports:** copra, wood, fish, palm oil.

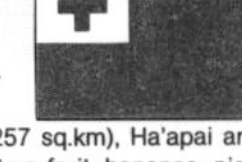

ONGA

ngdom of Tonga, area 699 sq.km, population 92,000 (1978), **kingdom, member of the Com-**
onwealth (King Taufa'ahau Tupou IV since 1965).
ıe archipelago called Friendly Islands is formed by 169 islets – 3 chief groups: Tongatapu (257 sq.km), Ha'apai and
ava'u. **Capital:** Nukualofa 17,400 inhab. – **Currency:** pa'anga. **Economy:** coconuts, copra, citrus fruit, bananas; pigs.

UVALU

ıe Tuvalu Islands, area 26 sq.km, population 6,500 (1977), **independent state, member**
the Commonwealth (Prime Minister Toalipi Lauti since 1978).
ıe archipelago was formerly called Ellice Islands. **Capital:** Funafuti 826 inhab. **Population:** Polynesians. – **Currency:**
ustralian dollar. **Economy:** coconuts, copra, fishing; raising of pigs and poultry.

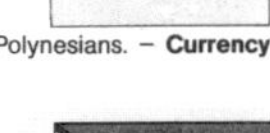

ANUATU

ea 14,763 sq.km, population 99,000 (1977), **independent state** (President Ati George Soka-
ani since 1980).
ıe archipelago was formerly called New Hebrides. Group of 40 islands; the largest are Espíritu Santo, Malekula,
ate, Eromanga. **Capital:** Vila (Efate I.) 16,604 inhab. (1975). **Economy:** coconuts 268,000 tons, copra, bananas, cocoa,
offee; cattle 115,000 head, pigs. Mining of manganese 19,100 tons (Efate I.). – **Exports:** copra, manganese.

RITISH TERRITORY:

ITCAIRN ISLAND – area 4,6 sq.km, population 65. Capital: Adamstown. **Economy:** fruit, vegetables.

RENCH TERRITORIES:

RENCH POLYNESIA – Polynésie Française, area 4,000 sq.km, population 137,382 (1977), **French overseas**
rritory (Governor Daniel Videau). **Administrative units:** 5 districts: Îles du Vent 1,173 sq.km (the largest Tahiti
042 sq.km) and Îles sous le Vent 474 sq.km called Îles de la Société (Society Is.); Îles Tuamotu and Îles Gambier
5 sq.km, Îles Marquises 1,274 sq.km, Îles Toubouai (Îs. Australes) 164 sq.km. **Capital:** Papeete 36,784 inhab.
onomy: mining of phosphates (Makatea); crops (1977, in 1,000 tons): coconuts 165, copra 23, sugarcane, cotton,
offee, tropical fruit; poultry, fish catch 3,000 tons. – **Exports:** copra, vanilla, citrus fruit. Tourism (92,000 visitors)
of importance (1976). The uninhabited island of **Clipperton** is under the authority of the Governor.
EW CALEDONIA and dependencies-Nouvelle-Calédonie, area 19,058 sq.km, population 136,000 (1977), **the Pacific**
erseas territory (Governor Jean-Gabriel Eriau); **dependencies:** Îles Loyauté (2,072 sq.km), Île des Pins, Îs. Huon, Îs. Bé-
p, Îs. Chesterfield, Î. Walpole. **Capital:** Nouméa 59,869 inhab. (1974). **Population:** French (45%), Polynesians 10%.
onomy: important mining of nickel 118,945 tons (1976, third world producer); deposits: iron ore, chromium, lead.
rops (1977): coconuts, coffee, bananas; raising of cattle, poultry, pigs.-**Exports:** nickel and nickel ore (98%), coffee, fruit.
ALLIS AND FUTUNA, area 255 sq.km, population 9,800 (1978), **overseas territory; seat of the administration:**
ata Utu (Î. Uvéa) 600 inhab. **Economy:** coconuts 10,000 tons, copra, fruit.

ERRITORIES OF THE UNITED STATES:

MERICAN SAMOA and dependency, area 197 sq.km, population 34,000 (1977), **unincorporated territory of the U.S.A.**
apital: Pago Pago 2,451 inhab. (on Tutuila), **seat of the Government:** Fagatogo 1,340 inhab. (1970). **Economy:**
oconuts, bananas, copra, cocoa; fishing. – **Exports:** canned fish, copra.
UAM, area 549 sq.km, population 104,000 (1977), **unincorporated territory of the U.S.A. – Capital:** Agana 2,199 inhab.
onomy – crops: maize (corn), bananas, coconuts, citrus fruit, sugarcane; fishing. The air and naval base.
IDWAY ISLANDS, area 5 sq.km, population 2,300 (1978), **overseas territory of the U.S.A.** under naval administration.
ORTHERN MARIANA ISLANDS, area 404 sq.km, population 18,000 (1977), 14 islands – the largest Saipan and
nian. **Capital:** Garapan. **Economy:** crops of maize (corn), tropical fruit, coconuts, fishing. Important tourism.
ACIFIC ISLANDS, The Trust Territory of the Pacific Islands, area 1,410 sq.km, population 110,000 (1977),
ust territory of the U.S.A. Includes 2,127 islands, islets and atolls of Micronesia (of which only 90 are inhabited).
dministrative units: 6 districts – Truk Is., Marshall Is., Ponape, Palau Is., Yap, Kosrae. **Capital:** Koror (on Palau Is.).
onomy: mining of phosphates and bauxite; crops: coconuts, copra, vegetables, tropical fruit; fishing.
AKE ISLAND, area 8 sq.km, population 2,000 (1977), **island under naval administration of the U.S.A.,** air base.

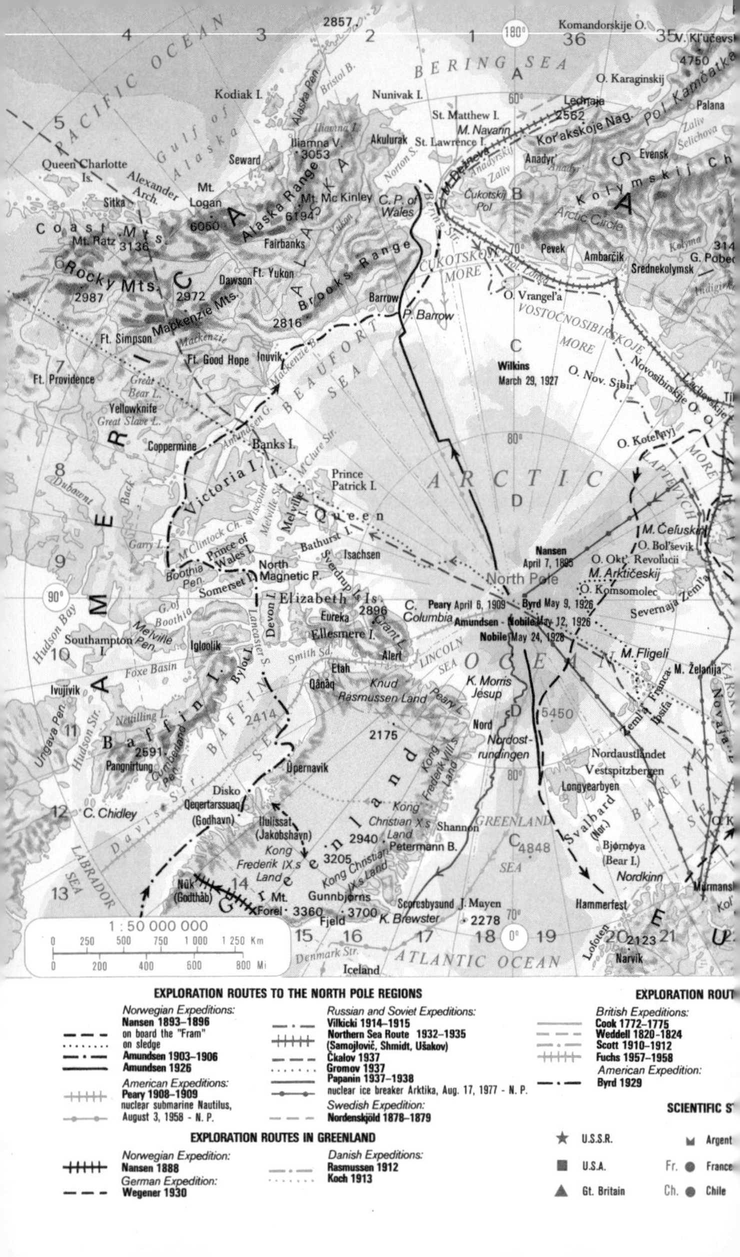

EXPLORATION ROUTES TO THE NORTH POLE REGIONS
Norwegian Expeditions:
Nansen 1893–1896
on board the "Fram"
on sledge
Amundsen 1903–1906
Amundsen 1926
American Expeditions:
Peary 1908–1909
nuclear submarine Nautilus, August 3, 1958 - N. P.
Russian and Soviet Expeditions:
Vilkicki 1914–1915
Northern Sea Route 1932–1935 (Samojlovič, Shmidt, Ušakov)
Čkalov 1937
Gromov 1937
Papanin 1937–1938
nuclear ice breaker Arktika, Aug. 17, 1977 - N. P.
Swedish Expedition:
Nordenskjöld 1878–1879
EXPLORATION ROUTES IN GREENLAND
Norwegian Expedition:
Nansen 1888
German Expedition:
Wegener 1930
Danish Expeditions:
Rasmussen 1912
Koch 1913
EXPLORATION ROUT
British Expeditions:
Cook 1772–1775
Weddell 1820–1824
Scott 1910–1912
Fuchs 1957–1958
American Expedition:
Byrd 1929
SCIENTIFIC S
U.S.S.R.
U.S.A.
Gt. Britain
Argent
Fr. France
Ch. Chile
1 : 50 000 000
0 250 500 750 1 000 1 250 Km
0 200 400 600 800 Mi
PACIFIC OCEAN
BERING SEA
ARCTIC OCEAN
ATLANTIC OCEAN
BEAUFORT SEA
BAFFIN SEA
LINCOLN SEA
GREENLAND SEA
LABRADOR SEA
ČUKOTSKOJE MORE
VOSTOČNOSIBIRSKOJE MORE
MORE LAPTEVYCH
Queen Elizabeth Is.
Greenland
North Pole
North Magnetic P.
Nansen April 7, 1895
Peary April 6, 1909
Byrd May 9, 1926
Amundsen - Nobile May 12, 1926
Nobile May 24, 1928
Wilkins March 29, 1927
Alaska
Alaska Range
Brooks Range
Rocky Mts.
Mackenzie Mts.
Coast Mts.
Mt. Mc Kinley 6194
Mt. Logan 6050
Kodiak I.
Nunivak I.
St. Matthew I.
St. Lawrence I.
Fairbanks
Barrow
P. Barrow
Inuvik
Victoria I.
Banks I.
Baffin I.
Ellesmere I.
Devon I.
Svalbard (Nor.)
Vestspitzbergen
Nordaustlandet
Longyearbyen
Bjørnøya (Bear I.)
Hammerfest
Narvik
Murmansk
Iceland
Nûk (Godthåb)
Qeqertarssuaq (Godhavn)
Ilulissat (Jakobshavn)
Upernavik
O. Vrangel'a
O. Nov. Sibir'
O. Kotel'nyj
M. Čeluskin
O. Bol'ševik
O. Okt'. Revolucii
O. Komsomolec
Severnaja Zemľa
Zemľa Franca-Iosifa
M. Želanija
Novaja Zemľa
Hudson Bay
Foxe Basin
Davis Str.
Denmark Str.
180°
90°
0°

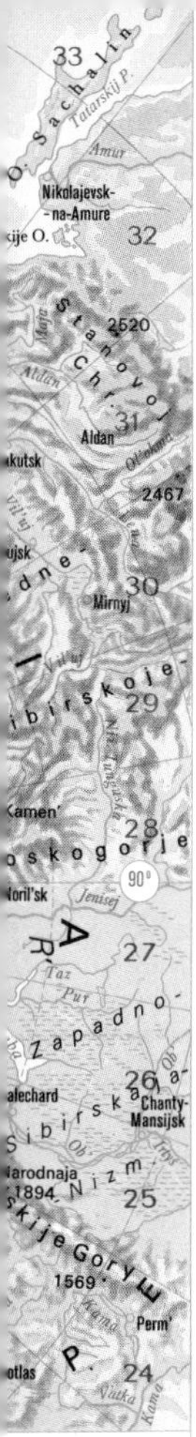

EGIONS
Soviet Expeditions:
Lazarev 1819–1821
n 1958–1959
Expedition:
958
xpedition:
0–1912

R. of S. Afr. ● Republic of South Africa
Pol. ● Poland

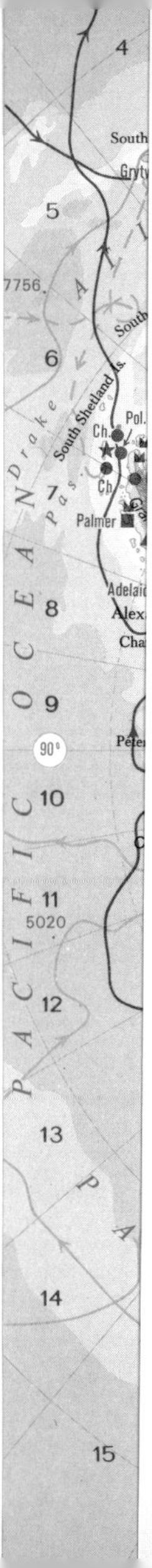

bacteria and fungi live below the surface of the weather-worn rocks. As regards its **fauna** there are two species of penguin as well as the skua, a predatory gull-like bird, though there are large numbers of animals in the sea. As far as **Man** is concerned the region is inhabited only by the staff of the scientific research stations. Great mineral wealth is assumed, but no mining is as yet in progress.

In 1773 Captain James Cook was the first to set eyes on South Georgia and the Sandwich Group during his voyage along the Antarctic Circle. The Russian explorers F. F. Bellingshausen and M. P. Lazarev discovered Peter Ist I. and saw Alexander I. in 1819–21; in 1819 the South Shetland Is. were discovered by W. Smith, but the coast of the Antarctic continent was not discovered until 27 February 1831 when John Biscoe reached the mountainous Enderby Land. J. C. Ross discovered Victoria Land, Ross I. and the Ross Ice Shelf in 1840–43. The first to stand on the continent was C. E. Borchgrevink (1894–95) near Cape Adare. Attempts to reach the South Pole date from the early 20th century; in October 1909 E. Shackleton reached Lat. 88°23′ – 180 km from the Pole–but he was forced to turn back. The first to stand on the Pole was Roald Amundsen on 14 December 1911, and he was followed by R. F. Scott on 18 January 1912. Aerial surveys of Antarctica began in 1928; in the years 1928–30 H. Wilkes explored Palmer Land, on 20 November 1929 R. E. Byrd reached the Pole and he led a total of 5 expeditions until 1947. During expeditions in 1934–38 L. Ellsworth crossed Western Antarctica by plane. The first research stations were set up in 1947, and during the International Geophysical Year 1957–58 alone there were 55 such bases. On 14 December 1958 a Soviet expedition reached a place known as the Pole of Relative Inaccessibility, 82°06′ S.Lat. and 54°58′ E.Long. On 1 December 1959 a treaty was signed for peaceful cooperation in scientific investigation and research and the demilitarization of Antarctica; the treaty came into force on 23 June 1961. Some countries lay claim to territory in Antarctica in the given **sectors** (south of 60° S.Lat.).

AUSTRALIAN ANTARCTIC TERRITORY 45° to 136° E.Long. and 142° to 160° E.Long., an **area** roughly **6.4 mn. sq.km** with Mawson as the largest station; incl. Macquarie Is. 176 sq.km, Heard I. and McDonald I. 258 sq.km.
BRITISH ANTARCTIC TERRITORY 20° to 80° W.Long., since 1962 a colony roughly **388,500 sq.km in area,** incl. the South Orkney Is. 622 sq.km, South Shetland Is. 4,622 sq.km and Graham Land. Argentina has laid claims to a part of the British sector from 25° to 74° W.Long. and Chile to 53° to 90° W.Long. and have set up their own stations.
TERRES AUSTRALES ET ANTARCTIQUES FRANÇAISES, since 6 August 1955 French overseas territory, roughly **395,500 sq.km in area.** Adélie Coast, about 388,500 sq.km with the Dumont d'Urville station, from 136° to 142° E.Long.; islands in the Indian Ocean: Îles Kerguélen 6,232 sq.km with the Pt-aux-Français research station, Îs. Crozet 476 sq.km, Î. Amsterdam 66 sq.km and Î. St. Paul 7 sq.km.
NORWEGIAN DEPENDENCY 20° W.Long. to 45° E.Long., known as Queen Maud Land. The Dependencies in the Atlantic Ocean: Bouvetøya 59 sq.km, Peter Ist I. 249 sq.km.
NEW ZEALAND ROSS DEPENDENCY between 160° E.Long. and 150° W.Long., an **area of about 453,000 sq.km,** mostly ice shelf (330,000 sq.km) with the Scott Station.
PRINCE EDWARD ISLAND and MARION ISLAND, area 255 sq.km, occupied by South Africa since 1947.
The SECTOR OF THE U.S.A., between 80° and 150° W.Long. has not been officially proclaimed. Of the four American bases in the Antarctic, McMurdo Station is the largest; it has an atomic power station. The Amundsen-Scott Station stands on the South Pole itself.
The **U.S.S.R.** has 6 permanent scientific research bases in Antarctica; the largest of these is Mirnyj

add page 182

ARCTIC REGIONS

The Arctic – the northern Arctic polar region – owes its name to the Ancient Greeks, for it lies below the northern constellation of the Great Bear, and the Greek for bear was "Arktos". The boundary of the Arctic region does not run only along the Arctic Circle (66°32' N.Lat.), making it 21.18 mn. sq.km in area; it is defined climatically by the 10°C July isotherm, which roughly coincides with the northern timber line and the range of tundra and taiga. This gives the Arctic an **area of 26.4 mn. sq.km**, of which 18.5 mn. sq.km are ocean and 7.9 mn. sq.km (30%) islands and continent. The point closest to the North Pole is Morris Jesup Cape on Greenland (83°40' N.Lat.). **The largest islands** are Greenland 2,175,600 sq.km and the Canadian Arctic islands 1,403,134 sq.km, of which Baffin I. 507,414 sq.km, Victoria I. 217,274 sq.km, Ellesmere I. 196,221 sq.km lie within the American Arctic; Novaja Zeml'a 82,180 sq.km, Svalbard 62,050 sq.km and Zeml'a Franca-Iosifa 16,100 sq.km are within the European Arctic, and Novosibirskije Ostrova 38,400 sq.km, Severnaja Zeml'a 37,560 sq.km and Ostrov Vrangel'a 7,270 sq.km are in the Asian Arctic. **The Arctic Ocean** proper covers an area of 13,950,000 sq.km with a maximum depth of 5,450 m; in the region of the North Pole it is covered with a thick layer of pack-ice and drift-ice, which extends southwards beyond 70° N.Lat. The depth of the ocean at the North Pole is 4,316 m.

Climate: The inner part of the Arctic Ocean is permanently covered with pack-ice (annual precipitation is 100–300 mm), and such is the intense cold that its effects spread far into the northernmost countries of the world. The warming effect of the ocean and its currents make it possible for the temperature to rise in the short summer on the North Pole to +1 to +3°C, while in the long winter it drops in January to about −35 to −40°C. Extreme winter temperatures spread to subarctic continental Sibir' (Siberia) (Ojm'akon −78°C), Canada (Fort Good Hope −78.2°C) and Greenland (−62°C). The ice shield covers 1,830,000 sq.km of Greenland and 155,000 sq.km of the Canadian Arctic islands. The land that is not icebound has permafrost with tundra **vegetation** (mosses, lichens, perennial plants, dwarf bushes) that passes into subarctic taiga. **Animal life** is limited in species and number (polar bears, polar foxes, hares, reindeer, caribou, waterfowl, etc.), and there are more species in the sea (cod, flatfish, walrus, seal, whale). Along the edge of the Arctic there are only sparse **settlements** of Eskimos, Lapps, Nenets, Yakuts, Chukchi, white immigrants and others, who make a living by fishing, trapping and reindeer farming and the extraction of minerals, which is on the increase. The northernmost settlement is Alert in Canada (82°30'N.Lat.) on Ellesmere I.. 3,431 inhabitants live on Svalbard (1975, of these 2,485 Russians) and 831,590 tons of coal was mined. Petroleum is being drilled in North Alaska, petroleum and natural gas are also found on the Arctic islands of Canada.

The Vikings, who settled in Iceland 870–930, were the first to explore the Arctic. Eric the Red discovered Greenland in 986, and around 1000 Leif Eriksson reached the American continent near Cape Dyer on Baffin I. Svalbard was discovered in 1194. The Russians explored northern Sibir' (Siberia) in the 17th century, and in 1648 S. Dezhnyov circumnavigated the eastern point of Asia. The North-East passage along the northern coast of Europe and especially Asia was first made by A. E. Nordenskjöld (1878–79), the North-West passage along the North American coast by R. McClure (1850–53) from the Bering Strait in the direction of the Atlantic Ocean, and in the opposite direction by Roald Amundsen (1903–06). The scientific exploration of the Arctic region began at the end of last century: F. Nansen and O. Sverdrup crossed the South of Greenland in 1888.

Robert E. Peary explored northern Greenland in 1892–1900 and he and M. Henson were the first people to stand on the North Pole (6 April 1909). Richard E. Byrd reached the Pole by air on 6 May 1926, two days later R. Amundsen, L. Ellsworth and U. Nobile did so in an airship; in 1937 the Russian polar explorers I. Papanin and O. Schmidt again used a plane; the Americans reached the Pole under the sea in their nuclear submarine Nautilus on 3 August 1958; Guido Monzino followed Peary's route with a dog team and reached the Pole on 19 May 1971. The Soviet atomic ice-breaker Arktika reached the Pole on 17 August 1977.

ANTARCTICA

Antarctica – the southern Antarctic polar region – derives its name from the Greek word for "opposite the Arctic", "opposite the North". The boundary of Antarctica is given by the Antarctic Circle (66°32' S.Lat.) as well as by the climatic 10°C January isotherm (the warmest), roughly concurrent with the southern timber line; within this boundary **the Antarctic measures 67.84 mn. sq.km** (incl. parts of South America). It comprises the Antarctic continent 14,108,000 sq.km in area, incl. the ice shelf, 13,209,000 sq.km without the ice shelf (i.e. 8.8% of land surface), with 75,570 sq.km of islands and the surrounding sea with the more distant subantarctic islands extending over an area of 13,198 sq.km. Antarctica lies at a distance of 4,000 km from Africa, 3,200 km from Australia and only 1,450 km from South America. The Ross Sea (off the Pacific Ocean) and the Weddell Sea (off the Atlantic Ocean) penetrate deep into the continent and divide it into the larger Eastern Antarctica on the side of the Atlantic and Indian oceans and the smaller Western Antarctica on the Pacific side, where the large Antarctic Peninsula stretches in the direction of South America. The permanent ice-cap, of mean thickness 2,500 m, leaves only insignificant parts of the coast and the highest rocks free of ice. The average altitude of Antarctica is 2,280 m. The South Pole lies on the South Polar Plateau at a height of 2,765 m above sea level and the thickness of ice there reaches 2,810 m. The highest peak, the Vinson Massif in the Ellsworth Mts. in Western Antarctica is 5,140 m high, Mt. Kirkpatrick in the Transantarctic Mts. in Eastern Antarctica is 4,528 m and the highest active volcano is Mt. Erebus, 3,795 m, on Ross I.

The region of the South Pole has a severe, harsh polar **climate,** for the mean summer temperature (January) is −28°C, the July temperature (winter) is about −50°C, with precipitation not exceeding 80 mm in the prevailing calm; by contrast, the Antarctic coast (e.g. Wilkes Land) has the following mean temperatures: January −0.2°C, July −25.8°C, mean annual temperature −13.9°C, and 340 stormy days (highest number in the world). Absolute minimum temperature was recorded at the Vostok station (3,488 m high): −88.3°C. The lowest annual mean temperature is found at the Pole of Cold: −57.8°C. **Flora:** there exist only 3 species of flowering plants in Antarctica, about 60 varieties of moss and lichen, and microbes, algae,

INDEX

The index contains in alphabetical order all geographical names used in the maps. Apart from towns and cities all names are marked with appropriate abbreviations (for example, R. = river, I. = island, L. = lake, etc.). Following each name there is a number signifying the map number, and a letter and a number indicating the section of the map in which the place is located. The letters (marked in red on the maps) refer to sections of latitude, and the numbers (also marked in red) to sections of longitude. For example, "Mamoré, R., 37 D 3" means that the river Mamoré appears on map 37 in square D 3.

Rivers are indexed under the names given them by the country of their source. Thus the Danube comes under Donau and the Rijn under Rhein. For major geographical names, the squares given are those in which the name actually appears. For example, England, 9 B-D 3-5.

In some maps the names of administrative units, states and their subdivisions have been replaced by numbers, which are explained in the legend for the map. In the index this number is given in brackets after the name of the administrative unit.

When there is more than one place with the same name, the country where each is situated is given in brackets. For example, Victoria (Canada) 33 D 7; Victoria (Hong Kong), 19a B 3; Victoria (U.S.A.), 34 F 7. In the case of rivers, the sea, lake or river into which they flow is given, e.g. Negro (Amazonas), R., 37 C 3; Negro (Atlantic Ocean), R., 37 F 3.

ABBREVIATIONS USED IN THE INDEX

Admin. U. = Administrative Unit
Arch. = Archipelago
B. = Bay
C. = Cape
Can. = Canal
Chann. = Channel
Depr. = Depression
Des. = Desert
Fs. = Falls
G. = Gulf
Glac. = Glacier
H., Hs. = Hill, Hills
I., Is. = Island, Islands
L. = Lake
Mt., Mts. = Mountain, Mountains
Pen. = Peninsula
Pk. = Peak
Pl. = Plain
Plat. = Plateau
R. = River
Reg. = Region
Res. = Reserve, Reservoir
S. = Sea
St. = State
Str. = Strait
Sw. = Swamp
Val. = Valley
Vol. = Volcano

ABBREVIATIONS USED IN THE TEXT

m = metre
km = kilometre
sq.m = square metre
sq.km = square kilometre
ha = hectare
kg = kilogramme
ton = metric ton (tonne) = 1,000 kg
cub.m = cubic metre
l = litre
hl = hectolitre
mn. = million
agglom. = agglomeration
R.S.F.S.R. = Russian Soviet Federal Socialist Republic

C

D

G

H

I

M

N

S

T

U

W

X

Y

Z